M000006131

Grounding Social Sciences in Cognitive Sciences

GROUNDING SOCIAL SCIENCES IN COGNITIVE SCIENCES

edited by Ron Sun

The MIT Press
Cambridge, Massachusetts
London, England

© 2012 Massachusetts Institute of Technology

All rights reserved. No part of this book may be reproduced in any form by any electronic or mechanical means (including photocopying, recording, or information storage and retrieval) without permission in writing from the publisher.

MIT Press books may be purchased at special quantity discounts for business or sales promotional use. For information, please email special_sales@mitpress.mit.edu or write to Special Sales Department, The MIT Press, 55 Hayward Street, Cambridge, MA 02142.

This book was set in Stone Sans and Stone Serif by Toppan Best-set Premedia Limited. Printed and bound in the United States of America.

Library of Congress Cataloging-in-Publication Data

Grounding social sciences in cognitive sciences / edited by Ron Sun.
p. cm.
Includes bibliographical references and index.
ISBN 978-0-262-01754-1 (hardcover : alk. paper) 1. Cognitive science. 2. Social sciences. I. Sun, Ron, 1960–
BF311.G7686 2012
302–dc23
2011044282

10 9 8 7 6 5 4 3 2 1

Contents

Preface

This book explores the cognitive (psychological) basis of the social sciences and the possibilities of grounding the social sciences in the cognitive sciences, broadly defined. The result is what I call cognitive social sciences (or cognitively based social sciences)—an integrative intellectual enterprise.

The cognitive sciences have made tremendous strides in recent decades. In particular, computational cognitive modeling (computational psychology) has changed the ways in which cognition and psychology are explored and understood in many profound respects. There have been many theoretical or computational models proposed in the cognitive sciences, leading to detailed understanding of many cognitive or psychological domains and functionalities. Empirical research has also progressed to provide a much better understanding of many psychological phenomena.

Given the advances in the cognitive sciences, can we leverage these successes for the sake of better understanding social processes and phenomena? More fundamentally, can the cognitive sciences (including computational psychology, experimental cognitive psychology, social-personality psychology, developmental psychology, psycholinguistics, philosophy of mind, cognitive neuroscience, and so on) provide a better foundation for important disciplines of the social sciences (sociology, anthropology, economics, political science, and so on)?

Thus far, although it is still very much neglected, there have been various efforts at exploring this topic. Some of the efforts have been computationally motivated. Others are more empirical or theoretical in nature. The present volume includes some of the major work in all of these directions, written by some of the best experts in various fields of the social sciences and the cognitive sciences.

The focus of this book is the unification of the social and cognitive sciences through "grounding" of the social sciences in the cognitive sciences, broadly construed and broadly inclusive. This book is not limited to computational approaches, or to any other specific methodology. It includes chapters on a range of topics, selected to capture issues in a wide selection of social science fields. Thus, for example, someone from behavioral economics could pick up the book to see what related work is being done in other social science fields. To achieve a proper balance between breadth and depth, each chapter aims to combine the rigor and depth of a research article with the breadth and appeal of a handbook chapter.

A summary of the key features of this book is as follows:

• A unique agenda: the broad exploration of the "grounding" of the social sciences in the cognitive sciences is unique and relevant.
• A comprehensive scope: this book is broader in scope than any other book on this or similar topics (possibly pointing to a new general direction for the social sciences as a whole).
• Multiple approaches: the book includes multiple approaches and perspectives, either theoretical, experimental, or computational, which may compete with or complement each other.
• Interdisciplinary interaction: the goal of the book includes the facilitation of the interaction of many relevant disciplines, including cognitive psychology, social-personality psychology, computational psychology, sociology, anthropology, economics, political science, philosophy, artificial intelligence, and so on.

The readership of this book may include academic researchers and graduate students in fields ranging from sociology, economics, political science, and anthropology, to cognitive psychology, cognitive modeling, and social psychology, and even further to artificial intelligence and philosophy. In particular, I have in mind readers from a social sciences background who are interested in incorporating the considerations of human cognition or psychology into their studies, as well as readers from a cognitive or psychological background who are interested in tackling social issues from cognitive or psychological viewpoints. In addition, the book may be of interest to policy makers and other practitioners, as well as to laypersons interested in the aforementioned directions.

The book may be suitable for graduate-level courses and seminars on this topic, but may also be extended to the advanced undergraduate level.

I would like to thank Phil Laughlin of the MIT Press for his support along the way. Thanks are due to Jon Kable, Herbert Gintis, Todd Hare, Joel Mort, Jesper Sorensen, Adrian Murzac, Kristen Monroe, Ilkka Pyysiäinen, Peter Bull, Bradd Shore, Norbert Ross, Kimberly Gross, and others for reviewing draft versions of the chapters of this book. Thanks are due to Paul Thagard, Colin Camerer, Paul Glimcher, Dan Ariely, Philip Tetlock, Bob Barsky, John Hibbing, Riccardo Viale, Darren Schreiber, and others for their suggestions or comments, which helped to improve the book. Thanks are due to Selmer Bringsjord for supporting teaching release, which made producing this book (as well as other books) possible. My work has been financially supported (in part) by ONR, ARI, and AFOSR (thanks are due to Paul Bello, Jun Zhang, and others).

I INTRODUCTION

1 Prolegomena to Cognitive Social Sciences

Ron Sun

1.1 Something Missing

On a chilly autumn day in 2001, I walked through the broad sidewalks of the strangely quiet streets of Chicago in the early morning, a then unfamiliar city to me, to get to a meeting on "new approaches to the social sciences." With an eager anticipation of being enlightened, I rushed through the ten-minute walk and arrived early at the grand and equally strangely quiet granite-clad University of Chicago Business School building in downtown Chicago for the meeting.

Sitting in the audience with the expectation of being intellectually stimulated, I soon discovered that something important was missing. The longer I sat there, the more I felt that way. This idea was gnawing at me.

What I was feeling missing from this otherwise interesting meeting was a particular type of explanation of social processes and phenomena—what I considered to be a fundamental type of explanation for social processes and phenomena. This type of explanation may be termed *psychological explanation* or *cognitive explanation* (in the broadest sense of the word *cognitive*). I prefer to refer to it as *cognitive explanation*, in recognition of the fact that many disciplines concerned with the human mind have come to be known, collectively, as the *cognitive sciences* (notice the plural form here). (I often would use *cognition-psychology* as a single term to highlight the inseparable nature of these two.)

In the evening, back in my hotel room, I continued my rumination. Looking out onto a 180-degree view from the floor-to-ceiling, wrap-around window of the corner room in the high-rise hotel, I could see a panorama of the city with its neon signs and flickering lights. I wondered whether it was indeed possible to explain a substantial part of social processes and phenomena from a cognitive-psychological point of view, whether correspondingly agent-based social simulation could be made more psychologically realistic, and whether the social sciences could thus be put on a more solid footing that was more "scientific" (but not necessarily more mathematical) in some way.

Evolutionary explanations have been popular in many segments of the scientific community, but they tend sometimes to provide only unverifiable "just so" stories. Mathematical models such as game theory are useful and well respected, but they are often too normative and fail to take into account real-world complexity. The social sciences are broad-ranging, varied, interesting, and stimulating, but they are also often chaotic and confusing. How do we make sense out of this chaotic, exciting scene?

1.2   Why Cognitive Sciences Are Needed

By any measure, the cognitive sciences (including computational psychology, experimental psychology, linguistics, cognitive neuroscience, and so on) have made tremendous strides in recent decades. In particular, computational cognitive modeling (i.e., computational psychology; see, e.g., Sun, 2008) has changed the ways in which cognition-psychology is explored and understood in many profound respects.

For example, there have been many detailed models of cognition-psychology proposed in the cognitive sciences (broadly defined, as mentioned above), leading to more in-depth, more mechanistic, and more process-based understanding of cognitive-psychological domains and functionalities. Empirical psychological research has also progressed to provide us with a better understanding of many phenomena, from "pure" cognition to social cognition and beyond.

Given such advances in the cognitive sciences, the question now is: can we leverage these successes for the sake of better understanding social processes and phenomena? More fundamentally, can the cognitive sciences provide a better foundation for important disciplines of the social sciences (e.g., sociology, anthropology, economics, political science, communication, as well as some more "humanity" related fields such as history, ethics, religion, law, literature, and so on)?

Thus far, although very much a neglected topic, there nevertheless have been various efforts at exploring this topic. Some of the efforts were computationally motivated; see, for example, *Cognition and Multi-Agent Interaction* (Sun, 2006). Other efforts were more empirical or theoretical in nature; see, for example, *Cognitive Dimensions of Social Science* (Turner, 2001).

Evidently, there are both theoretical and practical rationales for the establishment and development of "cognitive social sciences." Any social process occurs through the actions and therefore the minds of the individuals involved (DiMaggio, 1997; Turner, 2001; Sun, 2001, 2006). Whether in a specific context it is a deciding factor or not, taking cognition-psychology into serious consideration would be a reasonable step in trying to reach an in-depth, fundamental understanding of social phenomena. Some cognitive-psychological process details may turn out to be unimportant for a particular phenomenon, but this possibility cannot and should not be determined and declared a priori. Instead, it needs to be ascertained through empirical and theoretical work examining all factors involved, cognitive-psychological factors included.

To look at the issue in another way: the cognitive sciences may serve as a basis for the social sciences, in much the same way that physics provides grounding for chemistry or quantum mechanics provides grounding for classical mechanics. Social, political, and cultural forces, although perhaps "emergent" (as often claimed), act both *upon* individual minds and *through* individual minds. In that sense, minds, however complex or simple one conceives them to be, are the basis of social processes and phenomena. Macro-micro (social-psychological) interactions thus do exist and need to be understood. These two types of forces (macro and micro) interact with each other, giving rise to complex sociocultural and cognitive-psychological phenomena (e.g., Tetlock & Goldgeier, 2000; Sun, 2006).

The social sciences are facing their share of challenges, in terms of making significant breakthroughs, becoming more rigorous, connecting better with the physical sciences, and so on (see, e.g., chapter 14 by Mathew McCubbins and Mark Turner in this book). I would contend that the social sciences might find their future in the cognitive sciences, at least in part, which may well lead to a powerful, productive, and unified intellectual enterprise. Such a unification, or grounding as I called it (Sun, 2006, 2010), may provide the social sciences with imaginative research programs, novel paradigms and frameworks, new syntheses, hybridization, and integration, and so on, in addition to providing the cognitive sciences with new data sources and problems to account for.

Some sociologists (such as cognitive sociologists) and anthropologists (such as psychological and cognitive anthropologists), as well as social and cultural psychologists have been interested in socioculturally shaped cognition. That is, they are interested in how culture and social processes shape individuals' minds (see, e.g., Zerubavel, 1997; Cerulo, 2002; D'Andrade & Strauss, 1992). The other side of this equation—how cognition (human psychology) shapes, substantiates, and grounds social institutions, social structures, social processes, and culture—is largely underexplored (of course, with exceptions as always; see, e.g., Sperber, 1996). The fact that this issue has been underexplored makes it even more important a candidate for serious examination, in both theoretical and empirical ways.

Looking into the future, one can easily see how the understanding of human cognition-psychology and its relation to sociocultural processes may lead to better understanding of a wide range of important issues in the social sciences, ranging from religion and international relations to politics and economics (e.g., see the chapters on these topics in the current volume). These issues are important not only for academics, but also for policy makers and practitioners in many different fields. There have been some promising signs already from the nascent field of cognitive social sciences, as described by various chapters in this book. I will get back to the topics discussed in these chapters a little later. For now, let us look into a broad framework first, which justifies the aforementioned "grounding" (integration/unification).

1.3 Levels of Analysis and Links across Levels

As discussed in Sun (2006), one interesting but unfortunate characteristic of the current social and cognitive sciences is a relative lack of interaction and integration among disciplines (the kinds of collaboration reported in this volume are relatively rare). Each discipline tends to consider a particular aspect and more or less ignore the rest. They generally do not work together (although there have been calls for cooperation; see, e.g., chapter 15 by Herbert Gintis).

Instead of adhering to this relative isolation of disciplines from one another, we may adopt a broader perspective. For one thing, we may take a look at multiple levels of analysis. As we will see, these levels of analysis in the social sciences can be cast as a set of related disciplines, from the most macroscopic to the most microscopic. These different levels include the *sociological, psychological, componential*, and *physiological* levels. In other words, as has been argued in Sun, Coward, and Zenzen (2005) and Sun (2006), we may view certain different disciplines as different levels of abstraction for exploring essentially the same broad set of theoretical questions (with different emphases, of course).

The *sociological level* includes sociocultural processes, social institutions, structures, organizations, and inter-agent interactions, as well as interactions between agents and their sociocultural environments. These issues have been studied by sociology, anthropology, political science, and economics.

Next is the *psychological level*, which covers individual behaviors as well as concepts, beliefs, knowledge, and skills employed by individuals. Between this and the sociological level, the relationship of individual concepts, beliefs, knowledge, and skills with those of the society and culture, and the processes of change of these, independent of or in relation to those of society and culture, may be investigated.[1] This level examines human behavioral data, comparing them with models and with insights and constraints from the sociological level and more detailed information from the lower levels.

The third level is the *componential level*. This level attempts to understand the mind in terms of its components, applying the language of a particular theoretical paradigm. This level may involve conceptual, computational, and/or mathematical structural specifications, such as specifying computationally an overall architecture of the mind and the components therein (e.g., Newell, 1990; Sun, 2002). Essential processes within each component as well as essential connections among components may also be specified. Constructs and data from the psychological level—that is, the psychological constraints from above, which bear on the division of components and the processes within components—are among the considerations. This level may also incorporate biological and physiological notions regarding divisions; that is, it can incorporate constraints and ideas from the level below. This level results in mechanisms, though they may be computational-mathematical and thus somewhat abstract compared with the physiological level.

Although the componential level is essentially about intra-agent processes, conceptual, computational, or mathematical models developed therein may be used to capture processes occurring at higher levels, including interactions at the sociological level that involve multiple individuals (Sun, 2006). That is, we may construct agent models from a sub-agent level (the componential level), but go up from there to the psychological and sociological levels. For example, the CLARION cognitive architecture model specifies component mechanisms and processes and their interactions, and then moves up to higher levels to account for psychological and sociological data (Sun, 2002).[2]

The lowest level of analysis is the *physiological level*, which refers to the biological substrate (the biological implementation) of the mind's computation. This level has been the focus of a range of disciplines. Biological substrates may provide useful inputs as to what kind of computation is likely at a higher level and what a plausible architecture at a higher level should be like. Thus the utility of this level includes facilitating analysis at higher levels, using lower-level information to narrow down choices in determining the overall architecture as well as choices in describing componential processes.

Although theoretical or empirical work is often limited to within a particular level, this need not be the case: cross-level and mixed-level analysis and modeling could be enlightening, and might even be crucial (Sun, Coward, and Zenzen, 2005; Sun, 2006). These levels, as proposed above, do interact with each other (e.g., by constraining each other or grounding each other; more on this below) and may not be easily isolated and tackled alone. Moreover, their respective territories often lack clear-cut boundaries.

Normally, theories begin with the specification of units of analysis within a specific level, such as the sociological level. Theories that cross or mix levels subdivide such units and therefore may prompt deeper explorations (e.g., cognitive analysis of sociological issues). In relation to the theme of the present book, crossing and mixing levels of analysis constitutes the meta-theoretical foundation of cognitive social sciences, the integration of the cognitive and social sciences, which will be explicated in more detail below.

A key theoretical issue in this regard is the micro-macro link between society and individuals (see, e.g., Alexander, Giesen, Munch, & Smelser, 1987; Sawyer, 2003; Sun, 2001) or, more specifically, the micro-macro link between the social and the cognitive-psychological, crossing the first two levels (or more). The general questions regarding the micro-macro link are as follows: how do individuals affect collective processes and phenomena, and how do collective processes in turn affect individuals? In order to explore the questions at a sufficient depth, it is necessary to delve into individual cognition-psychology, because the cognitive-psychological processes of individuals are presumably the most important factors at the micro level. (Of course, one may choose to believe that individuals are just puppets of inescapable social forces, but in that case there is, practically speaking, no longer a question of the micro-macro link.) Hence crossing and mixing the sociological and psychological levels (as well as possibly other levels) is the prerequisite for a better understanding of social processes and phenomena (from the standpoint of the micro-macro link), as argued in the previous section.

Another key theoretical issue in this regard is downward versus upward causation across levels. This issue has been controversial (see, e.g., Wimsatt, 1997). In the present context, upward causation refers to influences from the micro to the macro (from individuals to society), and downward causation refers to influences from the macro to the micro (society to individuals). The precise nature of these two directions of causation, however, may be murky. For example, it is unclear whether downward causation from a macro state is supervenient on causation within micro states and, if so, whether it is meaningful to separate out downward causation (see, e.g., Kim, 2006; Craver & Bechtel, 2007). I will discuss this issue later in relation to the nature of cognition-psychology specifically (as opposed to pure philosophical argumentation).

The following sections look into specific cases of crossing and mixing levels with regard to analyzing sociocultural and psychological phenomena, while keeping in mind both upward and downward causation.

1.4   Grounding of Culture in Psychological Processes

We may first examine the relationship between culture and individual. In particular, the influence from culture to the cognitive-psychological, an instance of downward causation, has been emphasized in the literature in recent decades (e.g., Zerubavel, 1997; Shore, 1998; see also chapter 4 by Bradd Shore in this volume). However, in this relationship, besides downward causation, we also need to examine the importance of the cognitive-psychological to culture. Geertz (1973) claimed, "We are, in sum, incomplete or unfinished animals who finish or complete ourselves through culture." But, apparently—at least to some extent, and possibly to a very large extent—culture must function *through* the cognitive-psychological.

It seems fairly straightforward that culture is, at least in part, based on our innate cognitive-psychological capabilities and tendencies. As Richerson and Boyd (2005) argued:

Culture causes people to do many weird and wonderful things. Nonetheless, the equipment in human brains, the hormone-producing glands, and the nature of our bodies play a fundamental role in how we learn and why we prefer some [cultural] ideas to others. Culture is taught by motivated human teachers, acquired by motivated learners, and stored and manipulated in human brains. Culture is an evolving product of populations of human brains, brains that have been shaped by natural selection to learn and manage culture.

Chapter 10 (by Harvey Whitehouse) in the present book makes similar points about evolved psychology and culture.

As an example of this point, naive sociological classification reveals the relationship between cognitive capabilities and cultural categories (Sperber & Hirschfeld, 1999). Children tend to attend to surface differences in forming categories and interpret these categories in terms of these superficial features. For example, children sort people by gender before they sort them by political party affiliation. That is, children learn to pick out social groups that are visibly distinct earlier than they learn about other, less visually marked, groups (although exceptions exist). Those categories based on surface differences are more culturally salient as a result of this cognitive tendency.

Cultural tools or artifacts and their relationship to corresponding cognitive processes are another example. Lévi-Strauss (1971) suggested that the structure of oral narratives might reflect an optimal form for memory (a cognitive process) unaided by external inscriptions. Boyer (2001) suggested that counterintuitiveness of religious beliefs might serve mnemonic purposes (a cognitive process) as well.

Culture, furthermore, may even be defined in many ways by individual cognition, although on a collective scale. For example, if people of a particular culture collectively recognize that social castes are genuine social categories, they become genuine social categories. The formation and recognition of social groups relies heavily on verbal labels, communication (verbal and nonverbal), stereotyping (embedded in communication and action), and other cognitive processes. Furthermore, the supplement, or even displacement, of "natural" features of social group membership (or category membership in general) by cultural features makes possible the construction of novel social groupings (or novel categories in general). That is, the cognitive ability to construct abstract or otherwise less "natural" features leads to cognitive changes in individuals and to cultural changes on a collective scale, which may in turn effect cognitive changes in additional individuals. This process goes from the individual mind to the sociocultural, and then back to the individual mind.

The stability and transmission of culture are also critically dependent on cognitive-psychological processes and their characteristics. Sperber (1996) argued for an "epidemiological" approach to culture. According to this view, culture is the distribution of ideas, concepts, representations, and artifacts among a population. Specifically, social interactions may distribute similar mental representations and public productions (behaviors and artifacts) throughout a population; mental representations and public productions stabilized through such processes constitute the cultural.

However, cultural contents transmitted through a chain of individuals (through their cognitive-psychological processes) undergo changes, distortion, and decay. For one thing, imitation is not very reliable (e.g., one cannot easily imitate internal mental states, such as implicit skills, implicit memory, and so on; Sun, 2002). Research on memory and communication shows that both memory and communication often involve reconstruction rather than mere copying. So, what makes stable cultural contents possible? There are two types of explanations (see Sperber & Hirschfeld, 2004; see also chapter 9 by Ilkka Pyysiäinen in this volume): (1) the stabilizing role of psychological learning biases in transmission (favoring, for instance, prestige or conformity), (2) the stabilizing role of the predisposition to acquire knowledge structured in domain-specific ways (within domain-specific modules that embody biases). Both approaches recognize the importance of the psychological in explaining the macro-level phenomenon of culture.

At this point, it seems that we are focusing on telling only half the story (an often neglected half). Just as social processes must be understood in relation to cognitive-psychological processes (although they may not be fully determined by such processes), it seems that cognitive-psychological processes need to be understood in relation to sociocultural processes (e.g., Nisbett & Norenzayan, 2002). It goes both ways, it appears. Besides, we do not always know exactly what constitutes cognitive-psychological universals (as opposed to cultural specifics) in every case. For example, research has shown the importance of specific cultural models for individual cognitive and motivational processes (see, e.g., D'Andrade & Strauss, 1992 and Shore, 1998; see also chapter 4 by Bradd Shore in this volume).

In this regard, currently scientific research points to two fundamental ideas about the human mind: (1) the mind is the product of evolution (thus cognitive-psychological universals are likely; Carruthers, Laurence, & Stich, 2005); and (2) the mind is shaped by culture (even specific forms of culture; e.g., Shore, 1998). These two views of the mind are not mutually incompatible (Richerson & Boyd, 2005; Schaller, Norenzayan, Heine, Yamagishi et al., 2009). Furthermore, they are not incompatible with the idea of grounding social phenomena in cognition psychology in the quest to better understand social processes and phenomena. This is because cognitive-psychological universals resulting from evolution likely exist, and they likely affect culture and social processes (as variously argued by, e.g., Sun, 2001; Sperber & Hirschfeld, 2004; Carruthers et al., 2005; Semin & Echterhoff, 2010), so understanding social phenomena through cognition-psychology is useful.[3] It is likely that the mind may be "fine-tuned" by different cultures, but cognitive-psychological universals may not be completely displaced or eliminated (at least in most cases; Nisbett & Norenzayan, 2002; Carruthers et al., 2005; Semin & Echterhoff, 2010). Therefore, grounding sociocultural phenomena in cognition-psychology is theoretically possible despite the existence of cultural differences.

Furthermore, it is possible that culture, the collective phenomenon, is enacted at the individual level, at least in part, through shaping and fine-tuning innate cognitive-psychological capacities (cognitive-psychological universals) in individuals (Sun, 2001; Sperber & Hirschfeld, 2004), so that cultural influence on cognition-psychology may in fact be an instance of social phenomena enabled through psychology. The enactment of culture through shaping and fine-tuning innate cognitive-psychological mechanisms and processes (cognitive-psychological universals) of individuals may be understood in this way: culture is the distribution of ideas, representations, and so on in a population (e.g., according to Sperber, 1996), and culture-specific cognitive, motivational, and other representations are made possible on the basis of tapping into and fine-tuning innate psychological mechanisms and processes (Sun, 2001; Carruthers et al., 2005; Sun, 2009; Semin & Echterhoff, 2010).

Moreover, it is understood that cognitive-psychological universals may have evolved in specific historical sociocultural contexts (e.g., in hunter-gatherer societies) and hence take on characteristics of these contexts, which is another sense in which culture might be grounded—that is, in this case, engraved—in individual cognition-psychology.

At an individual level, mechanistically, it has been hypothesized that culture may manifest in individual minds as *schemas* (Fiske & Linville, 1980; DiMaggio, 1997; etc.). However, schemas can exist and be accessed (and thus affect behavior) only through innate cognitive capacities such as memory, reasoning, and decision making. This point seems to have been well established in the cognitive sciences, especially in computational psychology. Hence the importance of cognitive-psychological universals.

However, culture need not be understood as simply the vague notion of schemas, whatever that notion means in practice. It should be, and can be, more specifically pinned down in a mechanistic, process-based way, as computational cognitive modeling (computational psychology) would provide.

In the cognitive sciences, schemas are hypothesized slot-filler structures employed in the interpretation of input, guiding of action, and storage of knowledge in memory. Connectionist models, however, view them as resulting from excitatory and inhibitory connections among many simple processing units. Connectionist models posit that memory consists of a connected network of units, each of which can have a certain activation. Knowledge is distributed over the units and the connections (with weights) between them.

Furthermore, the more recent development of *cognitive architectures* leads to a more comprehensive, yet more detailed, understanding of the mind. The notion of (computational) cognitive architecture denotes generic, comprehensive computational models of the mind (Newell, 1990; Sun, 2002). These models capture mechanistic and process details of generic psychological functions, thereby producing realistic computational simulations in a qualitative or quantitative way (e.g., Sun, Slusarz, & Terry, 2005; Helie & Sun, 2010). For example, CLARION is such a cognitive architecture, one that notably centers on the distinction between implicit and explicit processes, as well as the interaction of multiple subsystems (modules; Sun, 2002).

With the cognitive architectures in mind, it is possible to take a broader view and make more general points about the mind, while at the same time being more specific and detailed. It may be claimed, in accordance with the ideas above, that culture (at the individual level) is the complex and specific patterns of interaction of an individual with its social and physical environments. In particular, in accordance with the CLARION cognitive architecture, it involves the implicit psychological processes underlying such interactions, as well as explicit beliefs and knowledge that may result from and impinge on such interactions (Sun, 2002).

Implicit (unconscious) processes that govern an agent's interaction with the world may, in some sense, be described as *schemas*. They often consist of relatively fixed patterns of interaction—specific actions for specific situations, similar situations leading to similar actions, and so on (in a statistical sense; that is, with certain variability and flexibility). However, such interactions are often rather direct—unmediated by explicit processes and unreflective (Heidegger, 1927/1962; Dreyfus, 1992; Sun, 2002).

Beyond immediate, unreflective direct interactions, explicit thinking (explicit reasoning, explicit reflection, and so on) in the human mind also affects an individual's interaction with the world. Explicit conceptual representations and processes are abundant in human cognition (Carey, 2009) and may, directly or indirectly, guide an individual's interaction with the world. They are actually more akin to the traditional notion of schema discussed earlier. The interplay and balance of implicit and explicit "schemas" (or, more generally, psychological processes) are yet to be fully understood (see, e.g., Sun, Slusarz, & Terry, 2005 and Reber, 1989).

When discussing culture and schemas, it is also relevant to take into consideration the role of essential, intrinsic motivation in human beings (a crucial psychological aspect; Sun, 2009; Dai & Sun, 2011; Murray, 1938). At least in part, culture may be determined by essential human motivation.

For example, a culture may suppress a certain aspect of human motivation and highlight some others, but it nevertheless has to be in accord with essential human motivation as a whole in some way. Therefore, in a sense, culture may be (in part) viewed as a manifestation of essential human motivation (e.g., as discussed in Sun, 2009). Different cultures, then, may represent different manifestations of essential human motivation. I would argue that, to fully understand culture, it is important to understand essential human motives and take them into consideration when theorizing about culture. It is not enough merely to look at cognition in the narrow sense (learning, memory, concepts, skills, and so on); cognition in a broader sense—all psychological processes and mechanisms—must be considered.

Armed with these conceptual tools, we may address issues and problems related to culture in terms of the psychological. For example, issues of culture fragmentation, schema aggregation, and other related phenomena—identified by DiMaggio (1997) as some of the most puzzling conceptual problems—may be examined with the perspective above, as discussed in depth in Sun (in press). In general, cultural studies seem to have taken a cognitive turn in exploring a wide range of topics.

To reiterate the points discussed thus far regarding the psychological underpinning of culture: first, as argued thus far, cognition-psychology is important to culture. Second, culture may consist of more than schemas psychologically. The notion of schema, although useful as a first approximation, seems inadequate for capturing the full complexity of culture. Third, both implicit and explicit psychological processes need to be taken into consideration. Fourth, the relation of culture to human motivation also needs to be taken into consideration, which moves us further away from the simple notion of schema as well as cognition in the narrow sense.

Finally, although psychological aspects of culture are only one part of the sociology of culture, it is impossible to ignore them and still have a good understanding of how culture works. Any understanding of the impact of culture on daily practice has to be based on an understanding of the psychology of culture (Sun, 2001). Understanding of culture needs to be connected with the understanding of psychology, because culture is grounded in actual human psychology. Work in cognitive psychology, social psychology, computational psychology, and so on provides useful tools for pursuing such understanding.

1.5 Collective Ramification of Psychological Particulars

Upward causation from individuals to society may be revealed, in part, through demonstrating how cognitive-psychological factors and parameters affect macro-level social processes and phenomena. This can be done through computational simulation, as well as through empirical work. Using such explorations, it is possible to discern the importance of the psychological to the social.

First, the significance of cognitive factors in the formation, adaptation, and maintenance of social institutions should not be underestimated (although some social phenomena may be cognition-independent, as some would plausibly argue). Some work has been done on this often-neglected issue.

One example along this line is an application of the CLARION cognitive architecture mentioned above to simulating survival strategies of tribal societies under various environmental conditions (Sun & Naveh, 2007). These simulations dealt with a world with randomly distributed food items and agents. There were harsh, medium, and benign environmental conditions, distinguished by the agent-to-food ratios. Agents had a limited life span that varied depending on energy. Agents looked for and consumed food in an effort to prolong life.

A tribe in which each agent relied on only its own resources was said to have adopted an individual survival strategy. A tribe in which resources were transferred from one individual to another adopted a social survival strategy. For example, the "central store" was a mechanism to which all agents in a tribe transferred part of their resources. The resources collected by the central store could be redistributed to members of the tribe (according to some formula).

The agents in this simulation were cognitively realistic. The CLARION cognitive architecture from which these agents were constructed captured a variety of cognitive processes in a psychologically realistic way (Sun, 2002). Therefore, simulating social survival strategies using CLARION could shed more light on the role of cognition and its interaction with social institutions and processes. The major objective, in fact, was to investigate that interaction (that is, the micro-macro link) through varying cognitive parameters of the agents.

Through extensive simulation, relationship was identified between various cognitive parameters and social processes, indicating, for example, that the social institutions and norms adopted (such as survival strategies) might have something to do with the cognitive abilities, tendencies, and characteristics of the agents involved. This relationship, which may be termed *social-cognitive dependency*, could have significant theoretical and empirical implications (amid other types of dependencies in different directions; more on this later). For example, some forms of social systems and social institutions might be suitable for certain cognitive characteristics, but unsuitable for others. Thus, one social system or institution might not be universally better or worse than another. Rather, a host of other factors—cognitive factors, in particular—might affect which social system or institution is best in each case. Sociocultural variability exists as the result of active human agency and human psychology, and not merely top-down inscription onto the minds of individuals. An earlier article, Sun (2001), contains a more substantial discussion of the close relationship between cognitive and social processes, and advocates the exploration of cognitive-psychological principles of sociocultural processes.

This work has since been extended into investigating motivational factors and their relevance to social institutions. Similar demonstrations via simulation of the role of psychological factors in social phenomena and processes have been described in domains ranging from organizational decision making (e.g., Sun & Naveh, 2004) to academic publishing (e.g., Naveh & Sun, 2006).

Aside from simulation studies, there have been other indications of upward causation in which psychology plays a crucial role. A case in point is the institution of religion, which has been shown to be shaped by cognitive-psychological factors.

In this regard, Atran and Norenzayan (2005) argued that "religion is not an evolutionary adaptation per se, but a recurring cultural byproduct of the complex evolutionary landscape that sets cognitive, emotional, and material conditions for ordinary human interactions." In relation to the supernatural aspects of religious beliefs specifically, Boyer (2001) argued that "minimally" counterintuitive concepts that violated a small number of intuitive expectations, such as a talking tree or an invisible man, were better remembered than either intuitive concepts or maximally counterintuitive concepts that violated a larger number of intuitive expectations. A number of empirical studies have found some support for this cognitively based explanation. It has also been hypothesized that better recall for "minimally" counterintuitive concepts was a consequence of evolutionary processes that resulted in a cognitive architecture in which such concepts were better remembered. Hence, such concepts likely won the competition among ideas so that their representations became widespread and cultural (cf. the epidemiology of beliefs of Sperber, 1996, discussed earlier). See also chapters 8 and 10 by Scott Atran and Harvey Whitehouse, respectively, in this volume.

Beyond memory advantages (which are cognitive in the narrow sense of the term), there are also other psychological (e.g., perceptual, motivational, or emotional) factors underlying religion. For example, the supernatural agent concept common to almost all religions involves the triggering of an innate agency detector (Atran & Norenzayan, 2005), the natural domain of which encompasses animate objects relevant to individual survival, such as predators, protectors, and prey, but may also extend to other things. Furthermore, in relation to essential human motivation (Sun, 2009), it has been argued that religion responds to some basic human needs and motives, such as the need for safety, the need for a superior authority, and so on, although some researchers dispute such arguments (Sperber & Hirschfeld, 2004). In particular, core religious beliefs often "minimally" violate ordinary notions about what the world is like, enabling individuals to imagine supernatural worlds (that are "minimally" impossible) that solve the most serious existential problems, including, in particular, the inevitability of death (Atran & Norenzayan, 2005). Here again, we see how large-scale social phenomena, including social institutions such as religion, may be explained, at least in part, by the cognitive-psychological particulars of individuals.

Culture, as discussed at length earlier, also heavily involves upward causation from cognitive-psychological characteristics to macro-level social phenomena, although the conventional wisdom has focused more on causation in the other direction. Because this topic has been discussed earlier, these points concerning how the psychological (at the micro level) affects culture (at the macro level) will not be repeated here.

It is also worth pointing out that some cultural input that are contrary to innate cognitive-psychological constraints and preferences may be rejected or transformed. An example is creole. When different linguistic communities are brought together, the linguistic consequence is the emergence of a pidgin (a cobbled language of which no individual is a native speaker). Sometimes, children are raised with a pidgin language. When pidgin utterances are input to the language-acquisition device in children, a creole, a natural and fully elaborated language, may be the output. Children transform an incomplete cultural form into a fully developed one. This can be attributed to the fact that they are equipped with an innate (evolved) cognitive device for acquiring language, which has its own preferences and constraints (Bickerton, 1990). This phenomenon, again, shows the importance of cognitive-psychological particulars for macro-level phenomena (such as language).

Despite the significance of individual psychological particulars, notably, people often overlook the importance of cognition-psychology when theorizing about social phenomena. For example, current work in social simulation tends to ignore the role of cognitive-psychological processes and to adopt simplified models of agents instead. The social sciences (and "social engineering" in practice) ignore cognition-psychology at their own peril. In history, there have been many examples of failure of social theories, social institutions, or social practices, due to the failure to take into account important factors of human psychology. For example, some socioeconomic theories failed because they failed to take into account human psychology and human nature (especially motivation, emotion, and so on). For another example, some religious doctrines have rarely been strictly obeyed in the modern world. Although it is almost a necessity that certain counterintuitive beliefs as well as other anomalies are instituted in religion, many practices that go against essential human nature have not been strictly followed, at least since modernity (Sun, 2006).

1.6   Two-Way Interaction

At this point, it would be helpful to examine some scenarios of simultaneous occurrence of both upward and downward causation and their interaction. It helps to examine specific ways in which downward or upward causation actually takes place, instead of abstract descriptions. Again, I want to emphasize the importance of the cognitive-psychological in grounding the social in this discussion. The existence and the importance of simultaneous micro-to-macro and macro-to-micro influences have been identified, described, and argued for, including computational modeling and simulation of such influences.

For example, in Sun (2001), "power asymmetry" in the two-way interaction was discussed: "Although the relationship between an individual agent and society is complex and goes in both ways, it has been recognized that, fundamentally, it is the influence from the society on individuals that matters more. ... Individuals find themselves already 'current in the community' (Heidegger, 1927/1962). Their cognitive processes and thus their behaviors are shaped by their social environments" (see also Bourdieu & Wacquant, 1992). Macro-to-micro downward causation takes place in many ways. In this process, individual psychological processes serve as the basis and the constraints for such shaping to take place, thereby grounding social practices, norms, culture, institutions, and a variety of other aspects in individual cognition-psychology.

Vygotskian "internalization," in particular, is important in this regard. Vygotsky (1986) emphasized social interaction as a major determinant in the development of thinking in individuals. One aspect of internalization lies in the genesis of verbal thoughts. According to Vygotsky, speech develops before the development of internal verbal thinking. It starts at the single-word level, which serves the function of labeling. Labeling itself is sociocultural, because it is based on the established convention of a sociocultural and linguistic community. However, when more complex speech develops, it directly serves a social function (e.g., to get someone to do something). When speech loses its social functions (e.g., when nobody responds to a child's request), it can be turned inward and thus become a request or command to oneself (in Vygotsky's term, an egocentric speech). Speech can thus be transformed from serving an interpersonal function to an intrapersonal function. Egocentric speech can be further turned inward and become internal verbal thoughts. Internal thinking, accomplished without overt utterances or actions, relies on internalized signs/symbols from sociocultural contexts. Internalized sociocultural signs/ symbols enable individuals to develop rich representations, including those formed socioculturally and historically (Sun, 2001). However, internalized signs/symbols are not innocuous: they carry with them particular sociocultural perspectives and biases. Through internalization, the thinking of an individual may be thereby mediated by externally given signs/ symbols, along with their associated perspectives and biases. Such internalization has implications for grounding the social in the psychological, because it is one specific way through which downward causation is made possible.

Based on this notion of internalization, detailed computational models may be developed. CLARION may be used as an example here. Internalization can be accomplished in CLARION through the "top-down" assimilation process (described in Sun, 2002), which matches well the phenomenological characterization of internalization. The direct acceptance of external symbols, rules, and so on into the explicit processes of CLARION captures the initial stage of internalization. The assimilation into the implicit processes, however, captures a deeper process by which external symbolic structures are meshed with implicit routines, reflexes, and behavioral propensities so that they can effectively affect an individual's comportment in the world. On the other hand, implicit learning also captures the internalization of sociocultural aspects through interacting with those aspects (Sun, 2002). Through internalization, according to CLARION, the behavior of an individual and the psychological processes underlying the behavior are mediated by the sociocultural world, including signs/symbols as well as associated perspectives and biases.

The other side of the coin is the process that gives rise to the sociocultural environment (from the interaction of individuals): although the influence of society on individuals is overwhelming, the influence in the other direction is also important. As emphasized by phenomenological sociologists, social reality is an "ongoing accomplishment" actively constructed through organized practices of everyday life by individuals. Social reality is, in some ways, an aggregate product of the actions, routines, skills, knowledge, decisions, and thoughts of individuals, each of whom has a direct, meaningful interaction with his or her world (Sun, 2001). Evidently, in this process, the cognitive-psychological inner working of individuals matters, because it is such inner working that leads to thoughts and actions by individuals. So, as long as we acknowledge the existence of the influence in an upward direction—that is, as long as we reject the notion that individuals' actions are completely, inescapably determined by external social forces—it is almost inevitable that we acknowledge the significance of individual cognition-psychology in affecting macro-level social processes, structures, and institutions (as demonstrated in the previous section by simulation examples and other illustrations). Even in internalization discussed earlier, individuals naturally gravitate toward those perspectives and biases provided by culture that strike a chord with their innate psychological propensities and prior learning and experiences. Individuals need not be consciously aware of this process; human instinct is often more powerful than conscious reasoning (see chapter 7 by Kristen Monroe).

Note that the micro-to-macro (individual-to-society) influence has been discussed, for example, by Schelling (1971), Axelrod (1984), and others. What has not been emphasized sufficiently is the role of individual cognitive-psychological processes in this influence. However, Sun (2001, 2006) emphasized this role, which went beyond the usual treatment of upward causation.

Going back to the two-way (micro-to-macro and macro-to-micro) interaction, let us revisit the tribal society simulation discussed earlier. In that simulation, on the one hand, there is the *social-cognitive dependency* alluded to earlier, which indicates that the social institutions and norms adopted might have something to do with the cognitive abilities and characteristics of the agents involved. However, on the other hand, some cognitive attributes may have been selected through evolution to work with certain social systems and cultural environments, which may be termed the *cognitive-social dependency* (Sun & Naveh, 2007). There are, of course, other types of dependencies: cultural practice, social institution, human psychology and behavior, and physical environment influence each other; together, these dependencies form a complex dynamic system of interwoven interactions. In such a dynamic system, it is important to understand not only direct effects of dependencies, but also indirect effects, which are not obviously related to their causes but are often crucial for discerning the functional structures of the system. A simulation of both upward and downward causation was described by Bravo (2009) in which micro-level processes influence macro-level institutions, which in turn influence micro-level processes.

Macro-level social structures often have demonstrable causal effects on individuals, even when those individuals are not consciously (explicitly) aware of them (Sun, 2001). However, individuals might, at least sometimes, come to explicitly recognize macro-level phenomena and explicitly alter their behaviors to take account of them. This may be termed *cognitive emergence* or *implicit-to-explicit explicitation* (Sun, 2002). This explicit recognition requires higher-level cognitive abilities with some understanding of wider social contexts. The importance of the cognitive-psychological to the social is evident in this process.

In CLARION, implicit perception, cognition, and action can be carried out, responding to environmental regularities and internal/external reward structures, taking (implicit) account of various types of macro-level structures and institutions. Explicit mental recognition of the macro-level structures may emerge through a variety of cognitive means, including "bottom-up learning" (explicitation) through turning implicit representations into explicit representations (Sun, 2001, 2002).

An illustrative case of both upward and downward causation (often used in the discussion of this topic) is as follows: individuals interact locally and move in a given spatial environment; they construct buildings that stand or fall depending on their usefulness to the individuals involved (upward causation). The space through which individuals move is defined by where buildings are. This is a macro-level structure that influences individual movements and the interactions among individuals (downward causation). The movements and the interactions of the individuals in turn influence the survival of the buildings, and therefore the structure of the space (upward causation).

In this case, the downward causal forces appear given and non-negotiable. That is, once a building is constructed, it seems to have an existence of its own, independent of the people who built it or are using it. It appears to exert causal influence on those who are using it, whether or not they are consciously (explicitly) aware of the exact nature of the structure or its impact on them. The theoretical question is whether there are non-physical social structures that operate like such a building; that is, whether there can be social structures as autonomous from individuals as a building can be once it has been built.

Some believe that there are. However, others (such as phenomenological sociologists or interpretive sociologists) disagree. Some theorists further argue that buildings are not completely autonomous either—what matters is how the buildings are perceived or interpreted (by architects, engineers, owners, occupants, caretakers, passersby, and so on). This argument not only leads to an individualist perspective (Sawyer, 2003; Sun, 2001), but also to a cognitive-psychological perspective, because individuals' psychological processes determine their perception, interpretation, and consequently action. Regardless of whether they are tangible or not (i.e., physical or non-physical), macro-level structures constrain individual behaviors only if individuals exist—macro-level structures are not independent of individuals in that sense. Furthermore, they matter only if individuals perceive, interpret, recognize, memorize, and react to them. So, *ultimately*, they exist in and through individuals and their cognition-psychology.

This point applies not only to macro-level structures such as political systems, norms, and other non-physical structures, but also to physical structures such as buildings, roads, bridges, and so on, because otherwise buildings and roads will not be buildings and roads, but piles of stones, bricks, concrete, steel, and so on. Any macro-level structure matters only if it exists in a perceivable and relevant way for individuals, which can then be taken into consideration in the actions by individuals. Macro-level structures matter only if they affect the actions of individuals. The upshot is that the cognitive-psychological perspective—that is, the perspective from the viewpoint of individuals—is important in this matter, and moreover, that the issue cannot be fully understood without the cognitive-psychological perspective.

Furthermore, Craver and Bechtel (2007) argue that all causation actually takes place within levels, and therefore there is really no cross-level causation. Across levels, there are constitutive relationships. So, what is regarded as "downward causation" (e.g., from the social to the individual) does not involve top-down causes, but only within-level causation plus constitutive relationships across levels (see also chapter 9 by Ilkka Pyysiäinen). This view is consistent with the multilevel analysis framework outlined earlier, and argues for the importance of psychological understanding of the social.

Along this line, some social science researchers have focused on individuals' internal psychological representations of macro-level social aspects. For example, Fiske & Linville (1980) and DiMaggio (1997) claimed that the notion of schema was especially relevant to the individual internal representation of culture (as discussed earlier). The idea that social structures exist simultaneously through individual internal mental representations and in concrete social relations was also central to Nadel's (1957) theory.

Regardless of whether there are "autonomous" macro-level structures, representations of macro-level structures (including their processes and mechanisms) need to be taken into consideration in theorizing and in agent-based social simulation (Sun, 2006). This is because such mechanisms, processes, and structures are pervasive in society (at least in the modern world), and their existence is readily felt from the perspective of individuals (such as a physical market, a currency note, a law, a bus route, a highway, and so on).

1.7 An Overview

The remainder of this book has been designed to capture issues in a wide selection of areas and fields in the social sciences. Chapters are divided into four major parts, focusing on culture, religion, politics, and economics, respectively, in addition to a final part that examines unifying perspectives in general.

Part II of this volume is concerned with culture and how it is related to cognition-psychology. The three chapters cover a range of issues, from psychological explanations of cultural differences to cognitive effects of cultural models. They cover the impact of cognition on culture, as well as the impact of culture on cognition. As discussed earlier, ultimately, culture might be instantiated through cognitive-psychological processes.

Chapter 2, by Paul Thagard, addresses methodological issues, including the relation between the cognitive and the social sciences. It rejects the view that the study of humanity must be a hermeneutic enterprise eschewing the concepts and methods of science such as psychology. But it also rejects the reductionist view that social phenomena should be directly derived from cognitive phenomena. The methods pursued in this chapter aim at providing explanations of social phenomena by drawing on models of human cognition.

Chapter 3, by Nobert Ross, asks the question of what the cognitive sciences can do for anthropology in studying different cultures. Multiple examples illustrate how methods and theories from the cognitive sciences can enrich anthropology in significant ways. It argues that the cognitive sciences and anthropology complement one another, and that together they can form new approaches for addressing important questions concerning the human mind and society.

Chapter 4, by Bradd Shore, focuses on the issue of perspective in cultural models, specifically the distinction between egocentric and allocentric cultural models. It takes this distinction, which has been studied in terms of spatial cognition, into less obvious areas of mental representation, and tries to take the discussion on culture and cognition a step further.

Part III is concerned with the cognitive-psychological basis of politics. The three chapters in this part together explore how politics may be better understood from a cognitive-psychological perspective. These chapters call for further work along this line to advance the understanding of the cognitive-psychological basis of political science.

Chapter 5, by Stanley Feldman, Leonie Huddy, and Erin Cassese, touches on an important issue. It argues for a fine-grained understanding of emotion in understanding political cognition. Different emotions may lead to different styles of cognition, which affect individual political opinions and the resulting political dynamics. To fully understand domestic political processes, international relations, and so on, a better understanding of the roles of emotion and motivation is needed (Lebow, 2008; Dai & Sun, 2011).

Chapter 6, by Peter Bull and Ofer Feldman, shows the relevance of a number of cognitive theories to understanding political discourse. The chapter does so in multiple cultural contexts. Contemporary politics is "mediated" politics: the communication skills of politicians play a crucial role. The cognitive theories discussed in this chapter lead to better understanding of political communication and political behavior, and are useful for framing future research.

Chapter 7, by Kristen Monroe, argues for a moral psychology that is appropriately constrained by the architecture of the human mind—its development, emotion, social psychology, and the limits of human capacity for rational deliberation. The chapter shows how details of human psychology help determine moral choices. This point has significant theoretical and practical implications for political science and beyond.

Part IV of this book includes three chapters on religion and the relationship between religion and psychology. These chapters show how religion may be understood (in part) through human cognition-psychology. All three chapters discuss the complex interaction between cognitive-psychological factors on the one hand and sociocultural factors on the other in the context of evolution.

Chapter 8, by Scott Atran, explores an array of factors leading to religion. It argues that religion, an interwoven complex of rituals, beliefs, and norms, arises from a combination of the mnemonic power of counterintuitive representations, the evolved willingness to adopt culturally acquired beliefs (e.g., from commitment-inducing devotions and rituals), and the selective effect of competition among societies and institutions. None of these (many of which are cognitive-psychological) evolved for religion per se, but they together, possibly along with other factors, give rise to the institution of religion.

Chapter 9, by Ilkka Pyysiäinen, specifically examines competing explanations of the persistence of religious beliefs. This persistence may be in part explained by various biases in cultural transmission; these biases include tendencies to do what the majority does, to imitate prestigious individuals, and to punish non-cooperators. But biases cannot operate in a cognitive vacuum; to fully account for them, cognitive considerations are required. This discussion points to multilevel explanations that do not necessarily entail reductionism.

Chapter 10, by Harvey Whitehouse, explores how ritualized behavior may be rooted in psychology, linked to the natural human propensity to imitate trusted others. The role of ritual in the formation and regulation of human societies is discussed (e.g., rituals may benefit group building). Anthropological research, including case studies, field research, and large-scale ethnographic surveys, has been conducted. These studies lead to the development of theoretical models and agent-based computational simulations.

Part V explores the cognitive-psychological basis of economics, including the debates and controversies that it engenders. These chapters show that, in general, it is highly beneficial to study cognitive-psychological factors when investigating economic issues and problems.

Chapter 11, by Don Ross, begins with a broad sketch of views toward psychology in the history of economics. It reviews the current state of theoretical modeling of the economic agent; in particular, it addresses a specific phenomenon—intertemporal discounting of utility. It points out that the cognitive sciences have not yet become a significant supplier of variables or parameters to economic models. Future interdisciplinary collaboration likely depends on better integration of cognitive models and multi-agent models of social interaction.

Chapter 12, by Joseph Kable, argues for the role of neuroeconomic research within the context of multiple levels of analysis (e.g., as discussed earlier in section 3). The field of neuroeconomics provides useful lessons regarding the promises and pitfalls of drawing links across the cognitive and the social sciences. This chapter highlights some of those lessons while providing an overview of neuroeconomics. It argues that neuroeconomics provides examples of how the social sciences can be grounded in the cognitive sciences—not just in psychology, but also in cognitive neuroscience.

Chapter 13, by John McArdle and Robert Willis, discusses frameworks used by psychologists and economists for studying the development of ability, knowledge, and skills over the human life cycle. Economists were largely unaware of the theory of fluid and crystallized intelligence in psychology, while psychologists were equally ignorant of the theory of human capital in economics. The chapter shows the parallel between the two theories and ways in which they may be integrated for studying practical issues.

Part VI contains two chapters on broad issues across fields: culture, religion, politics, and economics. How can the social sciences be grounded in the cognitive sciences? How can the social and cognitive sciences be more unified? These are fundamental metatheoretical questions.

Chapter 14, by Mathew McCubbins and Mark Turner, discusses what are believed to be important ideas offered by the cognitive sciences to the social sciences. In the past, the cognitive sciences have undermined confidence in some apparently unobjectionable assumptions held by many social scientists. This chapter instead offers some positive suggestions for the social sciences from the cognitive sciences.

Chapter 15, by Herbert Gintis, identifies a number of components for a unified social science: gene-culture coevolution, game theory, the theory of norms, the rational actor model, and complexity theory. Evidently, cognition-psychology plays an important role here. But is it emphasized sufficiently and properly? An even more important question is: What else is needed as part of the foundation for the future social sciences? These questions are yet to be answered (cf. Camerer, 2003; Sun, 2001).

The contributors of these chapters were asked to provide (a) an overview of a field, (b) an in-depth discussion of a research program, and (c) a broader discussion addressing a set of issues concerning the cognitive social sciences. It is useful to achieve a proper balance between breadth and depth.

The contributors were asked to address, among others, the following questions in their chapters:

• What are the relevant major open issues in your (social sciences) field? How does cognitive-psychological understanding shed new light on these open issues?

• What are the future potentials and possibilities in shedding more light on your field through cognitive-psychological investigation (or "grounding")?

Then, the contributors were asked to address the following broader questions:

• What general lessons have been learned in investigating cognitive-psychological factors in your field? What are the general benefits and pitfalls of such an investigation?
• Can such an investigation be generalized to other areas/fields within the social sciences? What are the possible ways of generalizing the approach to other areas/fields?

With these questions addressed to various extents in this book, I hope that the final product reasonably clearly, evenhandedly, and convincingly demonstrates the pros and cons, the general applicability, and the future prospect of the cognitive social sciences.

1.8 Final Remarks

To ground or not to ground the social sciences in the cognitive sciences: that is the fundamental question.

There have always been at least two schools of thought on this question: (1) cognition-psychology is an important factor in, or even the holy grail and the final frontier of, the social sciences; or, (2) cognition-psychology is largely irrelevant, in terms of being a major deciding factor, in social matters, compared to the strength of social forces. I have argued in favor of the first view all along. To add more support to this view, let me cite briefly some well-known authors and schools of thought.

While an adequate account of historical precedents of the cognitive social sciences would take far too long, it is worth mentioning some particularly relevant ones. Max Weber, for example, pointed out that, unlike the physical sciences, the social sciences need to gain an "empathetic understanding" of the "inner states" of social actors, and thus gain an understanding at both the level of causation and the level of "meaning" (that is, cognition/motivation of social actors). Alfred Schutz, for another example, attempted to understand the construction of social reality from the point of view of the individual in terms of meaningful actions, motivations, and a variety of different kinds of social relationships.

In a related fashion, psychoanalytic anthropologists have conducted their fieldwork and then used psychoanalytic techniques to analyze the generated materials, thereby grounding the social in the psychological. More recently, cognitive anthropology has drawn on insights from the contemporary cognitive sciences in its theories and analysis (such as applying the notion of schema, as discussed earlier).

Paul DiMaggio put it this way:

Cognitive aspects of culture are only one. ... part of the sociology of culture's domain. But it is a part that we cannot avoid if we are interested in how culture enters into people's lives, for any explanation of culture's impact on practice rests on assumptions about the role of culture in cognition. I have argued that we are better off if we make such models explicit than if we smuggle them in through the back door. (DiMaggio, 1997)

My epiphany from that cold autumn morning in Chicago seems to have been (at least partially) validated.

Finally, I shall add that the theme of the present book happens to be the "grounding" of the social in the cognitive-psychological, or to put it in another way, the importance of the cognitive-psychological to the social. However, this emphasis does not exclude influence in the other direction, nor other ways of unifying or structuring different disciplines, whether as different levels or not. Every discipline has its place (more or less). In particular, "once developed, a cognitive theory ... will not displace or dismiss social science, any more than the theory of evolution supplanted the local study of zoological phenomena in all their particularity" (Turner, 2001, p.12). It is impossible to emphasize everything under the sun in one volume; we have to be selective, emphasizing one aspect out of many. In this case, I have emphasized what I consider to be an important, indispensable aspect of unifying the cognitive and social disciplines.

Acknowledgments

This work has been supported in part by the ONR grant N00014–08–1-0068; thanks are due to Paul Bello. An accompanying workshop was supported by AFOSR; thanks are due to Jun Zhang. Herbert Gintis and Ilkka Pyysiäinen provided useful comments on the draft version of this chapter.

Notes

1. See Sun (2001) for a more detailed argument for the relevance of sociocultural processes to cognition-psychology and vice versa. More on this in the sections below.

2. The importance of this level has been argued for, for example, in Sun (2002) and Sun, Coward, & Zenzen (2005).

3. "Of course genetic elements of our evolved psychology shape culture—how could it be otherwise?" (Richerson & Boyd, 2005).

References

Alexander, J., Giesen, B., Munch, R., & Smelser, N. (Eds.). (1987). *The micro-macro link*. Berkeley, CA: University of California Press.

Atran, S., & Norenzayan, A. (2005). Religion's evolutionary landscape: Counterintuition, commitment, compassion, communion. *Behavioral and Brain Sciences, 27,* 713–730.

Axelrod R. (1984). *The evolution of cooperation*. New York: Basic Books.

Bickerton, D. (1990). *Language and species*. Chicago: University of Chicago Press.

Bourdieu, P., & Wacquant, L. (1992). *An invitation to reflexive sociology*. Chicago: University of Chicago Press.

Boyer, P. (2001). *Religion explained: The evolutionary origins of religious thought*. New York: Basic Books.

Bravo, G. (2009, September). The evolution of institutions for commons management: An agent-based model. Presentation at the European Social Simulation Association conference, University of Surrey, Guildford, England.

Camerer, C. Loewenstein, G. & Rabin, M. (Eds.) (2003). *Advances in behavioral economics*. Princeton, NJ: Princeton University Press, Russell Sage Foundation.

Carey, S. (2009). *The origin of concepts*. New York: Oxford University Press.

Carruthers, P. Laurence, S., & Stich, S. (2005). *The innate mind: Vol. 2. Culture and cognition*. Oxford: Oxford University Press.

Cerulo, K. (2002). *Culture in mind: Toward a sociology of culture and cognition*. Oxford: Routledge.

Craver, C., & Bechtel, W. (2007). Top-down causation without top-down causes. *Biology and Philosophy, 22,* 547–563.

Dai, D. Y., & Sun, R. (2011). Where is the unity of attention, representation, and performance? In D. Y. Dai (Ed.), *Attention, representation, and human performance: Integration of cognition, emotion, and motivation*. London: Psychology Press.

D'Andrade, R. G., & Strauss, C. (Eds.). (1992). *Human motives and cultural models*. Cambridge: Cambridge University Press.

DiMaggio, P. (1997). Culture and cognition. *Annual Review of Sociology, 23,* 263–288.

Dreyfus, H. (1992). *Being-in-the-world.* Cambridge, MA: MIT Press.

Fiske, S. T., & Linville, P. W. (1980). What does the schema concept buy us? *Personality and Social Psychology Bulletin, 6,* 543–557.

Geertz, C. (1973). *The interpretation of cultures.* New York: Basic Books.

Heidegger, M. (1962). *Being and time.* New York: Harper and Row. (Original work published 1927)

Helie, S., & Sun, R. (2010). Incubation, insight, and creative problem solving: A unified theory and a connectionist model. *Psychological Review, 117*(3), 994–1024.

Kim, J. (2006). Emergence: Core ideas and issues. *Synthese, 151,* 547–559.

Lebow, R. N. (2008). *A cultural theory of international relations.* Cambridge: Cambridge University Press.

Lévi-Strauss, C. (1971). *L'homme nu.* Paris: Plon.

Murray, H. (1938). *Explorations in personality.* Oxford: Oxford University Press.

Nadel, S. F. (1957). *The theory of social structure.* New York: Free Press.

Naveh, I., & Sun, R. (2006). A cognitively based simulation of academic science. *Computational and Mathematical Organization Theory, 12,* 313–337.

Newell, A. (1990). *Unified theories of cognition.* Cambridge, MA: Harvard University Press.

Nisbett, R. E., & Norenzayan, A. (2002). Culture and cognition. In D. Medin & H. Pashler (Eds.), *Stevens' handbook of experimental psychology: Vol. 2. Memory and cognitive processes* (3rd ed.). New York: John Wiley.

Reber, A. (1989). Implicit learning and tacit knowledge. *Journal of Experimental Psychology. General, 118*(3), 219–235.

Richerson, P.J. & Boyd, R. (2005). *Not by genes alone: How culture transformed human evolution.* Chicago: University of Chicago Press.

Sawyer, K. (2003). Artificial societies: Multiagent systems and the micro-macro link in sociological theory. *Sociological Methods & Research, 31*(3), 2003.

Schaller, M. Norenzayan, A., Heine, S. J., Yamagishi, T., & Kameda, T. (Eds.), (2009). *Evolution, culture, and the human mind.* London: Psychology Press.

Schelling, T. C. (1971). Dynamic models of segregation. *Journal of Mathematical Sociology, 1,* 143–186.

Semin, G. R. & Echterhoff, G. (Eds.) (2010). *Grounding sociality: Neurons, mind, and culture.* London: Psychology Press.

Shore, B. (1998). *Culture in mind: Cognition, culture, and the problem of meaning.* New York: Oxford University Press.

Sperber, D. (1996). *Explaining culture: A naturalistic approach.* Oxford: Blackwell.

Sperber, D. & Hirschfeld, L. (1999). Culture, cognition, and evolution. In R. Wilson & F. Keil (Eds.), *MIT encyclopedia of the cognitive sciences.* Cambridge, MA: MIT Press.

Sperber, D. & Hirschfeld, L. (2004). The cognitive foundations of cultural stability and diversity. *Trends in Cognitive Sciences, 8*(1), 40–46.

Sun, R. (2001). Cognitive science meets multi-agent systems: A prolegomenon. *Philosophical Psychology, 14*(1), 5–28.

Sun, R. (2002). *Duality of the mind.* Mahwah, NJ: Lawrence Erlbaum.

Sun, R. (Ed.). (2006). *Cognition and multi-agent interaction: From cognitive modeling to social simulation.* Cambridge: Cambridge University Press.

Sun, R. (Ed.). (2008). *The Cambridge handbook of computational psychology.* Cambridge: Cambridge University Press.

Sun, R. (2009). Motivational representations within a computational cognitive architecture. *Cognitive Computation, 1*(1), 91–103.

Sun, R. (Ed.). (2010). *Proceedings of the Workshop on Cognitive Social Sciences.* New York: RP Press.

Sun, R. (In press). From cognition to culture and back: Lessons and potentials. In B. Kaldis (Ed.), *Mind and society: Cognitive science meets the philosophy of social sciences* (Synthese Library Series). Dordrecht, The Netherlands: Springer.

Sun, R., Coward, L. A., & Zenzen, M. J. (2005). On levels of cognitive modeling. *Philosophical Psychology, 18*(5), 613–637.

Sun, R., & Naveh, I. (2004). Simulating organizational decision-making using a cognitively realistic agent model. *Journal of Artificial Societies and Social Simulation, 7*(3).

Sun, R., & Naveh, I. (2007). Social institution, cognition, and survival: A cognitive-social simulation. *Mind and Society, 6*, 115–142.

Sun, R., Slusarz, P., & Terry, C. (2005). The interaction of the explicit and the implicit in skill learning: A dual-process approach. *Psychological Review, 112*(1), 159–192.

Tetlock, P., & Goldgeier, J. (2000). Human nature and world politics: Cognition, identity, and influence. *International Journal of Psychology, 35*(2), 87–96.

Turner, M. (2001). *Cognitive dimensions of social science*. Oxford: Oxford University Press.

Vygotsky, L. (1986). *Thought and language*. Cambridge, MA: MIT Press.

Wimsatt, W. C. (1997). Aggregativity: Reductive heuristics for finding emergence. *Philosophy of Science, 64*, S372–S384.

Zerubavel, E. (1997). *Social mindscapes: An invitation to cognitive sociology*. Cambridge, MA: Harvard University Press.

II CULTURE

2   Mapping Minds across Cultures

Paul Thagard

2.1   Introduction

Anthropology is the study of the physical, social, and cultural development of humans. It has been considered part of cognitive science since that enterprise was organized in the 1970s (Gardner, 1985; D'Andrade, 1995), but anthropology is also classed as a social science because of its concern with groups and cultural interactions. Ironically, cognitive anthropology has waned as a practice within the community of anthropologists under the influence of postmodernism, at the same time that there has been an explosion of research by cognitive and social psychologists concerned with culture (e.g., Kitayama & Cohen, 2007). These divergences make anthropology an excellent field in which to examine the relevance of cognition in individuals to the operation of societies.

In contemporary social science, the two most prominent accounts of the relation between the social and the psychological are *methodological individualism*, the reductionist view that everything social is caused by the actions of individual people; and *postmodernism*, the holistic view that reality is a matter of social construction. Methodological individualism remains dominant in economics and political science, where social events are viewed as arising from individual actions determined by rational self-interest, as in game theory. In contrast, many researchers in anthropology, sociology, and history have adopted the postmodernist view that the individual can be largely ignored in favor of attention to social processes such as discourse and power. Neither of these approaches is adequate to explain complex social phenomena such as culture.

Cultural psychologists have adopted a richer view of the dynamic interdependence of self systems and social systems, arguing that the psychological and the cultural mutually constitute one another and must be analyzed and understood together (Markus & Hamedani, 2007, p. 3). From the perspective of the natural sciences, the idea of mutual constitution is highly puzzling, because constitution in physical systems is a unidirectional, asymmetric part-whole relation. Particles such as protons make up atoms, which make up molecules, which make up cells, which make up tissues, which make up organs, which make up organisms, which make up species. Analogously, in accord with methodological individualism, it would seem that people make up social groups, not vice versa. Can sense be made of the idea that social groups constitute people?

Here is the key insight to resolve this seeming conundrum: the actions of groups result from the actions of individuals who think of themselves as members of groups. What makes a group a group is not the sort of physical bonding that makes a group of cells into an organ. Rather, social bonds are largely psychological and arise from the fact that the individuals in the group have mental representations, such as concepts, that mark them as members of the group. The bonding process is not purely psychological, however, as it can also include various kinds of physical interactions that are social, linguistic, or both, such as participating in rituals and legal contracts, or even just making eye contact. These interactions tie people together into groups when they result in mental representations (affective as well as cognitive) through which individuals come to envision themselves as part of the group. Without such envisioning, the group cannot continue to function collectively. For example, by going together to church, school, and work bees, church members reinforce the beliefs and attitudes that mark them as members in contrast to other groups. To take a simpler example, a marriage is not simply a legal arrangement, but is also a social group that depends on the development and maintenance of emotional bonds through ongoing emotional interactions (Gottman, Tyson, Swanson, Swanson, & Murray, 2003).

This account does not, however, reduce the social to the psychological, because the psychology of individuals cannot be understood without appreciating the centrality of the social to the self. Many psychologists have observed that group membership and distinctions between in-groups and out-groups are an important part of self-identity (Tajfel, 1974; Brewer & Yuki, 2007). What individuals do is greatly affected by how they think of themselves in relation to ongoing interactions with various groups. Individuals are social in that a person's thinking and behavior depends substantially on representation of and interaction with other people. It is an exaggeration to conclude from such interactions that the social constitutes the individual in the way that individuals constitute groups; for a rigorous analysis of such constitution, see Findlay and Thagard (2011).

Nevertheless, the mutual relevance of the social and the psychological shows the need for an alternative to both simplistic individualism and obfuscatory holism.

This chapter proposes such an alternative: the method of multilevel interacting mechanisms. Historical and cultural explanations need not be restricted to psychological and social levels, but can also benefit from incorporating neural and molecular mechanisms. The value of considering multilevel mechanisms can be shown by considering explanations of the surprising and disturbing frequency of suicide in aboriginal communities in Canada and elsewhere. This chapter shows how explanation of suicide in aboriginal communities can fruitfully operate on multiple levels, with factors including social forces, cognitive-affective structures, and underlying neural and molecular operations.

New tools are required to analyze the mental structures that underlie social interactions. Hence, this chapter applies a new technique called *cognitive-affective mapping* (Thagard, 2010b; Findlay & Thagard, in press). A cognitive-affective map is a diagram that displays not only the conceptual structure of people's views but also their emotional nature, showing the positive and negative values attached to concepts and goals. This technique is based on HOTCO (an abbreviation of hot coherence), a model of how people make decisions and other inferences through emotional coherence (Thagard, 2006). Cognitive-affective maps can be used not only to describe cultural thought, but also to explain resulting actions by members of cultural groups.

From the perspective of multilevel interacting mechanisms, alleviating social problems such as aboriginal suicide requires change at multiple levels. Some of the changes needed are conceptual, such as the replacement of historically inaccurate and pejorative terms like "Indian" by more historically accurate and reputable ones like "aboriginal," "native American," and "First Nations." Hacking (1999) described the *looping effect of human kinds,* in which adding new ways of categorizing social groups can contribute to changes in the ways in which those groups interact. This chapter introduces the idea of *revalencing,* changing the emotional values of the concepts used to guide social interactions. Hacking's looping effect and revalencing can be understood as the result of multilevel interacting mechanisms, ranging from the social to the molecular.

Another tool helpful for understanding the relation between the individual and social is computational modeling of group interactions. Agent-based modeling has become a standard technique in social science and artificial intelligence (Sun, 2006, 2008) to explain the operations of groups in terms of the computationally intelligent behavior of individuals. This chapter presents a new model, HOTCO 4, that extends agent-based modeling to consider a higher level of group dynamics.

This chapter concludes by addressing core methodological issues about culture, meaning, and the relation between the cognitive and social sciences. It rejects the postmodernist view that the study of humanity is a hermeneutic enterprise eschewing the concepts and methods of sciences such as psychology. But it also rejects the reductionist view that the social sciences can be grounded in the cognitive sciences by simply deriving social phenomena from cognitive phenomena. The method of multilevel interacting mechanisms is neither reductionist nor antireductionist, but instead aims to show how explanations of social phenomena (such as aboriginal suicide) and historical events (such as Canadian Mennonite migration) can profitably draw on new ways of understanding human behavior.

2.2   Multilevel Interacting Mechanisms

Individual-social interactions can be clarified by adopting the method of multilevel interacting mechanisms that has been advocated for in relation to the study of creativity (Thagard & Stewart, 2011) and the self (Thagard & Wood, 2011). Adapting ideas developed by philosophers of science such as Bunge (2003) and Bechtel (2008), we can define a system as a quadruple of environment, parts, interconnections, and changes, or EPIC for short. Here the parts are the objects (entities) that compose the system, and the environment is the collection of items that act on the parts. The interconnections are the relations among the parts, especially the bonds that tie them together, and the changes are the processes that make the system behave as it does. The multilevel interacting mechanism method is a specification of a general approach to cognitive and social explanations that attempts to incorporate and develop previous insights about the architecture of complexity (Simon, 1962), the epidemiology of representations (Sperber, 1996), explanatory pluralism (McCauley & Bechtel, 2001), systems thinking (Bunge, 2003), cognitive social science (Turner, 2001) and multilevel cognitive modeling (Sun, Coward, & Zenzen, 2005).

Thagard and Wood (2011) argue that the self can best be understood as a multilevel system operating at social, individual, neural, and molecular levels, which are the levels that can be used to explain emotions, consciousness, and other important aspects of thinking (Thagard, 2006, 2010a). At the first or social level, the set of parts consists of individual persons. The social parts are influenced by an environment that includes all the objects that people causally interact with, including natural objects, artifacts, and social organization. The interconnections at the social level consist of the myriad relations among people, ranging from mundane perceptual ones such as one person being able to recognize another, to deeper bonds such as being in love, to larger group relations such as belonging to the same sports team. Changes at the social level consist of the many kinds of human interaction, ranging from talking to playing games to sexual intercourse.

At the second or individual level, the self consists of particular behaviors and the many mental representations, such as concepts and rule-like beliefs, that people apply to identify themselves and others. Behaviors are properties of individuals, and mental representations can be considered as parts of them if one replaces a commonsense, dualist view of the mind with a scientific one that takes minds to be information-processing brains (Thagard, 2010a).

For the neural subsystem—the third level—the most important parts of the brain are neurons, which are cells that also exist in related parts of the nervous system such as the spine. The interconnections of the neural system are largely determined by the excitatory and inhibitory synaptic connections between neurons, although glial cells in the brain and hormonal processes are also relevant (Thagard, 2006, chapter 7). The environment of the neural system is better described at a smaller scale than the level of whole objects appropriate for the psychological and social levels. For example, photons of light stimulate retinal cells and initiate visual processing in the brain, and sound waves affect the structure of the ear and initiate auditory processing. Thus, the environment of the neural system consists of physiological inputs that influence neural firing. Finally, the changes in the neural subsystem include alterations in firing patterns resulting from excitatory and inhibitory inputs from other neurons, as well as alterations in the synaptic connections.

Moving down to the fourth level, molecular mechanisms are highly relevant to understanding neural and psychological aspects of the self. Neurons are cells consisting of organelles such as nuclei and mitochondria, and the firing activity of neurons is determined by their chemical inputs and internal chemical reactions. Aspects of the self such as personality are influenced by biochemical factors, including genes, neurotransmitters, and epigenetic factors that modify the expression of genes. Genetic effects on behavior are shown by studies that find higher correlations between some features in identical twins than in nonidentical ones, for example in

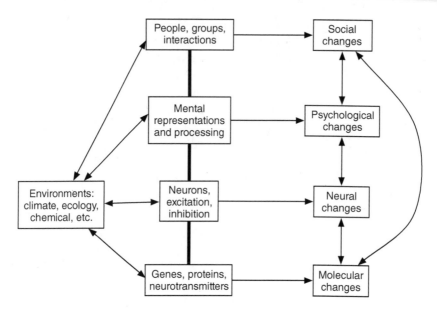

Figure 2.1
Diagram of the self as a multilevel system. Lines with arrows indicate causality. Thick lines indicate composition.

tendencies toward mental illnesses such as schizophrenia. Humans have variation in genes that determine the receptors for more than fifty different neurotransmitters that affect neuronal firing. In sum, a self is a system consisting of subsystems at four levels—social, psychological, neural, and molecular—each of which includes environment, parts, interconnections, and changes.

The resulting view of the self treats it neither as reducible from psychological and biological processes, nor as a mere social construction. Figure 2.1 presents the self as a complex system of multiple interacting mechanisms. Note that causality runs both up from biological processes and down from social processes.

Figure 2.1 provides a schema for filling out explanations of the kinds of social phenomena described later, such as why suicide is so common among aboriginals, why the Mississaugas sold southern Ontario to the British, and why Mennonites migrated to the Waterloo region. The individual, psychological level that is captured in part by the cognitive-affective maps described below in section 3 is an important part of the story, but full-blown cultural explanations should also pay attention to social, neural, molecular and environmental factors. Like people in general, social scientists have a preference for single-factor explanations, but the complexity of the cultural world demands viewing it as a complex system with many interacting causal influences. Computer models such as those described in section 6 can also be a valuable tool for representing such complexity.

2.3 Cognitive-Affective Mapping

As a case study of the interpenetration of cognitive and social explanations, consider the history of Waterloo, Ontario. It is a prosperous city of 120,000 people with two universities, several insurance companies, and many high-technology companies, including Research in Motion, where the Blackberry was invented. Europeans first settled the region in the early 1800s, when a group of Mennonite farmers from Pennsylvania purchased land from British speculators, who had bought the land from Joseph Brant, leader of the Six Nations (Iroquois) tribes. More than a thousand square miles of land along the Grand River had been given to the Six Nations by the British, whom the tribes had supported during the American War of Independence, in compensation for loss of traditional Iroquois land in upper New York State. In order to make this gift, the British bought the Grand River valley from the Mississaugas, a native group that had settled southern Ontario after replacing the original Iroquoian people. In a deal of dubious legality, the British had also bought the land now occupied by the city of Toronto from the Mississaugas. These developments raise many fascinating historical questions, including:

1. Why did the Mississaugas sell much of southern Ontario to the British?
2. Why did the Mennonite settlers move from Pennsylvania to Waterloo region?

Answering such questions requires attention not only to cognitive processes operating in individual minds but also to social processes of the sort that have been studied by anthropologists and sociologists.

To understand the actions of people from groups as culturally different as Mennonites and Mississaugas, we need a way of grasping the structure of their concepts, beliefs, and attitudes. The technique of cognitive-affecting mapping originated as a way of understanding and facilitating conflict resolution (Findlay & Thagard, in press), but also has direct relevance to fostering cross-cultural communication. Researchers in psychology, computer science, and other fields have used the technique of *cognitive maps* (also known as conceptual graphs, concept maps, or mind maps) to illustrate the conceptual structures that people use to represent important aspects of the world (e.g., Axelrod, 1976; Novak, 1998; Sowa, 1999). But such maps fail to indicate the values attached to concepts and other representations such as goals, and therefore are inadequate to capture the underlying psychology of conflicts and other important social phenomena. They lack an appreciation of *affect*, which is the complex of emotions, moods, and motivations that are crucial to human thinking.

A cognitive-affective map is a visual representation of the emotional values of a group of interconnected concepts. It employs the following conventions:

1. Each concept is represented by a node (vertex). Favorable nodes are represented by circles. Unfavorable nodes are represented by hexagons. Neutral nodes are represented by rectangles. Degree of favoring and disfavoring is represented by thickness of lines and darkness of color, which is optional.

2. Each link between concepts is represented by a line. Supportive links are represented by solid lines. Conflictive links are represented by dotted lines. The strength of support or conflict is represented by thickness of lines.

Figure 2.2 schematizes this kind of representation.

Concept maps have long been used by cognitive anthropologists to understand the concepts, schemas, and connectionist underpinning of thoughts across cultures (see, e.g., D'Andrade, 1995), but the resulting models ignored the emotional character of the underlying thinking. Cultures differ not only in having different concepts and connections between those concepts, but also in attaching different values and emotional reactions to concepts. Cognitive-affective maps are intended to depict concep-

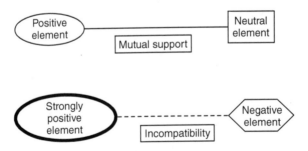

Figure 2.2
Schema for a cognitive-affective map. Use of color is optional depending on the medium used.

tual and emotional mental states of individuals who are typical of members of a group.

Cultural differences can be made clear by cross-cultural examples drawn from Ontario history. No information is available to map the attitudes of the Mississaugas in the early 1800s, when they signed treaties with the British that ceded large tracts of land that even then were much more valuable than the goods exchanged for them (a thousand British pounds each for the Grand River valley and Toronto areas; see Bellegarde, 2003; Gibson, 2006). But it is clear from historical investigations that the general conceptual scheme of the Mississaugas was very different from that of the British administrators with whom they were negotiating. The Mississaugas had spiritual beliefs similar to those of other speakers of the Anishinabe languages, including people variously known as Ojibway, Algonquin, and Chippewa. A cognitive-affective map can be a useful tool for displaying just how much the Anishinabe views differed from the more familiar Christian mythology of the British.

Figure 2.3 is a highly simplified exhibition of some of the key concepts of Anishinabe spirituality, based largely on Johnston (1995). The central concept is *manitou*, which refers to a supernatural spirit, or mystery.

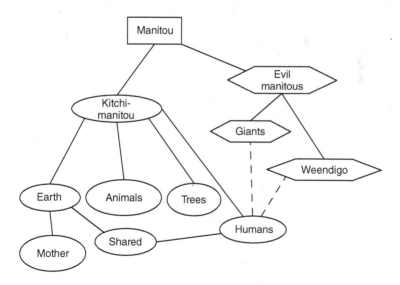

Figure 2.3

Cognitive-affective map of Anishinabe view of the world. Emotionally positive concepts are in circles, negative concepts are in hexagons, and squares indicate neutrality. Solid lines between concepts indicate mutual support, whereas dotted lines indicate conflict.

Johnston (p. xv) writes: "According to tradition, Kitchi-Manitou (the Great Mystery) created the world, plants, birds, animals, fish and the other *manitous* in fulfillment of a vision." This makes Kitchi-Manitou analogous to the Christian creator, but otherwise the Anishinabe ontology is very different from the Christian one. For Christians, only humans are transcendental beings with immortal souls, but for the Anishinabe all animals, plants, and the earth itself are *manitous*. Like many indigenous peoples, the Anishinabe did not employ the European concept of land as private property, so the value of land derived from its shared use rather than from ownership (Gibson, 2006; Rogers & Smith, 1994).

Of course, figure 2.3 is only the beginning of an explanation of why the Mississaugas agreed to sell cheaply two valuable parts of what is now southern Ontario. Because the Mississaugas viewed the land as shared rather than owned, they were not giving up ownership if it. Other important factors include the Mississaugas' acquired dependence on European manufactured goods such as rifles and cloth, and the fraudulent nature of some of the agreements. According to the Indian Claims Commission of Canada, the first Toronto transaction of 1787 was recorded only by a blank deed, and the British Crown may have failed in 1805 to disclose to the Mississaugas the invalidity of the 1787 surrender and the increased territory involved in the new purchase (Bellegarde, 2003). In January, 2010, the Canadian government offered the Mississaugas $145 million to settle the long-standing land claims.

Hence, fraud and colonial oppression figure large in explanation of why the Mississaugas sold their land, but in the background has to be a question about their understanding of land ownership. Their spiritual beliefs emphasized shared use over private ownership, and in any case the Mississaugas may have been aware that the relevant areas had previously been inhabited by Iroquoian tribes such as the Hurons and Senecas. The Mississaugas had only moved into the Toronto area around 1690 (City of Toronto, 2010).

Compare the alleged sale of Manhattan to the Dutch for twenty-four dollars. According to one source:

The "sale" of Manhattan was a misunderstanding. In 1626 the director of the Dutch settlement, Peter Minuit, "purchased" Manhattan for sixty guilders worth of trade goods. At that time Indians did everything by trade, and they did not believe that land could be privately owned, any more than could water, air, or sunlight. But they did believe in giving gifts for favors done. The Lenni Lenape—one of the tribes that lived on the island now known as Manhattan—interpreted the trade goods as gifts given in appreciation for the right to share the land. We don't know exactly what

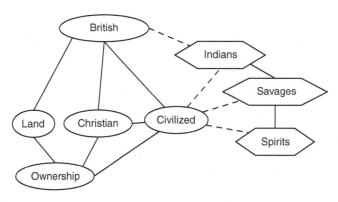

Figure 2.4
Cognitive-affective map of the British view of themselves and indigenous peoples.

the goods were or exactly how much a guilder was worth at that time. It has been commonly thought that sixty guilders equaled about twenty-four dollars. But the buying power of twenty-four dollars in 1626 is not known for sure. To Europeans, ownership of land was synonymous with wealth, power, and prestige. To purchase land meant that the purchaser had the exclusive right to own and use it. The Lenni Lenape did not realize that the Dutch meant to hold the land for their exclusive use. (National Museum of the American Indian, 2007, p. 36)

Similarly, it is possible that the Mississaugas viewed the transactions with the British very differently than did the British. Cognitive-affective maps can help to highlight the relevant cultural differences.

Figure 2.4 is a rough attempt to capture the attitude of the British colonials toward land and the natives, who were commonly viewed as savages. A crucial part of the cognitive-affective mapping in intergroup relations is the presence of positive in-group concepts and negative out-group concepts. We will see later that such representations are one reason why the relation between social and cognitive explanations is not reductionist. Section 6, below, will discuss the revaluing of concepts involved in replacing the term "Indian," which was based on Christopher Columbus's erroneous view that he had arrived in India. In figure 2.4, it might seem odd that there is a link between the concepts *Christian* and *ownership,* but recall that one of the Ten Commandments is "Thou shalt not steal."

Cognitive-affective maps can also be useful for comprehending why Mennonite settlers made the very difficult overland trek to the Waterloo region from Pennsylvania in the early 1800s. According to Epp (1974), their motivations were not just the availability of fertile land, but also freedom to practice their core religious beliefs, which included avoidance

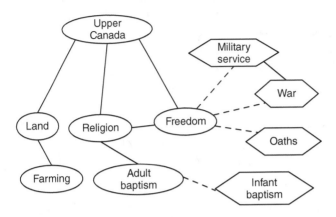

Figure 2.5
Cognitive-affective map of Mennonites motivated to migrate in pursuit of religious freedom.

of military service and oaths of allegiance, as well as the adoption of baptism for adults rather than infants. Like the Amish, the Mennonites were Anabaptists who had been attracted to Quaker-led Pennsylvania by promises of religious freedom, but the American revolution introduced new pressures toward military service. Upper Canada under the British offered renewed possibilities of freedom, as shown in figure 2.5. Not shown are the positive in-group attitudes toward Mennonites and other Anabaptists, in contrast with negative attitudes toward nonbelievers.

This section has shown how cognitive-affective maps can be used to portray the concepts and values that are part of the cognitive explanation of human actions, including ones involving social groups. Such maps are not merely a diagramming technique, since their components correspond directly to a computational model of emotional cognition called HOTCO, for "hot coherence" (Thagard, 2000, 2006). HOTCO is a neural network model in which the nodes have emotional valences as well as degrees of activation. It can also be implemented in more biologically realistic fashion by distributed neural networks corresponding to brain areas recognized for emotional processing, as part of the EMOCON (emotional consciousness) model (Thagard & Aubie, 2008; Thagard, 2010a).

The translation between cognitive-affective maps and HOTCO networks is simple: concepts become nodes in HOTCO networks, supportive connections become excitatory links, conflictive connections become inhibitory links, positive values become positive valences, and negative values become negative valences. A software tool, Empathica, is currently under development to provide an easy way to draw cognitive-affective maps and produce HOTCO models (Thagard, 2010b). Even without that tool, such maps can be produced using any drawing program via the following method (Findlay & Thagard, in press):

1. Identify the main concepts, beliefs, and goals of the person being modeled.
2. Identify these elements as emotionally positive or negative, and accordingly represent them by ovals or hexagons.
3. Identify relations between elements that are either complementary (solid lines) or conflicting (dashed lines).
4. Solicit feedback on the resulting cognitive-affective map from other knowledgeable people to see if it captures their understandings of the person and situation, and revise if needed.

The discussion in this section may misleadingly suggest that the goal of mapping minds across cultures is to provide an exclusively psychological explanation of social phenomena. Attention to a more recent phenomenon, the prevalence of suicide among Canadian aboriginal peoples, shows the need for explanations that operate at multiple levels.

2.4 Explaining Aboriginal Suicide

Since Durkheim's (1951) classic study, suicide has been a major topic for sociology, but it has also become a central concern for clinical psychologists (e.g., Joiner, 2010). Sadly, the historical development of native peoples in Ontario, the rest of Canada, the United States, and many other countries has been much less fortunate than that of the prospering Mennonites (Coates, 2004). In particular, the suicide rate among Canadian First Nations (aboriginals) is three times that of Canadians in general (Chandler et al., 2003). The Canadian Mental Health Association States:

Aboriginal people experience a broad range of health issues, and have the poorest health levels in the country. Aboriginal people have shorter life expectancies, experience more violent and accidental deaths, have higher infant mortality rates and suffer from more chronic health conditions.

Aboriginal people are also more likely to face inadequate nutrition, substandard housing and sanitation conditions, unemployment and poverty, and discrimination and racism, all important factors in maintaining health and wellness. ...

Many mental health problems of Aboriginals arise from a long history of colonization, residential school trauma, discrimination and oppression, and losses of land, language and livelihood. Many families were deeply affected by the government's residential school policy. Children were taken from their homes and sent to residential schools, where some experienced violence and abuse and many others lost their language and connection to their traditions, culture and community.

Rates of mental health problems, such as suicide, depression, and substance abuse, are significantly higher in many Aboriginal communities than in the general population. ...

Suicide and self-injury were the leading causes of death for Aboriginal youths. In 2000, suicide accounted for 22 percent of all deaths among Aboriginal youth (aged 10 to 19 years) and 16 percent of all deaths among Aboriginal people aged 20 to 44 years. Suicide rates of Registered Indian youths (aged 15 to 24) are eight times higher than the national rate for females and five times higher than the national rate for males. In 2005, there were 24 completed suicides in Nishnawbe Aski Nation territory, one of the highest rates in Canada. (Canadian Mental Health Association, 2010)

The causes of suicide operate at many levels, from the social and psychological to the neural and molecular. The individual, psychological level can be captured by providing a cognitive-affective map of the kind of negative self-identity that encourages a sense of hopelessness that leads to suicide. Chandler et al. (2003) report that in British Columbia there is a striking difference between First Nation groups that have maintained cultural continuity with traditional beliefs and those that have abandoned them: efforts by Aboriginal groups to preserve and promote their culture are associated with dramatic reductions in rates of youth suicide.

The difference is evocatively captured by an Ojibway writer, Richard Wagamese (2010), in an article ironically subtitled "What it means to be an Indian." He describes the many negative terms applied by the dominant white culture to natives: savage, redman, slow, awkward, lazy, shiftless, stupid, drunken, welfare bum, and so on. He eloquently writes: "You learned that labels have weight—incredible, hard, and inescapable. You learned to drink so that you wouldn't have to hear them" (2010, p. 9). But Wagamese then describes a transformation deriving from more positive self-identification: "But when you found your people you became Ojibway. You became Anishinabe. You became Sturgeon Clan. You became Wagamese again and in that name a recognition of being that felt like a balm on the rawness where they'd scraped the Indian away." When Richard Wagamese reconceptualized himself as a Anishinabe rather than a generic, downtrodden Indian, the change was behavioral and social as well as mental: he adopted different interactions, such as native cultural rituals.

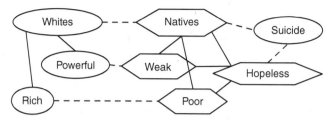

Figure 2.6
Cognitive-affective map of the appeal of suicide to aboriginal Canadians.

Figure 2.6 displays a small part of a conceptual structure and emotional attitude that can make suicide appealing to aboriginals. A crucial aspect of it is the contrast between the natives as an inferior (emotionally negative) group and whites as a superior (emotionally positive) group. The shift to a more positive identity such as Anishinabe removes this stark contrast by providing natives with a legitimate location in the universe.

The network shown in figure 2.6 is explanatory, not just descriptive, because of the underlying HOTCO model. The node for suicide gets positive activation and valence because of the inhibitory links it has with the negative nodes for natives and hopeless. In psychological terms, the despondent youths see suicide as a solution to their dismal self-identities as natives and their hopeless situations. That is why the figure includes dashed lines from the suicide node to the natives and hopeless nodes: suicide becomes desirable because it conflicts with two states that are undesirable. Note that the explanation of suicide sketched by figure 2.6 is psychological at the individual level and is also social, as people represent themselves both by membership in their own native group and by exclusion from the opposing group of white people. Section 5 will discuss this double aspect in terms of Hacking's idea of the looping effect of human kinds.

However, cognitive-affective maps only begin to provide an explanation of suicidal behavior by showing some of its social and psychological causes. They can at best capture some of the current psychological state of depressed aboriginals, neglecting the long history of oppression that contributed to feelings of hopelessness. A fuller historical explanation would need to cite other factors, such as the residential schools to which generations of aboriginal students were sent for cultural indoctrination, often accompanied by physical and sexual abuse. According to a leading researcher on the psychology of suicide, the desire for death usually results from a combination of two psychological states: the perception that one is a burden and the sense that one does not belong (Joiner, 2010). Aboriginals in Canada and other countries have been strongly sent the message that they do not belong, whereas the reintegration with native culture described by Wagamese can restore a sense of belonging.

The explanation of suicide need not operate only on the social and psychological levels, because there is abundant evidence of relevant mechanisms operating at the neural and molecular levels. Concepts and emotional reactions such as those shown in cognitive-affective maps can naturally be interpreted in terms of neural mechanisms (see, e.g., Anderson, 2007; Smith & Kosslyn, 2007; Thagard, 2010a). There is accumulating evidence that concepts are patterns of activation in populations of neurons, and that emotional reactions can also be understood as patterns of activation in multiple brain areas, combining cognitive appraisal and physiological perception (Thagard & Aubie, 2008). Hence the negative self-representation and feeling of hopelessness shown in figure 2.6 can naturally be understood as a neural process operating in the brains of suicidal individuals.

According to Jamison (2000), 90% of people who commit suicide have a diagnosable psychiatric illness such as bipolar disorder (manic depression), which is known to have underlying molecular mechanisms involving neurotransmitters such as dopamine and serotonin (Goodwin & Jamison, 2007). Other molecular factors involved in suicide concern abuse of substances such as alcohol, drugs, and inhalants; sniffing of glue, gasoline, and other noxious fumes is endemic among youths in some Canadian native reserves and among other disadvantaged populations. Substance abuse has a clear molecular basis: psychopharmacology has identified how substances such as alcohol have behavioral effects through neurochemical mechanisms involving neurotransmitters such as dopamine and GABA (Meyer & Quenzer, 2005).

Also relevant to suicide on the molecular level is research showing a genetic component to suicide (e.g., Jamison, 2000; Lemonde et al. 2003; Mitchell, Mitchell, & Berk, 2000), although there does not seem to have been much genetic research relevant to aboriginals or Canadian Mennonites. However, there have been very interesting findings among the Amish of Pennsylvania, who are Anabaptists like the Mennonites. In general, the Amish have suicide rates much lower than the general American population. One study found that four families accounted for 73 percent of all suicides, even though they represented only 16 percent of the total Amish population (Ginns, 1996). Specific associations between suicide and genes for neurotransmitter receptors are under investigation.

2.5   The Looping Effect of Human Kinds

Wagamese (2010) mentions negative terms such as "wagon burner" that have been applied to North American natives, whose increasingly common preference is to be known by a specific tribal name, in his case Ojibway and Anishinabe. Although the term "Indian" is still in use, for example in the Government of Canada's archaic Indian Act, Canadian usage now largely replaces it by the general term "First Nations," accompanied by specific tribal names like "Mississauga." In the United States, "Indian" is being supplanted by "Native American." These linguistic shifts are analogous to changes in terminology about other disparaged social groups, leading to adoption of terms such as "African American," "gay," and "mentally handicapped."

Some commentators dismiss such linguistic changes as puling political correctness, but the shifts can be better appreciated from the perspective of cognitive-affective maps combined with what Hacking (1999) calls "the looping effect of human kinds." Hacking differentiates between human kinds and more straightforwardly physical ones, such as molecules and galaxies, that are not affected by what they are called. In human societies, merely introducing a new term can lead to changes in the society that is being described. How people are characterized affects their behavior and hence changes society, so linguistic changes can have social consequences, which in turn can lead to further linguistic changes.

Such looping effects of categorization influence both cognition and emotion, because concepts serve both to describe parts of the world and also to evaluate them. If over time a concept acquires negative emotional valence, then that concept can have negative effects on behavior, including social contagion that can contribute to substance abuse, and even suicide, when the negative self-evaluations of one disadvantaged person spread to another. Section 4 described findings that native tribes in British Columbia who maintained cultural continuity with traditional values did not have the elevated suicide rates found in many Canadian aboriginal communities. Hence, altering conceptual schemes, by replacing concepts with negative emotional valences (such as "Indian") with more positive concepts (such as "Anishinabe"), can be a part of social change that has desirable consequences.

The novel term *revalencing* can be used to refer to the kind of conceptual change that takes place when the introduction of new concepts is accompanied by emotional change. The shift from "Indian " to "First Nations" is a kind of positive revalencing, replacing an emotionally negative concept with a positive one as part of an effort to bring about desirable social change, but there are also negative kinds. Propaganda is full of negative revalencing, in which groups are increasingly stigmatized by the introduction of new terms with negative emotional baggage, such as "feminazi," "raghead," and " camel jockey."

Hacking's looping effect of human kinds can be understood as an interaction between social mechanisms, such as interpersonal communication, and psychological mechanisms of the sort that are reflected in cognitive-affective maps. Revalencing changes such maps by introducing new concepts, by deleting old ones, and most importantly by replacing emotionally negative values with ones that are positive or at least neutral. Given the evidence that emotion is a multilevel process that includes neural and molecular changes (Thagard, 2006), we can infer that the looping effect operates at these levels in addition to psychological and social ones.

Revalencing can be a powerful force in group-group and intercultural relations. Marriage counselors sometimes tell conflicted couples to think in terms of "us" rather than "I" and "you." Similarly, cultural conflicts such as those between Israelis and Arabs can benefit from using "human" as a universal concept that ought to transcend ethnic and national divisions. Perhaps cognitive-affective maps will serve to reduce human conflicts not only by helping people to recognize differences in how members of diverse groups represent each other, but also by prompting efforts to revalence representations in ways that ease conflict through Hacking's looping effect.

2.6   Modeling Groups and Individuals: HOTCO 4

In physics, mechanisms are usually characterized by systems of differential equations, whereas in biology they are usually represented by a combination of words and diagrams. In cognitive science, computer programs have provided an effective way of representing mechanisms and applying them to psychological phenomena. In a traditional artificial-intelligence simulation of cognitive processes, the parts (mental representations) are modeled by data structures in a programming language, and the interactions between parts are captured by algorithms that are applied to the data structures to produce changes in them. The environment is usually limited to inputs provided by the programmer, but in a robotic system physical inputs can be generated by sensors such as cameras. Neural network models are a bit different, because mental representations are usually treated as the result of interactions of much simpler neuron-like parts, whose characteristics can also be represented by data structures in a programming language. For both cognitive and neural modeling, building a computer program provides a way of specifying the nature of the mechanisms thought to underlie thought, as well as an effective way of determining whether the program behaves in accord with the psychological and neural phenomena that are to be modeled and explained. On the methodology of computer modeling, see Thagard (2012, chapter 1).

Social models have become increasingly common in artificial intelligence and the social sciences under such rubrics as agent-based modeling and distributed artificial intelligence (Sun, 2006, 2008). Current models, however, lack the ability to explain group behavior in terms of intergroup interactions that include the representation of groups by individuals. HOTCO 4 is an initial attempt to model how the behavior of groups can depend on the behavior of individuals who represent groups.

HOTCO is a series of computational models of how cognition and emotion interact to produce important inferences such as decision making. The original version extended traditional connectionist constraint satisfaction techniques to model emotional constraints (Thagard, 2000). In artificial neural networks, neuron-like units typically have a property called *activation* corresponding to the firing rate of a real neuron. The activation of a unit representing a concept can then represent the degree to which the concept applies to the current input situation. HOTCO assigns to each unit an additional quantity called a *valence,* which represents the emotional value—positive or negative—of the concept. For example, the concepts *baby* and *ice cream* have positive valence for most people, whereas the concepts *death* and *vomit* have negative valence. In the original HOTCO, activations affect valences but not vice versa, in accord with the normative expectation that probabilities should affect utilities but not vice versa.

Psychologically, however, it is common for people's emotions and motivations to influence their beliefs, so HOTCO 2 was developed to model cases such as the O. J. Simpson trial, for which it is plausible that valences had an impact on activations (Thagard, 2006, chapter 8). Inference, however, is not simply an individual matter, since decisions and other conclusions are often reached in social contexts. Accordingly, HOTCO 3 provided a model of emotional consensus in the form of a multiagent system in which the inferences of multiple simulated people are shaped in part by interactions with other people and involve forms of emotional communication, including contagion, altruism, and empathy (Thagard, 2006, chapter 5). HOTCO 3 involves a single group of interacting individuals, but pays no attention to the interaction of groups.

The latest version, HOTCO 4, is an attempt to take groups even more seriously. It was inspired by an analysis of the Camp David negotiations of 1978, which produced a rare and dramatic breakthrough in Middle Eastern international relations—an enduring peace agreement between Israel and Egypt following a quarter-century of conflict (Findlay & Thagard, in press). The negotiations involved some interactions between individuals of the sort that HOTCO 3 models, but there were also group-group interactions at two levels. The main negotiators were Menachem Begin for Israel, Anwar Sadat for Egypt, and Jimmy Carter for the United States. Each leader, however, was assisted by a small team of high-level advisors, so the Camp David negotiations amounted to an interaction among three groups of negotiators. Just as important, the negotiators operated not only with representations of each other, but also with representations of the larger groups to which they belonged. Begin, for example, viewed himself as an Israeli and a Jew, in opposition to Egyptians, Americans, Muslims, and Christians.

The actions of the negotiators as they moved toward a groundbreaking accord cannot be understood in terms of some simple game-theoretic matrix. Findlay and Thagard (in press) use cognitive-affective maps to chart the mental changes in Sadat and especially Begin that led to agreement. These maps include representations that the negotiating leaders had of each other, as well as of their own and other groups. Begin's decisions at the beginning and end of the negotiations have been simulated using HOTCO 2, but there is insufficient knowledge about the internal dynamics of the Israeli, Egyptian, and American delegations to indicate how to produce a HOTCO 4 simulation of group-group interactions.

So HOTCO 4 has been applied to a much simpler kind of group-group interaction involving two couples who need to make a joint decision about where to go to dinner. It is not uncommon for established couples to think of themselves *as* a couple, and also to think of other established couples as more than just a pair of individuals. Hence a joint decision involving two couples can involve all of the following for each individual:

• representation of self, which social psychologists call the self-concept;
• representation of the other individuals, including both the individual's partner and the members of the other couple;
• representation of the couple to which the individual belongs;
• representation of the other couple;
• representation of the potential actions to be performed;

• representation of the goals to be accomplished, which can altruistically include not just individual goals but also the goals of each couple.

All of these representations include a cognitive judgment about how well they apply to what is represented (corresponding to the activation of a HOTCO unit), as well as an emotional evaluation (corresponding to the valence of a HOTCO unit). Clearly, these kinds of representations make the bonds between human individuals more complex than the instinctual ones found in animal groups such as wolf packs.

To be more concrete, consider two couples, Alice and Bob, and Cathy and Doug, deliberating about whether to go together to a Chinese restaurant or a steakhouse. The simplest kind of decision process would be to vote and go with the majority preference, but a more collegial process would involve discussion and interactions in which each individual takes into account the goals of other individuals as well as the groups. The group structure is crucial because the style of interaction within the couple is likely to be different from that between couples: that is, the groups really do matter. For example, Alice may communicate differently with Bob than she does with people with whom she is not partnered, and she may deal with Cathy and Doug in a way that reflects in part her conception of them as a couple.

HOTCO 4 operates by building on the emotional consensus process of HOTCO 3. Each of the four individuals in the two couples reaches a preliminary judgment about what he or she prefers to do—Chinese or steakhouse. This judgment reflects not only the person's own valences for particular outcomes but also those of the other individuals and groups. If no consensus exists, then further deliberation takes place involving both cognitive and emotional communication. Cognitive communication includes introducing new information about the extent to which actions actually do facilitate goals; this process is analogous to the change in information that takes place in the CCC model of scientific consensus (Thagard, 2000, chapter 7). Just as important, emotional communication takes place through mechanisms such as contagion, where one person picks up on the emotional reactions of another through physiological processes of mimicry and neural mirroring. Other mechanisms of emotional communication include altruism, where one person adopts the valences of another who is cared about, and means-ends information, where one person points out to another how actions accomplish goals. HOTCO 4 and the example described here have been implemented in the programming language LISP. Not yet implemented in HOTCO is empathy, a kind of analogical thinking that provides a means by which one person can see what it's like to be in another's shoes.

What makes the groups significant in HOTCO 4 is when individuals change their preferences based on what they see as the preferences of their own and other groups. For example, Alice will be reinforced in her inclination to eat at a Chinese restaurant if she notices a consensus with her partner Bob, because they are part of the same group. Whether there is a similar impact on her views deriving from a perceived consensus in the other couple will depend on her cognitive-affective representation of the other couple, for example whether they are valued as a fine couple and good company. In this way, the decision of the general group, consisting of the two couples, depends on group decisions within couples, as well as a whole network of mutual representations.

A key question here is: what makes individuals into a group? As section 2.1 of this chapter argued, the key bonds for social groups such as couples are representational and emotional: Alice and Bob form a couple because they think of each other and the group they form with positive emotions. If the emotions are negative, then the bond is much weaker and prone to dissolution, even in the face of legal bonds, such as marriage, that may also exist. Similarly, a nation or ethnic group is constituted partly by historical background, but more importantly by emotional representations, such as Begin's being proud to be a Jew and an Israeli.

Obviously, the interactions of two couples in the HOTCO 4 simulation under discussion are much simpler than the interactions of whole cultures examined earlier in this chapter. But they serve to work out in greater detail the kinds of social-cognitive-affective mechanisms that are required for broader cultural developments. For the couples, as for cultural groups such as the Mississaugas and Mennonites, there need to be cognitive and affective representations of both in-groups and out-groups. HOTCO 4 provides a start at seeing how the actions of groups (in this case couples) can result from the actions of individuals who represent themselves as members of groups. Changes in such representations can have important effects on the behaviors of individuals and groups.

2.7   General Discussion

As section 1 described, much work in current social science is dominated by two inadequate methodological approaches: the methodological individualism that prevails in much of economics and political science in the form of rational choice theory; and the postmodernism that prevails in much of anthropology, sociology, and history in the form of vague discussions of discourse and power relations. The cognitive sciences, especially psychology and neuroscience, can provide a powerful third alternative, but not simply by reductively explaining the social in terms of the psychological. Rather, the method of multilevel interacting mechanisms shows how to integrate the social and the cognitive sciences non-reductively, displaying both psychological effects on social processes, and social effects on psychological processes. Cognitive-affective maps are a useful technique for depicting some of the conceptual and emotional structures that are relevant to both psychological and social explanations of culture.

Like meaning in general, culture is not *just* in the mind, but it would be wrong to say that culture just is not in the mind at all. Meaning and culture are multidimensional processes involving interactions among minds, the physical world, and social relations. This *tri-relational* view of meaning and culture is shown in figure 2.7, which depicts the interactions of systems rather than reduction of one system to another. Mental representations such as concepts can be understood as patterns of firing in populations of neurons (2010a). Neural populations interact with the physical world both by perceptual processes such as vision that carry information to the world into the brain, but also by physical actions that originate with brain activity and result in changes in the world, such as, for example, when a person moves objects around. For meaning and culture, an especially important aspect of the physical world is the social interaction among people who use language and other forms of communication to exchange information (Strauss & Quinn, 1997).

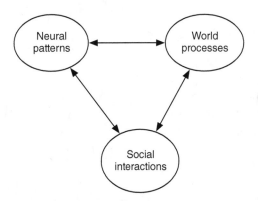

Figure 2.7
Meaning and culture as tri-relational interactions among minds (brains), the world, and other people.

Many philosophers misleadingly think of meaning as a kind of *content* (thing, possession) that representations have. Rather, we should construe the meaning of a concept such as *hat* as a complex relational process involving the neural populations that encode the concept, events in the world such as seeing hats and putting them on, and social interactions of people talking about and using hats. Similarly, cultural practices such as wearing different kinds of hats on different occasions (e.g., cowboy hats at rodeos) require a combination of neural activity, perception and manipulation of the world, and communication with other people involved in the occasion. The shift from thinking of meaning as a thing-like content to a relational process is analogous to similar conceptual changes that have taken place in physics, such as the shift from viewing the mass of an object as a simple quantity to viewing it as a relation with other objects in accord with the law of gravitation.

Further development of a tri-relational theory of meaning will contribute to an integrated understanding of psychological and social processes. This chapter has displayed the relevance of the cognitive sciences to the social sciences using several innovations: the method of multilevel interacting mechanisms, the technique of cognitive-affective mapping, the psychological interpretation of the looping effect of human kinds, and the HOTCO 4 computational model of group-group relations. These techniques have been applied to a range of interesting social phenomena, including the British acquisition of land from the Mississaugas, Mennonite migration from Pennsylvania to Ontario, and disturbingly high rates of aboriginal suicide. There are other promising and complementary approaches to integrating social and psychological phenomena, such as in neuroeconomics and sociolinguistics. Hence, the prospects are excellent for developing cognitive social science.

Acknowledgments

This research was funded by the Natural Sciences and Engineering Research Council of Canada. I am grateful to Bob McCauley, Ben Nelson, and two anonymous referees for comments on a previous draft.

References

Anderson, J. R. (2007). *How can the mind occur in the physical universe?* Oxford: Oxford University Press.

Axelrod, R. (Ed.). (1976). *Structure of decision: The cognitive maps of political elites.* Princeton, NJ: Princeton University Press.

Bechtel, W. (2008). *Mental mechanisms: Philosophical perspectives on cognitive neuroscience.* New York: Routledge.

Bellegarde, D. J. (2003). *Mississaugas of the New Credit First Nation inquiry.* Ottawa: Government of Canada Indian Claims Commission.

Brewer, M., & Yuki, M. (2007). Culture and social identity. In S. Kitayama & D. Cohen (Eds.), *Handbook of cultural psychology* (pp. 307–322). New York: Guilford Press.

Bunge, M. (2003). *Emergence and convergence: Qualitative novelty and the unity of knowledge.* Toronto: University of Toronto Press.

Canadian Mental Health Association. (2010). About mental health: Aboriginal people/First Nations. Retrieved Dec. 6, 2010, from http://www.ontario.cmha.ca/about_mental_health.asp?cID=23053.

Chandler, M. J., Lalonde, C., Sokol, B., & Hallett, D. (2003). Personal persistence, identity development, and suicide: A study of native and non-native North American adolescents. *Monographs of the Society for Research in Child Development, 68*(2), 1–130.

Coates, K. S. (2004). *A global history of indigenous peoples: Struggle and survival.* Houndmills, England: Palgrave Macmillan.

D'Andrade, R. (1995). *The development of cognitive anthropology.* Cambridge: Cambridge University Press.

Durkheim, E. (1951). *Suicide* (J. A. Spaulding & G. Simpson, Trans.). New York: Free Press.

Epp, F. H. (1974). *Mennonites in Canada, 1786–1920.* Toronto: Macmillan.

Findlay, S. D., & Thagard, P. (2011). How parts make up wholes: The dynamics of composition and emergence. Unpublished manuscript, University of Waterloo.

Findlay, S. D., & Thagard, P. (in press). Emotional change in international negotiation: Analyzing the Camp David accords using cognitive-affective maps. *Group Decision and Negotiation.*

Gardner, H. (1985). *The mind's new science.* New York: Basic Books.

Gibson, M. M. (2006). *In the footsteps of the Mississaugas.* Mississuaga, Canada: Mississauga Heritage Foundation.

Ginns, E. I. (1996). A genome-wide search for chromosomal loci linked to bipolar affective disorder in the Old Order Amish. *Nature Genetics, 12*(4), 431–435.

Goodwin, F. K., & Jamison, K. R. (2007). *Manic-depressive illness: Bipolar disorders and recurrent depression* (2nd ed.). Oxford: Oxford University Press.

Gottman, J. M., Tyson, R., Swanson, K. R., Swanson, C. C., & Murray, J. D. (2003). *The mathematics of marriage: Dynamic nonlinear models.* Cambridge, MA: MIT Press.

Hacking, I. (1999). *The social construction of what?* Cambridge, MA: Harvard University Press.

Jamison, K. R. (2000). *Night falls fast: Understanding suicide.* New York: Vintage Books.

Johnston, B. (1995). *The manitous: The spiritual world of the Ojibway.* New York: HarperCollins.

Joiner, T. (2010). *Myths about suicide.* Cambridge, MA: Harvard University Press.

Kitayama, S., & Cohen, D. (Eds.). (2007). *Handbook of cultural psychology.* New York: Guilford Press.

Lemonde, S., Turecki, G., Bakish, D., Du, L., Hrdina, P. D., Bown, C. D., et al. (2003). Impaired repression at a 5-hydroxytryptamine 1A receptor gene polymorphism associated with major depression and suicide. *Journal of Neuroscience, 23*(25), 8788–8799.

Markus, H. R., & Hamedani, M. G. (2007). Sociocultural psychology: The dynamic interdependence among self systems and social systems. In S. Kitayama & D. Cohen (Eds.), *Handbook of cultural psychology* (pp. 3–39). New York: Guilford.

McCauley, R. N., & Bechtel, W. (2001). Explanatory pluralism and the heuristic identity theory. *Theory & Psychology, 11,* 736–760.

Meyer, J. S., & Quenzer, L. F. (2005). *Psychopharmacology: Drugs, the brain, and behavior.* Sunderland, MA: Sinauer Associates.

Mitchell, B., Mitchell, D., & Berk, M. (2000). The role of genetics in suicide and the link with major depression and alcoholism. *International Journal of Psychiatry in Clinical Practice, 4,* 275–280.

National Museum of the American Indian. (2007). *Do all Indians live in tipis? Questions and answers from the National Museum of the American Indian.* New York: Harper.

Novak, J. D. (1998). *Learning, creating, and using knowledge: Concept maps as facilitative tools in schools and corporations.* Mahwah, NJ: Lawrence Erlbaum.

Rogers, E. S., & Smith, D. B. (Eds.). (1994). *Aboriginal Ontario: Historical perspectives on the First Nations.* Toronto: Dundurn Press.

Simon, H. (1962). The architecture of complexity. *Proceedings of the American Philosophical Society, 106,* 467–482.

Smith, E. E., & Kosslyn, S. M. (2007). *Cognitive psychology: Mind and brain.* Upper Saddle River, NJ: Pearson Prentice Hall.

Sowa, J. F. (1999). *Knowledge representation: Logical, philosophical, and computational foundations.* Pacific Grove, CA: Brooks Cole.

Sperber, D. (1996). *Explaining culture: A naturalistic approach.* Oxford: Blackwell.

Strauss, C., & Quinn, N. (1997). *A cognitive theory of cultural meaning.* Cambridge: Cambridge University Press.

Sun, R. (Ed.). (2006). *Cognition and multi-agent interaction: From cognitive modeling to social simulation.* Cambridge: Cambridge University Press.

Sun, R. (2008). Cognitive social stimulation. In R. Sun (Ed.), *Cambridge handbook of computational psychology* (pp. 530–548). Cambridge: Cambridge University Press.

Sun, R., Coward, L. A., & Zenzen, M. J. (2005). On levels of cognitive modeling. *Philosophical Psychology, 18,* 613–637.

Tajfel, H. (1974). Social identity and intergroup behavior. *Social Sciences Information. Information Sur les Sciences Sociales, 13,* 65–93.

Thagard, P. (2000). *Coherence in thought and action.* Cambridge, MA: MIT Press.

Thagard, P. (2006). *Hot thought: Mechanisms and applications of emotional cognition.* Cambridge, MA: MIT Press.

Thagard, P. (2010a). *The brain and the meaning of life.* Princeton, NJ: Princeton University Press.

Thagard, P. (2010b). EMPATHICA: A computer support system with visual representations for cognitive-affective mapping. In K. McGregor (Ed.), *Proceedings of the workshop on visual reasoning and representation* (pp. 79–81). Menlo Park, CA: AAAI Press.

Thagard, P. (2012). *The cognitive science of science: Explanation, discovery, and conceptual change.* Cambridge, MA: MIT Press.

Thagard, P., & Aubie, B. (2008). Emotional consciousness: A neural model of how cognitive appraisal and somatic perception interact to produce qualitative experience. *Consciousness and Cognition, 17,* 811–834.

Thagard, P., & Stewart, T. C. (2011). The Aha! experience: Creativity through emergent binding in neural networks. *Cognitive Science, 35,* 1–33.

Thagard, P., & Wood, J. V. (2011). The self as a system of multilevel interacting mechanisms. Unpublished manuscript, Department of Philosophy, University of Waterloo, Waterloo, Canada.

Toronto, City of. (2010). Toronto culture—Exploring Toronto's past—Natives and newcomers, 1600–1793. Retrieved Dec. 6, 2010, from http://www.toronto.ca/culture/history/history-natives-newcomers.htm.

Turner, M. (2001). *Cognitive dimensions of social science*. Oxford: Oxford University Press.

Wagamese, R. (2010). One native life. *Canadian Dimension, 44*(Mar/Apr), 8–9.

3 Cognitive Basis of Culture and Cultural Processes

Norbert Ross

3.1 Introduction

The question of the cognitive basis of culture and cultural processes is, of course, a broad and complex topic that would require a much wider scope than could possibly be dealt with in this chapter. An examination of it would have to explore, among other topics, the evolution of the brain, the evolution of social complexity, and the origin of language, as well as the more general human capacity to deal with symbolic systems. While desiring to draw attention to the wider topic in order to make palatable the notion that cognition is an important ingredient in cultural processes, the goals here are much more humble: namely, to discuss some of the cognitive underpinnings responsible for shared cognitive models within human populations. Yet even for such a humble goal, the chapter falls short, since it lacks important accounts from the neurosciences, studies on infant communication, or a discussion of theory of mind. Instead, this chapter focuses on what are often termed individual cognitive processes that help shed some light on culture and, more specifically, the acquisition and production of shared knowledge. After a short comparison of the standard approaches in anthropology and the cognitive sciences, underlying understandings of culture and related approaches in studying culture are discussed. This leads to a discussion of knowledge acquisition as knowledge production and in relation to cultural change. After exploring different case studies, the chapter is summed up by suggesting a synergetic effort between the cognitive sciences and anthropology, mutually enriching both with respect to theory and methods.

Anthropologists usually take "learning by copying" for granted, ignoring the impossible task this puts on both individuals as well as "copying" as a mechanism. As a consequence of this misconception, anthropologists very often take agreement as the status quo, thereby defining cultural change as a "problem" to be understood. This chapter takes the opposite view. Problematizing cultural learning, this chapter argues that agreement, in fact, needs to be explained in cultural systems, in which change seems to be the inherent modus operandi.

From the outset, perceived differences between anthropology and cognitive psychology couldn't be greater. Anthropologists are known for going off the beaten path and canvassing the world for exotic customs and traditions. Participating in the life of (mostly) non-Western people has become the trademark of the anthropologist, who socializes over long periods of time with "natives" to better understand what they are doing and why. This stands in stark contrast to the cognitive scientist, who works out of her laboratory, conducting experimental research with the utmost scientific rigor, most often with the help of a team of students and research assistants. Here, the subject pool consists mostly of university students, about whom the researcher has little or no knowledge. This approach is most often justified by referring to the quest for discovering universal processes (e.g., how the mind works), again a stark contrast with the anthropologist's search for knowledge content and accurate description of the culturally unique life-worlds of those she studies. At times, one cannot help but wonder whether the universals encountered by cognitive scientists are limited to controlled lab settings and not necessarily reflective of real-world cognitions. Finally, anthropologists are more interested in group processes (i.e., culture, cultural change, and globalization) and hence pay less attention to processes taking place in the individual brain (i.e., cognition, categorization, or conceptual change).[1]

If one buys into these kinds of stereotypical differences, one could think of the two disciplines as complementary and, in fact, at times some of these differences have been prescribed as a healthy division of labor. In this division, anthropology would describe the context and content of thought—what people think—while the cognitive sciences explore the processes involved, or how people think (D'Andrade, 1981). However, upon closer inspection, several problems turn up immediately. If the cognitive scientist is interested in the universalities of the human mind, the question becomes whether such universals can be detected within the confines of the laboratory, artificial stimuli (think geometric forms on a computer screen), and a very limited participant pool (in terms of age, culture, and education). Furthermore, such a division of labor only works under the assumption that context, content, and process of thought can actually be separated and dealt with independently.

This assumption, however, seems not to be warranted (Ross & Medin, 2011). An increasing number of researchers now explore cognitive processes outside the laboratory and in diverse cultural settings. Indeed, the emerging understanding seems to be that US students are not at all representative of the many features measured and tested in the psychology lab (Henrich, Hein, & Norenzayan, 2010). While these findings are important, they tell us next to nothing about the causal factors that create the described differences in cognitive processing. Still, the data are sufficient to show that the social context within which the human mind develops plays a crucial part in its workings. While this should not come as a surprise, we are just beginning to understand the ways in which the social and physical environment habituates and primes the developing mind (see, for example, Inagaki, 1990; Ross, Medin, Coley, & Atran, 2003; Waxman, Medin, & Ross, 2007; Shenton, Ross, Kohut, & Waxman, 2011). As a result, engaging with participants from other cultures, understanding what people learn and know as well as the context and forms within which real-life learning takes place may be important steps for the cognitive sciences.

On the other side of the academic divide, things don't look much better. If anthropologists are interested in group processes and shared knowledge, the question becomes how these group processes are triggered, how they play out, and how knowledge becomes shared knowledge. To put it differently: while cognitive scientists struggle with a lack of understanding of (and interest in) how the cultural context influences the developing mind, anthropologists lack an understanding of (and interest in) how cultural context and knowledge is influenced by the developing mind. Interestingly, while most anthropologists see "culture" as entirely social, very few anthropologists—and those few only recently—are actually concerned with how cultural knowledge is acquired (Hirschfeld, 2002).

3.2 Concepts of Culture

The aforementioned lack of anthropological interest in exploring the acquisition of cultural knowledge is a consequence of a series of misconceptions in the field, dating from its inception as an academic endeavor.

First, anthropologists initially set out to study the life of human groups outside industrial societies. These groups—in the mind of the anthropologist and her readers—were best represented by the more traditional (i.e., more "exotic") sector of the societies. This sector usually consisted of the elders, who are often considered a community's encyclopedia. Within this framework, younger people know less of the cultural system, are less interesting to the anthropologist's endeavor, and might also already be "corrupted" by encroaching Western societies. Several problems emerge from this framework, but this chapter only wants to draw attention to a set of assumptions directly relevant to the present argument. First, the approach seems to assume that there is only one correct compilation of cultural knowledge (to be discovered by the anthropologist), and that this compilation of knowledge is held by the older people (ideally, spiritual leaders, etc.) of a community. Second, the approach assumes that knowledge is automatically acquired with age (since older people know more than youths) and that the acquisition of said knowledge basically represents a flawless reproduction of existing knowledge and models (see Quinn & Holland, 1987). Of course, if these assumptions were correct, the acquisition by learning of cultural models could basically be ignored as trivial.

The first set of assumptions is reflected in most definitions of culture. For example, Edward Tylor defined culture as "that complex whole which includes knowledge, belief, art, morals, law, custom, and any other capabilities and habits acquired by man as a member of society" (Tylor, 1871/1920, p. 1). Ward Goodenough took the next step, defining culture as "what is learned ... the things one needs to know in order to meet the standards of others" (1971, p. 19). To be clear, Goodenough—himself interested in cognitive processes—realized the consequence of his definition: "If culture is learned," he wrote, "its ultimate locus must lie in individuals rather than in groups. If we accept this, then cultural theory must explain in what sense we can speak of culture as being shared or as the property of groups at all, and it must explain what the processes are by which 'sharing' arises" (1971, p. 20). Note that Goodenough did not talk about how learning takes place, but about how "sharing arises," highlighting a process more complicated than the very often assumed "sunburn model," whereby exposure to culture makes for enculturation (see Quinn & Holland, 1987, for a critique of this model). At this point, it should be made clear that such sharing is not an either/or situation. While not explored sufficiently in cognitive anthropology, general anthropological literature has certainly shown that competing models exist within a population, and that the "correct meaning" is very often a contested issue, rather than a simple truth to be discovered by the researcher.

3.3   The Acquisition of Knowledge and Cultural Change

Unfortunately, while many anthropologists might subscribe to Goodenough's definition of culture, his call to understand the production of shared knowledge went largely unheard. Instead, copying of knowledge by younger generations was assumed automatic and flawless, and hence uninteresting for the anthropologist. In fact, these processes seemed to be so uninteresting that, more often than not, the acquisition of cultural knowledge is not even mentioned in anthropological accounts. However, looking at some of the topics studied by anthropologists sheds an interesting light on this issue. A central focus of anthropology—after describing cultural systems—has always been the issues of cultural change and acculturation as well as, more recently, globalization. Of course, such a focus on explaining change only makes sense if the opposite—that is, cultural stability and learning of shared knowledge over time—is the assumed default condition (see Ross, Timura, & Maupin, in press. In this scenario, only change needs explanation, not stability.

However, if learning and knowledge production are not explored, how could we possibly understand change? How could we possibly know whether and why cultural stability should be a default assumption? No one would reasonably argue that culture exists outside individuals' heads, aside from material expressions (Atran, Medin, & Ross, 2005). As a result, we need to account for the mental processes (within their specific context) that produce either agreement or disagreement among the individuals of a given population.

If cultural knowledge is shared knowledge, then change can take two basic forms: change over time in terms of model content, or change of agreement patterns (i.e., who agrees with whom). Content change might occur within individuals, who change their own models over time, or among kinds of individuals. In the latter case, specific groups of individuals at one point in time (say, all 30-year-old males) differ in their models from the same groups of individuals at another point in time. This latter process has generally received more attention among anthropologists (although not in such explicit terms). Exploring such patterns goes a long way toward understanding the specific history of cultural models and their content in a given population, underscoring that these patterns of culture and cultural change represent emerging products of individuals' cognitive behavior. In a sense, they represent the change of more or less agreed-upon models over time and across individuals. Such an approach calls for both an exploration of *what triggered change* and *how change took place*. It necessitates an exploration of the settings within which change might come about (think, for example, of schools as institutionalized engines for controlled conceptual change), as well as the individual cognitive processes involved, to understand the complex interactions of context, content, and process and emerging patterns of agreement and disagreement.

To elaborate on Goodenough: if the ultimate locus of culture lies in individuals rather than groups, then cultural theory must explain cultural processes as emerging products of individual cognitive processes within a specific historical and social context. It must also explain what the processes are by which "sharing" arises. In this view, culture and agreement cease to be default assumptions, the independent variables, but become something in need of explanation (see Medin, Ross, Atran, Burnett, et al., 2002; Ross, 2004).

This idea follows and elaborates on Dan Sperber's concept of the "epidemiology of representations," which problematizes the transmission of representations across different states (mental representations to public representations to mental representations) and actors (Sperber, 1996). A good illustration of this problem is the well-known telephone game, where a story told in a chain of individuals naturally undergoes significant transformations rather than being transported in high fidelity among the participants. While the findings of this game already cast some doubt on the feasibility of cultural learning through observation and copying alone, real life, of course, is not a telephone game. It appears that, in contrast to the telephone game, in real life a great deal of important cultural knowledge is not readily accessible for high-fidelity transmission. Or, said differently, in real life much of cultural knowledge is not openly available for learning that consists entirely of copying. Instead, learning seems to imply a good deal of knowledge production, based on limited input. If this is the case, we must understand the conditions under which this knowledge production takes place, taking into account the relevant cognitive and social factors. In this approach, culture is the emergent outcome of interacting social and cognitive factors (Sperber, 1996), with our foci set on understanding the history of agreement and disagreement within and across populations and the emerging contents of shared models.

Following this line of thought, the really interesting questions are not so much *why* cultural models change, but rather why some cultural models change so slowly, and how cultural models change in terms of content.

3.4 Culture and Expertise

So far we have only dealt with culture as "shared knowledge." However, the picture is further complicated by issues that are usually not related to culture. For example, a good deal of cognitive research has been dedicated to understanding the role of expertise in cognitive processes. Research by Boster and Johnson (1989) explored the effects of different levels of expertise on categorization. Specifically, they examined knowledge and sorting patterns (categorization) among expert fishermen and novice college students. These researchers noted that while morphological information about fish (provided on stimulus cards in the form of pictures) should be available to novices and experts alike, access to more specific information related to functional and utilitarian aspects requires expertise. Results indicated that the shift from novice to expert in fish categorization is not a shift from an incoherent to a coherent model, but instead represents a change from a readily available default model to a newly acquired model based on different goals and information. Clearly, then, the acquisition of knowledge (here specifically expertise) represents a conceptual shift, rather than a simple adding-on of new knowledge. Note how these findings also undermine the assumption of a continuous learning wherein the experts of a culture represent "better informants."

The point is further complicated by findings from Medin, Lynch, Coley, & Atran (1997). Here the researchers showed that tree experts with different types of expertise apply different categorization schemes to local trees. The authors point toward the acquisition of expertise (i.e., knowledge) as a process of modifying conceptual structures to attune to different kinds of information and goals, resulting in new categorization schemes.

While domain-specific expertise will very likely differ across cultures— think US undergraduate student versus tropical rainforest agriculturalist— and hence produce cultural differences, the phenomenon is also at work within populations.

In her study of curers and non-curers in the Purepecha town of Pichataro, Mexico, Linda Garro explored whether members of the two groups shared a common model of health and disease (Garro 1986; 2000). Using a question-answer frame technique, Garro was able to show that this was indeed the case. However, in her study curers agreed more with one another than with non-curers, or than non-curers with one another. This higher agreement implies more knowledge with respect to the culturally appropriate beliefs (Garro 1986; 2000).

Ross, Timura, and Maupin (in press) extended this research by revisiting the community after 30 years and by including biomedical personnel—a new group of experts—who arrived in Pichataro shortly after the study conducted by Garro. Despite a plethora of changes that had occurred in the community, the authors found that curers and non-curers still agreed with one another (and with the models from 30 years ago), but disagreed systematically with biomedical staff working in the community clinic and pharmacies (Ross, Timura, & Maupin, in press). It is important to keep in mind that expertise is a relative term, and a nonexpert adult in a Mexican community may have a great deal more knowledge of medicine and medicinal uses of plants than US undergraduates. All the Pichataro non-curers interviewed by Garro or Ross, Timura, and Maupin had substantial folk medical knowledge. They were adult women, many of whom had already cared for several children.

Let's take a closer look at this to understand the issue at hand. From an anthropological perspective, a description of this case would include a detailed ethnographic description of the 30 years that have passed and most likely focus on changes in ways of life. How did people live before versus now, and what are the conditions within which people make their treatment decisions? The study would conclude that despite strong attendance records at the clinic, the overall cultural model has not changed toward one held and disseminated by the clinic personnel. The clinic personnel, often members of the community themselves, would very likely be described as agents of change, with clearly limited success, which might be ascribed to a resistance in the population. After all, people might attend the clinic for convenience only, since the Mexican government often uses incentives (free food supply and cash payments) to increase use of medical services.

Note how such a description is completely devoid of individuals or individual minds, and how it does not tell us much about the actual processes involved. As a consequence, resulting theories of persistence of cultural knowledge or change therein do not represent theories of how change and persistence happen, but are descriptions of either of them within a specific historical setting. This becomes clear when we add a set of the following new questions related to the cognitive sciences and individuals' mental processes:

1. How and when were the specific folk medical models targeted by the initial study acquired? How and under what conditions did the adult generation of the first study acquire their folk medical models?

2. What were the conditions within which the contemporary adult generation (subjects for the second study) acquired their models? What was the input information available in both cases? What information was provided, how and by whom? How did this information interact with pre-existing models, and what did these pre-existing models look like?

Adding these questions to our exploration forces us to reframe our perspective in the above sense: change and persistence of cultural models over time are emergent outcomes of complex interactions. They both seem to be subject to similar underlying processes—the acquisition of knowledge by individuals. Hence, we need to understand the conditions that influence knowledge acquisition in their specific cultural and historical settings. On the level of theory, this calls for a detailed description and understanding of the specific processes involved, as well as the conditions within which they occur.

Of course, this is not always feasible. The project in Pichataro described above enrolled participants around the age of 50, meaning that at least parts of the acquisition of the relevant knowledge (while ongoing) probably took place years, if not decades, before the study. The same is of course true for the initial study conducted by Garro in 1979. The elicited knowledge in that study was clearly developed prior to the study and was probably building up (and changing) over quite some time.

This example should furthermore make it clear that within this context persistence of cultural knowledge across time and individuals cannot be described as a default outcome. Very rarely do the overall conditions of knowledge acquisition stay the same over a long period of time. Much more research is needed to establish a clear theoretical understanding of what aspects of these overall conditions interact, and in what ways, with the processes involved in knowledge acquisition. Cases where cultural change did or didn't occur are abundant. That our explanations consist mainly of post hoc reasoning makes it clear that there is a huge gap in our theoretical understanding.

In research among Lacandon Maya, clear intergenerational differences with respect to folk ecological models, values, and behavior among the adults of the two living adult generations were found (Ross, 2001, 2002). Differences were such that a continuum of ongoing learning (younger adults were still in the process of acquiring expert knowledge) could be ruled out. While the author was not able to directly observe the actual process of knowledge acquisition (members of the two groups were adults between the ages of 25 and 80 years old), the differences could be tentatively (yet, again, post hoc) attributed to a different upbringing of the members of the two generations. Adults of the older generation lived most of their lives in dispersed households close to their agricultural fields and apart from other families, moving their homes with the agricultural cycle. This lifestyle was interrupted when the area was declared a bioreserve and land rights were given to the Lacandon Maya families under the condition that they move into fixed communities. As a consequence, adults of the younger generation grew up in these newly founded communities, distant from the agricultural fields but close to other families. Within this new setting, a new set of values developed that distanced the community from the forest both in terms of physical distance to an important source of information (observation) as well as in terms of values and interest. While people lost a certain amount of interest in the forest, other topics became important in Lacandon life, providing new possibilities for thought and communication.

Again, while the outcome of these processes is clearly measurable, the specific processes remain obscure. What kind of input created what kind of knowledge? How did new shared knowledge arise? Similar effects of intergenerational change were reported for Itzá Maya, second-generation adults living in close proximity to the northern Guatemala rainforest (see Atran, Medin, Ross, & Lynch et al., 1999; Atran, Medin, Ross, & Lynch et al., 2002; and LeGuen, Iliev, Atran, Lois et al., 2010).

To get a better handle on proximate causes for change in knowledge acquisition, Shenton et al. (2011) explored the acquisition of plant knowledge in two different locales of one Tzotzil Maya municipality of Chiapas, Mexico. The two locales were picked to represent two extremes in terms of modernization in the community. The first was a municipality center, a small-scale urbanized setting, while the other locale was a small hamlet approximately 90 minutes away. Children and adults in the two locales differ in their exposure to and interaction with the environment; these differences translate into systematic differences with respect to the acquisition of plant knowledge (Shenton et al., 2011). (See Zarger & Stepp, 2004, for a study of children's plant knowledge over a 30-year time period in a Highland Maya community). While this study relates differences in patterns of knowledge acquisition to differences in exposure to the environment, the scope of the data is somewhat limited as it only examines the acquisition of plant names. Obviously, plant names can only be learned through a process of copying, and hence the research does not allow a window into the more generative aspects of knowledge acquisition. It is this latter aspect that will be attended to next.

3.5 Categorization, Reasoning, and Latent Knowledge

The intergenerational change of folk biological knowledge was briefly described above for both the Lacandon Maya of southern Mexico (Ross,

2001, 2002) as well as the Itzá Maya of northern Guatemala (Atran, Medin, Ross, Lynch et al., 2002). These differences already seem to rule out the idea of younger people (successfully) copying their elders, the experts of their community. Furthermore, in both studies the data sets consisted of individual interviews, in which each individual was queried with respect to about eight hundred specific animal-plant interactions. While this is definitely a high number of questions for a single interview, the actual number of species involved (fifty-six animals and plants) represents only a very small fraction of the species known to study participants. As a consequence, it seems implausible that individuals copy each other's responses or memorize all the possible interactions (of any given domain) in any formal or informal way. First of all, this information is not publicly available—in the sense that people talk about it. While people talk about plants and animals, these conversations are usually limited to certain species. Individuals do not—and cannot—include all possible species of the forest in their conversations. Furthermore, while learning all possible species interactions by copying would already create an insurmountable learning and memory problem, forest life is about much more than simple species interactions. As a result, the study tasks not only involved only a small fraction of potential interactions occurring in the rain forest, but in addition, such species-interactions constitute only one aspect of complex folk ecological knowledge. Finally, ecological knowledge is only one of a large number of cultural knowledge domains, making copying as the only or even the primary mode of knowledge acquisition extremely implausible.

Within this context, then, the question of how knowledge is acquired or generated, and how within this process agreement across individuals is achieved is anything but trivial. Once we reject the notion of cultural learning by copying alone and step away from the possibility of learning culture by sheer exposure, the problem of shared knowledge becomes an interesting issue in need of study.

How then does agreement across individuals emerge? How do people learn and know cultural knowledge? While much work remains to be done, several data sets exist that allow us a window into a potential answer to this puzzle.

Above, research indicating that categorization is influenced by goals and values as well as expertise was reviewed. Categories have often been described as the building blocks of thought (Smith & Medin, 1981), as they allow us not only to chunk information for easier storage and communication (D'Andrade, 1995), but, more importantly, because they organize our knowledge in ways that allow us to reason about the specific members of categories (Ross, Maupin, & Timura, 2011). This type of reasoning is termed category-based induction.

Probably the best-known theory of category-based induction is the Osherson, Smith, Wilkie, Lopez et al. (1990) similarity-coverage model. Three phenomena associated with the theory have received the most attention: *similarity*, *typicality*, and *diversity*.

The *similarity* principle of induction describes the fact that two kinds seen as similar (more closely related in terms of their taxonomic distance) are more likely to share a previously unknown (and invisible) property or characteristic, compared to two kinds that are taxonomically more distant. For example, informants usually judge mice and rats as more likely to share some unknown property than mice and penguins.

The *typicality* principle describes the fact that the more typical members of a category are more likely to have features common to all the category members than less typical ones. For example, if informants are told that sparrows have some protein x inside them and that penguins have some protein y inside them, they judge that it is more likely that all birds have protein x than protein y.

Finally, the *diversity* principle describes the fact that individuals are usually more likely to ascribe a property to the whole category when told that two taxonomically distant category members share that property, than when told that two taxonomically similar category members share a property. A projection from mice and cows to all mammals is stronger than a projection from mice and rats to all mammals.

Such processes might provide the basis for the independent production of shared knowledge. If categories provide the basis for reasoning (here, finding the solution to a novel problem) and if a category structure is shared, then we might reasonably assume that the outcome of category-based reasoning is shared across the individuals of a population. In this scenario, categories and category-structure constitute an important *latent knowledge*, with category-based induction providing one mechanism through which people generate new (explicit) shared knowledge. Is there some evidence that this is indeed the case?

In the already mentioned plant-animal interaction task, Atran, Medin, Ross, et al. (2002) compared the responses of Itzá Maya from two adult generations. The authors found that young Itzá Maya differed from their elders in some responses, yet still agreed with one another. This suggested that agreement can emerge in the absence of direct observation or teaching and learning (see Ross 2002 for a similar result from the Lacandon Maya). Exploring the cases in which young Itzá Maya agreed, but systematically differed from the responses of their elders, it became clear that young Itzá Maya use certain animal-plant interactions as base models and extend these responses to interactions with similar species. While on average this might provide a reasonable response (as similar animals might affect similar plants in similar ways), there are some cases where it doesn't, opening a window into the underlying mechanics of active knowledge production. For example, older Itzá Maya make a distinction between the way howler monkeys and spider monkeys affect the ramon tree. The ramon tree is well known in the area and both the leaves as well as the fruits are used. The fruit is fairly large and contains a large, hard seed. In the view of the elders, howler monkeys help the tree because they are big enough to swallow the entire seed and disperse it throughout the forest. This contrasts with the much smaller spider monkeys that hurt the tree—according to the elders—by cracking the seed open before swallowing it. Young Itzá Maya, however, overextend the category-based similarity between the two monkey species and attribute the same effect on the ramon tree to both species. Several other examples point in the same direction, supporting the idea that young Itzá indeed use their (shared) taxonomy to generate agreed-upon (albeit wrong) responses (see Atran, Medin, Ross, Lynch et al., 2002)[2].

Clearly, then, the category structure and category-based induction strategies provided our study participants with the ability to generate new knowledge on the spot in ways that produced shared knowledge without a teacher-learner environment or the ability to learn from direct observation.

Ross, Maupin, and Timura (2011) have further evidence for this from research on folk medical models among Mexican migrants residing in Nashville, TN. Employing different tasks to explore the knowledge organization with respect to nineteen illnesses, the authors found that these illnesses were divided into two real categories and two ad hoc groups. Real categories are marked by their potential for category-based induction, while ad hoc groups are formed on the spot for different purposes. Such ad hoc groups share certain features with real categories (see Barsalou, 1983), yet do not allow inference-making (think all "things red"). In their data, the authors found strong evidence that only the real categories provided informants with the ability to make inferences (Ross, Maupin, & Timura, 2011).

These studies point at taxonomic knowledge as important *latent knowledge* that is available to produce new knowledge when needed. As mentioned above, category-structure changes in expertise (both in level and kind) lead to important group differences both within and across populations. The studies cited above among Lacandon, Itzá, and Tzotzil Maya all included a design in which expertise changed either over time or across study populations. While some of the tested knowledge might be directly affected by hands-on expertise (e.g., learning plant names as in Shenton et al., 2011), the actual group differences are exponentially larger given the potentially infinite number of different inferences to be made.

This leads to the question of the role of expertise and culture.

3.6 Expertise, Culture, and Cultural Change

If, as argued above, expertise differences create group differences, then the question becomes whether expertise differences within culture create disagreement beyond group membership. Asked differently, do experts and non-experts of a group agree more with one another than experts and non-experts of two different cultural groups? As already described, at least in the Purepecha community of Pichataro, curers and non-curers continue to share a common model of health and disease (Garro 1986, 2000; Ross, Timura, & Maupin, in press. In fact, curers and non-curers of the community agree more with one another than the curers agree with another set of medical experts, biomedical providers.

The same has been found in research among Menominee and Euro-American (nonprofessional) fish experts of Wisconsin. Comparing fish experts and novices of the two cultural groups living adjacent to each other, the authors found that each group's experts agreed more with the novices of their group than with the experts of the neighboring community (Medin, Ross, Atran, Burnett et al., 2002). Clearly, then, expertise is culture-specific. Following up on this initial finding, the authors explored the differences among fish experts in the two cultural groups in more detail.

Both groups shared a general overall agreement with respect to the sorting of fish (categorization), with Menominee experts having a sub-model not shared with Euro-Americans. For each group, a combined model was analyzed using multidimensional scaling. In order to represent the model of the Euro-American experts, two dimensions were needed, correlating with desirability and size of the fish. For the Menominee data, three dimensions were needed to achieve a fit for their sorting data. This third dimension—not found among Euro-Americans—correlated with what the authors termed an ecological dimension (for example, sorting by habitat).

This task was followed, months later, by several different but related tasks. First, participants were asked to describe fish-fish interactions for all possible pairs of twenty-one fish species (e.g., "Does the largemouth bass affect the river shiner, or the river shiner affect the largemouth bass?"). Notably, this task was conducted within about 1.5 hours, and at a fast pace (necessarily so, since the questions addressed about 420 potential relations).

As in the categorization task, the authors found cross-group consensus, and again Menominee experts held a clear submodel not shared by Euro-American experts. Menominee experts reported more relations overall and more reciprocal relations. In addition, Euro-American fishermen mainly reported interactions involving adult fish of the kind, "a musky will eat a northern." Menominee experts reported these relations, too, but also added relations characterizing whole life cycles (e.g., "a musky will eat a northern, northern fry hatch out about two weeks before musky fry, and northern fry will eat musky fry"), as well as relations other than food-chain (e.g., noting that a certain fish destroys the nesting place of another). The small set of relations reported by Euro-Americans but not Menominee fishermen all appeared to be overgeneralizations. They typically involved food-chain relations between predator and prey fish that are rarely, if at all, found in the same waters (a fact that was typically mentioned by Menominee participants).

Had the studies stopped here, it would seem natural to conclude that the authors were observing cultural differences in knowledge, and they might have been able to weave a story about why Euro-American fishermen mainly learn about adult fish. Readers in anthropology might have liked the authors' conclusion with respect to the intricate knowledge of indigenous people. However, here is where ethnography and spending plenty of time with fishermen outside of a formal interviewing situation were helpful, since what made the results puzzling for the authors was that they had heard more than one Euro-American expert mention that northern fish hatch in the spring a few weeks before muskies do. Judging from informal encounters with them, it also seemed implausible that Euro-American fish experts would have so little ecological knowledge.

In order to test this idea, the authors followed up with a second experiment exploring the knowledge individuals hold about fish habitats. In this task, participants sorted the fish according to different habitats. Two findings are important here. First, members of both groups did not differ in their responses and, second, Euro-American experts correctly described fish as not sharing a habitat, for which in the previous task they had described a "big eats small" relationship. These data reinforced the notion that cultural differences were triggered in part by the tasks and did not represent a simple cultural difference with respect to existing ecological knowledge.

Instead of representing differences in knowledge per se, perhaps the cultural differences encountered in this study were more about *knowledge organization* and related *accessibility*? The sorting task suggests that Menominee fishermen make use of an ecological organization, which might facilitate answering questions about fish-fish interactions. On the other hand, if the Euro-American experts focus more on taxonomic relations, it may take more time and effort to retrieve information about ecological relations. To test this idea, several months later, the authors repeated the fish-fish interaction task, but this time reduced the number of probes from 441 to 35, while still allowing an hour for the task. If their analysis was correct, they should have found that 1) the cultural differences would disappear, 2) Euro-American experts should start to answer in terms of the full life cycle of fish, and 3) Menominee experts would be relatively less affected by the pace of the task. All three predictions received strong support (Medin, Ross, Atran, Cox et al., 2006).

The authors take these data to indicate two things. First, experts of the two groups share a good amount of base knowledge with respect to ecological relations and fish habitats. This should not be surprising given the fact that, on average, experts in the two groups had fished for over 40 years. Arguably, knowing where fish can be found represents an important piece of information within fishing. Second, this ecological knowledge is not equally accessible. The authors find systematic group differences (that one might be tempted to call cultural), yet these differences are not in knowledge, per se, but in access to this knowledge.

This allows elaboration on the role of categories. Previously it was argued that processes of category-based induction allow for the independent production of shared knowledge, if the underlying category-structure is shared. As a consequence, cultural differences in categorization will lead to cultural differences in the production of shared knowledge. Research by López, Atran, Coley, Medin et al. indicates that this is indeed the case. In this study the differences in reasoning between US students and Itzá Maya could be predicted based on differences in the underlying category structure (López, Atran, Coley, Medin et al., 1997). Data from the Menominee study allow further elaboration. Here, the category structure influenced accessibility of kinds of knowledge. While the expert members of the two groups clearly share a lot of base knowledge, that knowledge was not equally accessible when producing new knowledge—here, answers to questions about species interaction (Medin et al., 2006). In a sense, one could argue that the *relevancy of certain information* differed for members of the two cultures, leading to stable differences in cross-group agreement patterns (Sperber & Wilson, 1986).

In parallel research with children of the same populations, Ross, Medin, Coley, and Atran found these differences to emerge at a rather young age (2003). Menominee children (six years of age) already show greater attention to ecological relations than their rural Euro-American counterparts (Ross, Medin, Coley, & Atran, 2003). This should not be too surprising, given the finding that Menominee experts agree more with Menominee non-experts than experts from a neighboring community. However, it makes it clear that anthropologists need to pay more attention to the acquisition of cultural knowledge and explore the nature of cultural differences in more detail. In this specific case, it seems clear that the question is not so much how Menominee children learn specific ecological knowledge, but more how a specific ecological orientation gets passed on. This question will be returned to below.

3.7 Acquisition of Cultural Knowledge

A disclaimer is in order. Much more research is needed to better understand the processes, channels, and units by which cultural knowledge is acquired and transmitted. From the above, it seems safe to assume that learning by copying is only part of the story and that neither is there a guarantee of faithful copying, nor is it possible that copying is the main process of learning. Of course, the idea is not rejected that observation, instruction, and play are important aspects of a child's enculturation (see Zarger 2002). However, what is observed and how it is integrated into existing conceptual webs are two very different issues.

In this respect, it is worthwhile to look across another academic divide into the educational sciences. Strike and Posner (1992) argue that in order to understand conceptual change in educational settings, it is critical to understand the student's *conceptual ecology*, the collection of concepts and relations among them that is implicated in understanding experience (including class lessons) and generating knowledge from lessons. In the classroom, it is often the case that the very same existing conceptions (and misconceptions) that are the target of instruction provide the children's framework to integrate new information (see Vosniadou & Brewer, 1992, for a good example related to the concept of gravity). Note how this notion of learning positions the learner as an active participant in the process of knowledge generation. Knowledge is not achieved by exposure alone, but represents a complicated mental process. Note, furthermore, how this approach does not regard the developing mind as an empty container to be filled with knowledge (through observation or copying), but how existing knowledge gets reshaped and reformulated through processes of conceptual change.

If this perspective is correct, we need to better understand the conceptual ecology that is invoked when acquiring new knowledge. While preexisting notions might not be directly targeted and changed by cultural models, they clearly provide one input for the generation of cultural models. These preexisting models themselves might be either cultural in nature, or innate.

For example, humans tend to conceive of species of organisms in essentialist terms (see Gelman & Wellman, 1991; Wellman & Gelman, 1992). Stated briefly, humans seem to assume by default that living things belong to natural categories by virtue of a shared essence, which is passed from parent to offspring. This assumption generates obvious traits that are seen as indicative of an individual thing, and maintain its identity through time. Such concepts color everyday perception of living things and aid dramatically in the generation of new knowledge that goes beyond firsthand experience (for example, through the aforementioned process of category-based induction). In the educational setting, this tendency seems to lead to misconceptions of evolution as described in research on evolution education (e.g., Evans, 2000; Poling & Evans, 2002; Samarapungavan & Wiers, 1997), but also primes the way people think about race and ethnic groups (Hirschfeld, 1996; Gil-White, 2001).

Waxman, Medin, & Ross (2007) found that Menominee as well as rural and urban Euro-American children ascribe an essence to living species. In this view, species are what they are because of their essence. On top of this general framework, Menominee children—drawing on salient discourse about ethnic/racial identity in terms of blood quantum—may tend to identify blood as a carrier of essence. Employing the common "adoption paradigm," the authors asked children whether an animal adopted at birth by an animal from another species (such as pig and cows) would grow up to be a member of the species of the birth parents or of the adoptive parents. In a series of tasks, several conditions were described (such as "always snuggling up with the adoption mother," "eating the same food as adoption mother," "drinking the milk of the adoption mother," etc.).

Only in the case of a total blood transfusion with the blood from the adoptive mother were some Menominee children (and only they) willing to change their predictions concerning the kindhood of the adopted animal when fully grown.

Even the cryptic description of the task makes it clear that these results are based on questions an individual will probably never encounter outside the research context. Does this make the questions useless? No. First, most cultural knowledge is implicit. Second, responses to our questions are fairly systematic across individuals, indicating that children used similar strategies and extracted similar background information when producing new knowledge (answers to our questions). The result is agreement on a topic they had probably never thought about before, one that presumably reflects community discourse concerning blood quantum.

In the above case, children use existing discourse regarding their immediate environment to make sense of a presumably innate framework theory. This, of course, is different from the question of how children might acquire the ecological orientation mentioned above for the Menominee.

While the authors are far from being able to answer the question in detail, they think that direct interactions as well as an indirectly transmitted general epistemological framework are responsible. Bang, Medin, and Atran (2007) asked Menominee and Euro-American children and adults about the nature and frequency of their outdoor practices. They found that Euro-Americans were much more likely to engage in practices in which nature is backgrounded (e.g., playing baseball) and much less likely to engage in practices in which nature is foregrounded (e.g., berry-picking).

In related research, Unsworth (personal communication) asked Euro-American and Menominee adults to describe the last encounter they had with a deer. In addition to the content of the stories, she also recorded the gestures used. The two groups did not differ in the overall likelihood of using gesture, but they showed a very large effect of perspective when gesturing about deer. Euro-American adults would "place" the deer in some location using their hands, while a significant proportion of Menominee adults "became" the deer in gesture. That is, they were reliably more likely to take the deer's perspective in gesture than were Euro-American adults. Much more evidence is needed, yet we propose that these observations will hold the key to understand children's learning, reasoning, and discourse concerning nature.

To put it another way, perspective-taking might be related to the differences in how Menominee and Euro-American hunters see the world. In a sense, "becoming the deer" takes into account the surrounding world from a perspective other than direct, human egocentric interaction with the environment. Rather than being the center of the universe, where nature becomes a backdrop for hunting and fishing, Menominee seem to explore nature, at least in part, through multiple lenses and perspectives.

There is some preliminary evidence that this is also true for Tzotzil Maya of Chiapas. In a study that originally targeted the presumed connection between spatial language and spatial cognition (see Levinson, Kota, Haun, & Rasch, 2002; Li & Gleitman, 2002), the authors found that Tzotzil Maya were much more likely to engage in perspective-taking when recreating a spatial scene after a 180-degree rotation. Rather than arranging objects after rotation based on the individual's left and right, Maya participants arranged them in such a way that the object that used to be on their right side on a stimuli table was placed on their left side on the recall table (and vice versa). While these findings have previously been described as being caused by language use (Levinson, 2003), the data seem to contradict these claims. First, in a language production task, Tzotzil speakers used language opposed to what the findings in the rotation task would suggest. Second, objects that on the stimuli table faced Maya participants were placed on the recall table in such a way that they now faced away from the partici-pants—maintaining the direction in which they looked before. Third, similar results were found with speakers of different languages living in the same environment. Based on this study, it appears that Tzotzil-Maya take on the perspective of the observed objects (and the objects' left-right) rather than putting themselves at the center of attention. The authors are currently exploring this theory further by looking at an interaction of kind of stimuli and culture.

3.8 Discussion: What Can the Cognitive Sciences Do for Anthropology?

The examples above should illustrate how implementing methods and theories from the cognitive sciences can enrich anthropology. Let us first start with methods. Clearly, experimental methods can provide the anthro-pologist with insights difficult to obtain via more traditional techniques of informal interviews and participant observation. For example, it appears that in order to understand processes of change or persistence, we need to understand patterns of agreement among participants. It is hard to see how traditional anthropological methods can provide the kind of data needed to compare and calculate individual agreement and disagreement in statistical ways. Calculating consensus as well as submodels held by groups of informants—as described for various studies above—requires a controlled experimental setting that not only allows comparison of detailed responses across individuals, but also provides data suitable for statistical analyses. Experimental methods also often allow a better understanding of the problem at hand. For example, the differences between Menominee and Euro-American fish experts only became evident when comparing the results from different tasks. In this particular research, cultural differences were revealed only when participants responded under time pressure. This is an important finding, since it pointed the authors toward the nature of these differences: rather than being a difference in base knowledge, the difference seems to be in accessibility of knowledge. Again, it is hard to see how qualitative methods would have allowed the detection of these subtle but important cultural differences.

One of anthropology's strengths has always been its practice of talking with real people about real-world concerns. This is, of course, a striking difference when compared to most research in the cognitive sciences. Many of the questions raised in the experiments described above were probably entirely new to the study participants. However, this does not make these questions worthless. In fact, it might represent one of the strengths of the experimental approach, since it allows for (1) independently tested hypotheses developed from observations, and (2) direct exploration of how people create new knowledge. To be sure, the argument is not that anthropologists should give up their toolkit and replace it with experimental research conducted by cognitive scientists. However, the preceding examples have shown that adding experimental techniques will dramatically enhance the quality of anthropological research. Of course, the same holds true for the cognitive scientist as well. Experimental research needs to be grounded in a solid ethnographic understanding, both in design as well as the interpretation of data. Only by wedding these different methodological approaches will we be able to overcome severe limitations in the two fields.

While it is certain that applying more experimental methods will enhance anthropological research, the bigger gain will come from implementing research questions and theories from the cognitive sciences into anthropological research.

Again, the above discussion has hopefully made it clear that the difference between anthropology and the cognitive sciences seems to be one of scale. Anthropology explores group processes, often ignoring the fact that these group processes emerge out of the complex interaction of individuals and individual minds. To overcome this deficiency, they clearly need the cognitive sciences, which enable anthropologists to ground the group-based processes they strive to describe in cognitive processes taking place in individual minds. From this perspective, anthropological theories of change over time have to be seen and studied as part of a theory of acquisition of knowledge under changing conditions. In this view learning, the acquisition and generation of new knowledge, must be explored in terms of its cognitive bases (learning mechanisms as well as mechanisms of concept formation and conceptual change) and its social bases (learning input and context). Of course, neither of these components are simple or trivial, and neither are their interactions. As such, it is important to realize that to understand group processes it does not suffice to simply add up individual cognitive processes across all members of a population. As elaborated above, cognitive processes are embedded in their social context, and hence we can expect feedback as well as interaction across such processes. However, while such an exploration will require tools to model complex dynamic systems, it is clear that the solution has to come from both anthropology and the cognitive sciences.

To be clear, while anthropology has a lot to gain from integrating cognitive science methods and theories, the opposite is just as true. A focus on research in the sterile lab, with experiments taking place within a very narrow participant pool devoid of any social context, will at best create artificial results. Becoming an expert in a laboratory task (such as, for example, categorizing geometrical figures) has usually next to nothing in common with the time, engagement, and activities needed to become a Menominee fish-expert. As a result, data produced in this kind of laboratory research are highly questionable in terms of their ability to replicate real-life situations. In this sense, it seems that the cognitive sciences and anthropology complement one another nicely; combined, they will form a new approach able to address important questions of the human mind and society.

Acknowledgments

The writing of this chapter was supported by NSF grants 0726107 and 0929457 to Norbert Ross and 0527707 to Norbert Ross and Tom Palmeri.

Notes

1. To locate these processes in the individual brain does not strip them from environmental influences.

2. Here the standard to qualify a response as "wrong" was the response agreed upon by the elders. While we tend to believe their observational skills, it is important to note that in this context it is not important which of the responses is objectively accurate, but rather that they differ in systematic ways.

References

Atran, S., Medin, D., & Ross, N. (2005). The cultural mind: Environmental decision making and cultural modeling within and across populations. *Psychological Review, 112*(4), 744–776.

Atran, S., Medin, D., Ross, N., Lynch, E., Coley, J., Ucan Ek', E., et al. (1999). Folkecology and commons management in the Maya Lowlands. *Proceedings of the National Academy of Sciences of the United States of America, 96*(13), 7598–7603.

Atran, S., Medin, D., Ross, N., Lynch, E., Vapnarsky, V., Ucan Ek', E., et al. (2002). Folkecology, cultural epidemiology, and the spirit of the commons. *Current Anthropology, 43*(3), 421–450.

Bang, M., Medin, D. L., & Atran, S. (2007). Cultural mosaics and mental models of nature. *Proceedings of the National Academy of Sciences of the United States of America, 104*, 13868–13874.

Barsalou, L. (1983). Ad hoc categories. *Memory & Cognition, 11*(3), 211–227.

Boster, J., & Johnson, J. (1989). Form or function: A comparison of expert and novice judgments of similarity among fish. *American Anthropologist, 91*, 866–889.

D'Andrade, R. (1981). The cultural part of cognition. *Cognitive Science, 5*, 179–195.

D'Andrade, R. (1995). *The development of cognitive anthropology.* Cambridge: Cambridge University Press.

Evans, E. M. (2000). The emergence of beliefs about the origins of species in school-age children. *Merrill-Palmer Quarterly, 46*(2), 221–254.

Garro, L. (1986). Intracultural variation in folk medical knowledge: A comparison between curers and non-curers. *American Anthropologist, 88*(2), 351–370.

Garro, L. (2000). Remembering what one knows and the construction of the past: A comparison of cultural consensus theory and cultural schema theory. *Ethos, 28*(3), 275–319.

Gelman, S., & Wellman, H. (1991). Insides and essences: Early understandings of the nonobvious. *Cognition, 38*, 213–244.

Gil-White, F. (2001). Are ethnic groups biological "species" to the human brain? *Current Anthropology, 42*(4), 515–554.

Goodenough, W. (1971). *Culture, language and society*. Reading, MA: Addison-Wesley.

Henrich, J., Heine, S. J., & Norenzayan, A. (2010). The weirdest people in the world? *Behavioral and Brain Sciences*, *33*, 61–83.

Hirschfeld, L. A. (1996). *Race in the Making: Cognition, culture and the child's construction of human kinds*. Cambridge, MA: MIT Press.

Hirschfeld, L. A. (2002). Why don't anthropologists like children? *American Anthropologist*, *94*(4), 946–947.

Inagaki, K. (1990). The effects of raising animals on children's biological knowledge. *British Journal of Developmental Psychology*, *8*, 119–129.

LeGuen, O., Iliev, R., Atran, S., Lois, X., & Medin, D. (2010). A garden experiment revisited: Inter-generational changes in the sacred and the profane in Petén, Guatemala. Society for Anthropology of Science meeting, February 17–20, Albuquerque, NM.

Levinson, S. (2003). *Space in language and cognition: Explorations in cognitive diversity*. Cambridge: Cambridge University Press.

Levinson, S., Kota, S., Haun, B. M., & Rasch, B. (2002). Returning the tables: Language affects spatial reasoning. *Cognition*, *84*, 155–188.

Li, P., & Gleitman, L. (2002). Turning the tables: Language and spatial reasoning. *Cognition*, *83*, 265–294.

López, A., Atran, S., Coley, J., Medin, D., & Smith, E. (1997). The tree of life: Universal and cultural features of folkbiological taxonomies and inductions. *Cognitive Psychology*, *32*, 251–295.

Medin, D., Lynch, E., Coley, J., & Atran, S. (1997). Categorization and reasoning among tree experts: Do all roads lead to Rome? *Cognitive Psychology*, *32*(1), 49–96.

Medin, D., Ross, N., Atran, S., Burnett, R. C., & Blok, S. V. (2002). Categorization and reasoning in relation to culture and expertise. *Psychology of Learning and Motivation*, *41*, 1–41.

Medin, D. L., Ross, N. O., Atran, S., Cox, D., Coley, J., Proffitt, J. B., et al. (2006). Folkbiology of freshwater fish. *Cognition*, *99*(3), 237–273.

Osherson, D., Smith, E., Wilkie, O., Lopez, A., & Shafir, E. (1990). Category-based induction. *Psychological Review*, *97*, 85–200.

Poling, D., & Evans, E. M. (2002). Why do birds of a feather flock together? Developmental change in the use of multiple explanations: Intention, teleology and essentialism. *British Journal of Developmental Psychology*, *20*, 89–112.

Quinn, N., & Holland, D. (1987). Culture and cognition. In: D. Holland & N. Quinn (Eds.), *Cultural models in language and thought* (pp. 3–42). Cambridge: Cambridge University Press.

Ross, N. (2001). *Bilder vom Regenwald: Mentale Modelle, Kulturwandel und Umweltverhalten bei den Lakandonen in Mexiko.* Münster, Germany: Lit Verlag.

Ross, N. (2002). Cognitive aspects of intergenerational change: Mental models, cultural change, and environmental behavior among the Lacondon Maya of southern Mexico. *Human Organization, 61*(2), 125–138.

Ross, N. (2004). *Culture and cognition: Implications for theory and method.* Thousand Oaks, CA: Sage Publications.

Ross, N. & Maupin, J. & Timura, C. A. (2011). Knowledge organization, categories, and ad hoc groups: Folk medical models among Mexican migrants in Nashville. *Ethos, 39,* 165–188.

Ross, N., & Medin, D. (2011). Culture and cognition: The role of cognitive anthropology in anthropology and the cognitive sciences. In D. B. Kronenfeld, G. Bennardo, V. C. de Munck, & M. D. Fischer (Eds.), *A Companion to Cognitive Anthropology.* Oxford: Wiley-Blackwell.

Ross, N., Medin, D., Coley, J., & Atran, S. (2003). Cultural and experiential differences in the development of folkbiological induction. *Cognitive Development, 18*(1), 25–47.

Ross, N., Timura, C., & Maupin, J. (in press). Stability in emergent cultural systems: Globalization and cultural resiliency in folk medical beliefs. *Medical Anthropology.*

Samarapungavan, A., & Wiers, R. (1997). Children's thoughts on the origin of species: A study of explanatory coherence. *Cognitive Science, 21*(2), 147–177.

Shenton, J., Ross, N., Kohut, M., & Waxman, S. (2011). Maya folk botany and knowledge devolution: Modernization and intra-community variability in the acquisition of folkbotanical knowledge. *Ethos, 39,* 349–367.

Smith, E., & Medin, D. L. (1981). *Categories and concepts.* Cambridge, MA: Harvard University Press.

Sperber, D. (1996). *Explaining culture: A naturalistic approach.* Cambridge: Blackwell.

Sperber, D., & Wilson, D. (1986). *Relevance: Communication and cognition.* Oxford: Blackwell.

Strike, K. A., & Posner, G. J. (1992). A revisionist theory of conceptual change. In R. A. Duschl (Ed.), *Philosophy of science, cognitive psychology, and educational theory and practice* (pp. 147–176). Albany: SUNY Press.

Tylor, E. (1920). *Primitive culture*. New York: J.P. Putnam's Sons. (Original work published 1871)

Vosniadou, S., & Brewer, W. F. (1992). Mental models of the Earth: A study of conceptual change in childhood. *Cognitive Psychology, 24*, 535–585.

Waxman, S., Medin, D., & Ross, N. (2007). Folkbiological reasoning from a cross-cultural developmental perspective: Early essentialist notions are shaped by cultural beliefs. *Developmental Psychology, 43*(2), 294–308.

Wellman, H., & Gelman, S. (1992). Cognitive development: Foundational theories of core domains. *Annual Review of Psychology, 43*, 337–375.

Zarger, R. (2002). *Children's ethnoecological knowledge: Situated learning and the cultural transmission of subsistence knowledge and skills among Q'eqchi' Maya*. Athens, GA: University of Georgia Press.

Zarger, R., & Stepp, J. (2004). Persistence of botanical knowledge among Tzeltal Maya children. *Current Anthropology, 45*(3), 413–418.

4 Egocentric and Allocentric Perspective in Cultural Models

Bradd Shore

4.1 Introduction

Over the last several decades, cognitively oriented anthropologists have looked to developments in cognitive psychology and cognitive science to help them rethink the concept of culture (Gentner & Stevens, 1983; Johnson-Laird, 1983; D'Andrade, 1989, 1995; Holland & Quinn, 1987; Strauss & Quinn, 1997; Shore, 1991, 1996, 1998). One of the most fruitful approaches to a cognitively oriented anthropology is to conceive of culture as a socially distributed stock of models, at once mental and social. Cultural cognition is viewed as a set of relations between external social institutions and their cognitive analogs as culturally conditioned mental models. Viewed in this way, thinking and feeling become at once public activities, in the social world, and private activities, engaging individual minds (see Geertz, 1973, p. 45).

This chapter will present a basic outline of cultural models theory, and then illustrate this dimension of cognition by focusing in on how cultural models shape actors' point of view, both literally and more abstractly. The chapter will take up the classic distinction between allocentric and egocentric perspectives in spatial navigation and show how this distinction informs several kinds of orientational cultural models.

Using models as a unit of culture allows for the linkage between mental representations and social institutions. Modeling is understood as a key activity of both the human brain (i.e., the formation of neural networks) and of human societies (i.e., the creation of stable social institutions). Modeling is a simplification and patterning of empirical phenomena. Models have been stripped of some phenomenal dimensions, highlighting certain aspects of organization at the expense of others. Such simplifications afford perceptual salience, generalizability to numerous particular instances, and memorability of models within a community. Shared social models constrain individual mental representations such that experiences are likely to be interpreted in similar and easily communicable ways by members of a community. Experiences mediated by shared social and cognitive models often become the default readings of ordinary reality for members of a community.

Using models as the unit of culture conceives of culture as a socially distributed stock of schemas that mediate human knowledge and feeling. This approach to culture does not presume that it is a single bounded entity ("a culture"). It does not imply that sharing in a collective culture necessarily produces identical understandings. Nor does it suggest that cultural models are immune to change or contestation.[1] Though the idea of a shared culture does not preclude conflict between models or in individual interpretations, some significant degree of joint focus and social coordination is made possible by cultural models. Significant capacities for joint attention and social coordination are important adaptive mechanisms for social mammals with a high degree of mental and social flexibility.

Strauss and Quinn (1997, chapter 3) and Shore (1996, pp. 345–351) propose a computational view of the possible neural structure and behavior of cultural models using connectionist theory. Understanding cultural models through connectionist theory provides a theoretical framework that can account for their flexibility, their mutability, and their responsiveness to the contingencies of human experience. This conception of cultural models is a neurally grounded theory of social learning.

4.2 Types of Cultural Models

Cultural models have an inherently double life as mental representations, or mental models, and as publically available social institutions, or instituted models (Shore, 1991, 1998). As instituted models, cultural models appear in a great many different formats. Characterizing the symbolic and cognitive affordances for meaning-making of these different formats is an important goal for cognitive and phenomenologically oriented anthropologists interested in accounting for the shape and texture of human experience. What a cultural model means is linked with *how* it means.

The most popular kinds of models studied by anthropologists are linguistic cultural models. Linguistic models exploit the multidimensional modeling power of language. Language models reality for its users in sound-symbolic patterns; in lexical units (words or word parts); in syntactic patterns modeling ontological, temporal, spatial, or causal relations; and in higher-level discourse units such as narrative or conversation. A variety of linguistic model that has been extensively studied is metaphor. Metaphor models are linguistically formatted analogies that use relatively concrete images to conceptualize more abstract concepts. Metaphor has been proposed as a fundamental basis of human cognition (Lakoff & Johnson, 1980; Lakoff & Kovecses, 1987; Lakoff & Turner, 1989; Lakoff & Johnson, 1999; Lakoff & Nunez, 2001; Lakoff, 2002; Kimmel, 2002). The status of metaphors as an important instance of cultural model seems evident, but this association has not been made clear by Lakoff and his colleagues who, as linguists, tend to treat metaphor more as a general feature of human language practice and of human cognition (cf. Quinn, 1991).

The evolution of language has greatly enhanced the modeling capability of *Homo sapiens,* and the analysis of conceptual models embedded in human discourse has occupied considerable attention in cognitive anthropology. But language is far from the only way that cultural models are formatted. There is also a wide array of kinesthetic and gestural models (e.g., postures, hand gestures, facial expressions) in which the body becomes the prime modeling medium (McNeill, 2005). Other potent media exploit the modeling power of constructed forms and shapes involving the various senses and include color, sound (as in music), and smell (Howes, 2003).

In addition to their diversity of formats, cultural models vary in the kinds of work they do. From a functional perspective, we can distinguish the work of culture as orchestrating collective strategies of orientation, memory enhancement, task performance, and conceptual clarification. *Orientational models* function as connectors, mediating human orientations to time, space, and social relations. Orientational models include temporal models such as are embodied in clocks, calendars, and synchronization patterns; spatial models such as directional coordinates, maps, and conventional spatial layouts; personal space models; and social relations models such as naming systems, kin terms, codes of social etiquette, and speech and gestural codes conveying different degrees of intimacy and formality. *Task performance models* facilitate the learning and performance of complex routines and skills by simplifying and standardizing physical and mental operations. Examples of culturally modeled tasks range from basic routines for eating or grooming to food preparation recipes, musical scales and exercises, and directions for using appliances. *Mnemonic models* (e.g., songs, rhymes, checklists, aromatics) enhance the memorability of specific content. *Conceptual models* (e.g., theories, myths, ideas, stories, art forms, categories) provide the basis for shared or disputed ontological perspectives on reality. This list of functionally distinct models is intended to suggest only the most basic functions of cultural models and is not exhaustive. Moreover, these distinctions are not exclusive, since any particular cultural model can serve more than one function.

4.3 Cultural Thematicity

Understanding "culture" as a socially distributed stock of social and mental models with many formats and functions underscores the fact that a collective knowledge system is quite different from the notion of culture as a unitary, bounded entity. In Sperber's terms, distributed cultural systems, like gene pools, have an epidemiological rather than a categorical character (Sperber, 1985, 1996). But this very epidemiological character of culture viewed as models raises the important issue of what accounts for the perception of "family resemblance" among clusters of cultural models within a cultural community, an apparent resemblance that becomes more salient when cultural communities are contrasted with one another. What allows us to conceive of "a Japanese culture" that can be contrasted with "American culture?"

Such generalized cultural characterizations have gone under the labels of "culture," "ethos," "themes," and "worldview." They were the basis of the idea of dominant cultural patterns or configurations characterizing whole communities or historical epochs implicit in German idealist history and philosophy (as suggested by the term *Weltanschauung*: "worldview"). This holistic vision of cultural patterns was picked up by the emerging American school of cultural anthropology pioneered by Franz Boas and his students Ruth Benedict and Margaret Mead in the 1920s and 30s (Benedict, 1932, 1934; Mead, 1935/1971).

In the 1940s and 1950s, it reappeared with a psychological spin as "national character" and with a linguistic emphasis in the writings of Benjamin Lee Whorf, who proposed that contrasting global worldviews could be traced to the differences between grammatical patterns of languages (Gorer, 1948, 1950, 1955; Bateson, 1972; Carroll, 1956). More recent accounts of linguistic relativism have focused on experimentally derived differences in spatial orientation that are linked to key differences in linguistic resources and practice (Levinson, 2003; Levinson & Wilkens, 2006; Haun & Rapold, 2009; Majid, Bowerman, Kita, & Haun et al., 2004; Bennardo, 2009). The influence of this notion of cultural holism was extended into the latter years of the twentieth century by the early writings of Clifford Geertz (Geertz, 1973).

Several possible sources have been proposed for such holistic perceptions of what Strauss and Quinn call "thematicity" in culture (Strauss &
Quinn, 1997, pp. 118–120).

Strauss and Quinn offer a series of complementary answers to the
question of the origin of thematicity. First, thematic structures may be
the result of schemas learned very early in childhood. For example, the
American concept of love in adult marriage can be explained on the basis
of early childhood experiences of love (see Strauss & Quinn, chapter 7).
These experiences are then metaphorically adapted to the new context of
marriage, while retaining a great deal of those expectations learned in
infancy.

Second, such structures result from a schema being learned in a wide
variety of contexts. Taking the American schema of self-reliance as an
example, Strauss and Quinn (p. 119) argue that schemas are more likely
to be further generalized if they are initially learned in a number of contexts. American middle-class children learn the schema of self-reliance at
home, at school, throughout day care, and in extracurricular activities like
the Scouts and Little League. These institutions not only reinforce the
schema of self-reliance, they progressively broaden its definition, so that
it is eventually transformed into an internalized template for the child.
This means that the schema is progressively elevated to an ever-higher level
of schematicity. The schema is acquired on a higher level than any one of
the individual contexts would specify, a level of schematicity that spans
the contexts as underlying common denominator.

Third, there is a mechanism called "elective affinity" for new rhetoric
and products that appeal to a preexisting understanding. Elective affinity
is a selection process in which people screen new cognitive input for information that fits into an attitudinal stance toward which one is already
inclined. People pick out new information in order to corroborate their
prior assumptions. Therefore, elective affinity is an autodynamic process
that assimilates new information into preexisting schemas and stabilizes
them in the process (Kimmel, 2002, pp. 140-141).

Shore (1996) suggests several additional ways in which cultural models
become "thematic." Groups of cultural models appear to cluster in "families" of structurally analogous models. Since humans commonly employ
analogy in a generative and creative way, these similarities among models
can be traced to the fact that people will often employ one model as
the basis of other models (Gentner, 1983, 1989; Shore, 1996, 1998). For
example, since the 1970s airlines have used a hub-and-spoke model

for their route structures. This same structural model can be seen in the physical design of many airports, and also in the design of shopping malls. It should be possible to trace a history of design influence to see just how these special purpose schemas were derived from one another. Shore calls the shared underlying model a "foundational schema," a kind of master model that is the foundation for a family of specific models, and that lends a feeling of thematic and aesthetic linkage among a group of diverse models.[2]

One other potential source of thematicity in culture results from what might be called "cultural synecdoche," in which a particularly salient model comes to stand in for the culture as a whole. Strauss and Quinn (1997, pp. 145–152) call these "cultural exemplars." Many sports are played in the United States, but historically baseball has long had a special place in American experience as the quintessentially American game.[3] Communities often consciously select certain cultural models as a privileged way to represent themselves to outsiders and visitors. Usually such cultural exemplars are chosen because they are seen to represent some aspect of the community's sense of itself in a salient (and often flattering) way. Local dances (like the Hawaiian hula or the Argentine tango), styles of music (like French accordion playing), or ceremonies (like the Japanese tea ceremony) are used consciously by members of a community to stand in for the culture as a whole. Because they are deployed to produce a generalized image of a community, such exemplary models are frequently the basis of stereotypes that are as misleading as they can be evocative.

4.4 Modeling Perspective

Among the most basic functions of cultural models is coordinating orientation. Orientational models are general representations that individuals use as cognitive tools to orient themselves to their physical and social worlds. In this sense, orientational models are functionally equivalent to navigational devices such as compasses, GPS systems, and maps that Hutchins has called "cognitive artifacts" (Hutchins, 1995). Orientation is a complex and ongoing challenge for humans, since we must orient ourselves not only to the constantly changing physical world we inhabit (both natural and constructed), but also to the multilayered social world around us. The most obvious (and perhaps most primitive) navigational challenge humans face is ordinary spatial navigation. It is also the most studied and best-documented aspect of navigation. Since Edward Tolman's seminal 1948 essay "Cognitive Maps in Rats and Men," the most common metaphor for mental models has been the map, the most familiar of our spatial navigation tools (Tolman, 1948).

Spatial cognition has been discussed in relation to "frames of reference" in a number of disciplines such as information science, neuroscience, and linguistics (Avons, 2007; Friedman, 2005; Jackendoff, 1983, 1999; Levinson, 1996; Campbell, 1994; and Bennardo, 2009).[4] Spatial navigation for humans involves two distinct, neurally based reference frames that produce two complementary navigational perspectives, commonly referred to as egocentric and allocentric (or, more rarely, geocentric) frames of reference (Klatzky, 1998; Grush, 2000).

4.5 Egocentric Navigation

Egocentric navigation involves the mapping of dynamic visual (or haptic) relations between individuals and their environment as they move from place to place. The egocentric frame of reference generally engages sensorimotor representations and thus involves the representation of direct kinesthetic as well as visual experience. As Kelly and McNamara describe it:

The sensorimotor representation is thought to be used when performing body-defined actions, such as negotiating obstacles and moving toward intermediate goal locations like landmarks which can function as beacons. Because these behaviors typically rely on egocentrically organized actions, it makes sense that this sensorimotor representation should also be egocentrically organized, in order to maintain an isomorphic mapping between representations and response (Kelly & McNamara, 2008, p. 23).

It is commonly assumed in the literature that egocentric navigation engages transient representations ("snapshots") involving short-term memory representations subject to rapid fading, since any one snapshot while moving through an environment will quickly be superseded by an updated one (Kelly & McNamara, 2008; Mou, McNamara, Valiquette, & Rump, 2004). Researchers generally concur that the precise nature and degree of this perceptual transience is not well understood (Kelly & McNamara, 2008, p. 23).

It is not clear, for instance, what sort of memory processes are engaged in the case of navigation or direction-giving for the traversing of familiar paths, where visual frames for particular landmarks appear to have become stabilized mental representations subject to mental recall and linguistic translation (as in giving directions). In other cases, such as in locating one's space in a parking lot, or recalling the location of a house to which one travels frequently, such egocentric landmarks need to be stored in long-term memory. The difficulty in activating long-term memory in navigational tasks that routinely utilize short-term memory may have something to do with the mental effort and the special cues (colors, numbers, letters, totemic names for parking lot sections) often needed to remember the location of one's car in an unfamiliar parking lot.[5]

4.6 Allocentric Navigation

In contrast to egocentric navigation, allocentric navigation is often conceived as perspective-neutral. Allocentric navigation requires "objective" reference coordinates such as longitude and latitude marks or compass points, location-markers that are by their nature independent of any particular perspective. While allocentrically framed locations are translatable through abstract codes like words and numbers into perspective-neutral terms, it is also the case that it is not possible to represent directly (visually) a picture of space that fully transcends perspective.

While egocentric and allocentric navigation are both essential for human spatial navigation, the distinction has also been used to distinguish alternative general cognitive strategies shaping overall orientation to the environment. For example, Aginsky, Harris, Rensink, and Beusmans (1997) suggest that during spatial learning and development individuals acquire one of two dominant orientations to the environment. Those employing *visually dominated* navigation strategies "base their wayfinding decisions on visually recognizing decision points along a route: the decision points are not integrated into any kind of survey representation." In contrast, those employing *spatially dominated* navigational strategies "represent the environment as a survey map right from the start; that is, they do not pass through the landmark or route stages" (Aginsky et al., 1997, p. 317).

Allocentric navigation is usually visualized as "a bird's eye view" of a territory that allows viewers to locate themselves visually and conceptually (using reference-frame coordinates) at any point on the visible terrain. By projection of coordinates, allocentric navigation permits the extension of imagined navigation beyond the boundaries of the visible space. This simulation of a perspective-neutral view suggests that allocentric navigation cannot be derived directly from human visual experience, but represents an abstract and imaginative gathering together of all possible angles of vision into a single conceptual "panoptical" view.[6,7]

Wexler stresses the many cognitive advantages of allocentric perception of space over egocentric mappings in ordinary cognition.

Perceiving spatial information in an allocentric frame is perhaps the ultimate form of spatial constancy and has important advantages. An ecological advantage is that it allows an observer to see whether an object is moving at all, rather than moving in relation to the observer. An allocentric frame also has computational advantages. The world can be assumed to be stable in this reference frame, and representations of objects and spatial relations do not have to be updated as one moves about (Wexler, 2003, p. 340).

It is generally agreed by neuroscientists that the two kinds of basic framing for spatial orientation are grounded in different neural substructures (Brotons, O'Mara, & Sanchez-Vives, 2005; Hartley et al., 2003; Jordan et al., 2004; Wolbers, Weiller, & Buchel, 2004; Brese, Hampson, & Deadwyler, 1989). In 1971 O'Keefe and Dostrovsky proposed that the central brain structure that stored allocentric cognitive maps of the environment was the hippocampus, controlled by the firing of discrete pyramidal neurons called "place cells" (O'Keefe & Dostrovsky, 1971; Poucet, Lencke-Santini, Paz-Villagram, & Save, 2003).

Subsequent studies using rats navigating in water to a hidden platform and later studies of human navigation have suggested that allocentric mappings of the environment involve areas in the parahippocampal complex. Recent studies of brain activation in allocentric navigation suggest a more precise hippocampal locus of activation in humans as well as rodents: CA field 1 (Suthana, Ekstrom, Moshivaziri, & Knowlton, 2009).

In contrast with allocentric navigation, egocentric navigation requires additional inputs such as "internal cues, motor input vestibular and directional information" (Brotons et al., 2005, p. 60). All these sources of information are subject to coordination that researchers have termed "path integration," a process that involves both the hippocampus and the parietal cortex (Brotons et al., 2005; Whishaw, Hines & Wallace, 2001).

4.7 Linguistic Mediation of Spatial Orientation

Despite their ubiquity, spatial models have not been accorded the analytical attention by anthropologists that they have received from researchers in cognitive psychology and neuroscience. Rather than developing a theoretically robust understanding of the constraints governing the cultural modeling of spatial orientation, ethnographies have tended to describe spatial orientation in a more or less anecdotal way.

The most ambitious and detailed analysis of cultural modeling of spatial orientation has been the comparative study of spatial reference systems used by numerous language groups by Levinson and his colleagues in the

Language and Cognition group at the Max Planck Institute for Psycholinguistics at Nijmegen in the Netherlands. Levinson, who is a linguistic
anthropologist by training, is interested in the cognitive universals governing human spatial cognition and the possible semantic and cognitive
diversity in language use shaped both by cultural models of time and space
and variations in the grammatical resources of various languages. Levinson
writes:

The language of space becomes an important focus of research ... for a number of
reasons. First, it may help to reveal the underlying conceptual structures in human
spatial thinking, which may be harder to extract from an inarticulate species. Naturally, universals of spatial thinking should be reflected in universal conceptualizations in spatial language. Second, and contrastively, the very variability of language
promises an interesting insight into the cultural variability of spatial thinking. Third,
this reasoning presumes a close correlation between spatial language and spatial
thinking—essentially, a (possible partial) isomorphism between semantics and conceptual structure. Where we have linguistic universals, the correlation may be presumed to be driven by cognitive universals, But where we have cultural divergences,
languages may not so much reflect underlying cognition as actively drive it.
(Levinson & Wilkins, 2006, p. 1)

Levinson's research group has spent much of its time studying and
comparing the pragmatics of spatial orientation in a dozen highly diverse
language groups.[8] Detailed comparisons of the spatial semantics and pragmatics of these languages revealed "an extraordinary diversity in both the
underlying conceptualizations of spatial distinctions and the manner in
which they are coded in various languages" (Levinson & Wilkins, 2006,
p. 512).

In comparing the ways that these languages handle frames of reference,[9]
Levinson and Wilkens argue that spatial reference is governed by one of
three alternative frames: *absolute frames* (frame 1, based on "antecedent
fixing by a community consensus of arbitrary bearings"); *relative frames*
(frame 2), which depend on "mapping the bodily coordinates of the viewer
onto the scene"; and *intrinsic frames* (frame 3), which rely on "designating
facets of the grounding object" (Levinson & Wilkins, 2006, p. 512).

While the evidence is strong that this tripartite distinction represents
an important set of differences evident in the way languages can encode
the representation of an observer and an object in space, it is not clear that
the absolute/relative/intrinsic distinction is fundamental to all spatial perception or limited to the framing of observers' relation to large objects with
a clear front and back (such as a house or a person). The author suggests
that the triple distinction is really a subvariety of the more basic egocen-

tric/allocentric dichotomy, where we have one allocentric frame (frame 1), one egocentric reference frame (frame 2), and a third (frame 3) that is a "displaced egocentric frame" in which the orienting perspective is displaced from the viewer to the grounding object. In a sense, frame 3, the "displaced ego" frame, is a particularly interesting hybrid of egocentric and allocentric reference.

The egocentric-relative frame is the only one that was used by all language groups, leading to the conclusion that it may be the only universal reference frame in human languages, a view that is consistent with the fact that it seems to be a developmentally primitive and universal feature of human spatial orientation (Piaget & Inhelder, 1956; Johnston & Slobin, 1979; Tanz, 1982). Of particular note, however, is the discovery that basic directional specifications for locating places, people, and objects are not universally based on the familiar egocentric frame of reference that we use (as when we give directions using "left" and "right"). While it is the case that most of the world's languages do use egocentric reference frames for direction specification, there are numerous exceptions. There are a number of languages in which the dominant reference frame is an absolute rather than egocentrically relative frame. The most famous case studied is the Australian Aboriginal group of Guugu Yimithirr speakers from north Queensland, who use allocentric compass coordinates as their basic reference frame. Whereas English speakers reserve allocentric reference to specifying locations or directions of significant distance scale (e.g., "Montana is located *west of the Mississippi*" rather than "Montana is to the left of the Mississippi"[10]), localized direction giving seems to be "naturally" egocentric for English speakers. For Guugu Yimithirr speakers, however, the cardinal directions are the conventional way to refer to all spatial orientation, including localized orientation. For example, one would request someone to move "a bit to the east" or "west" rather than left or right (Levinson, 2003).

While most of the research cited above on spatial cognition aims at understanding human universals in spatial cognition, cultural anthropologists are also interested in documenting and accounting for the range of variation in spatial modeling suggested by differences in spatial models employed by cultural and linguistic groups.[11] Levinson and his colleagues have not attempted to determine any systematic causal patterns underlying the variations they have found in spatial reference. But their careful documentation of heretofore unrecognized variations in the use of allocentric and egocentric spatial reference frames demonstrates that, for humans, egocentric and allocentric spatial models can be employed quite differently in different cultural groups. Spatial cognition must be considered, like so much of human behavior, as the outcome of interactions between universal (innate) dimensions of spatial cognition and historically contingent cultural and linguistic models governing significant local variations in spatial orientation. An understanding of the particular cultural models governing spatial orientation is thus a necessary part of any account of human spatial cognition.

4.8 Cultural Models of Spatial Orientation

In addition to the linguistic analyses of spatial orientation discussed above, there are a number of interesting studies of cultural models of space from a cognitive perspective. A relatively early treatment of spatial models with cognitive implications is contained in a famous paper on dual organization in village layouts of the Bororo of Brazilian Amazonia by Lévi-Strauss. He proposes two different forms of dual-organization implicit in village spatial layouts. One, based on binary oppositions between groups, he calls "diametric dualism," while the other, based on a graded (i.e., continuous) opposition between center and periphery, he calls "concentric dualism" (Lévi-Strauss, 1967, 1976). With his structuralist perspective, Lévi-Strauss used this dichotomy to suggest the attempt by Bororo to symbolically reconcile two different notions: a static and synchronic conception of dualistic social structure based on rules of clan exogamy, and a dynamic and diachronic conceptualization of a constantly shifting social structure, a ternary structure disguised as a graded binary system, allowing for the perpetual reshaping of society by the historical contingencies of demography and marriage.

Such categorical and graded models of space are particularly powerful structures for modeling the relations of dynamic, egocentric perspectives and more static allocentric ones, and one would expect to find these distinctions developed extensively in cultural models. For example, Shore develops this distinction in relation to Samoan cultural models of village space, noting that diametric dualism bisects space into simple categories (front/back, sea/inland), using digital coding. By contrast, concentric dualism establishes a continuous gradient of status and power, with sacred space defined by the village center (the *malae*, or sacred village center) and increasingly secular space defined by movement outward (Shore, 1977, 1982, 1996). Shore notes that rather than simply providing two alternative readings of village space, the diametric and concentric mappings of village space model two very different perspectives on experience, with very different representational affordances:

Digital distinctions are the preconditions for generating categories. Classical categories have been defined in relation to clear boundaries. In terms of communication, their virtue is systematic coherence and clear intersubjective reference. What such digital structures cannot represent, however, are the gradations inherent in the individual's sensory and emotional apprehension of the world. Graded structures can better convey the shape of experience, yet they are incapable of formulating the discrete categories through which mutual orientation and reference are made possible. If graded coding underlies the subjective apprehension of experience as continua, so digital coding with its discursive capabilities facilitates intersubjectivity and social coordination (Shore 1996, p. 274).

While the Samoan case suggests how these two perspectives can provide resources for representing alternative perspectives within the same community, the literature on traditional navigational strategies for Micronesian and Polynesian sailors also suggests that these distinctions can account for general differences between dominant navigational models in different communities (Gladwin, 1972; Finney, 1994; Lewis, 1972; Hutchins, 1983, 1989).

David Lewis attempted to describe to a Micronesian navigator his understanding of the Micronesian navigation strategy that had been shown to him in an account of navigation between the islands of Ponape and Ololuk. The strategy included the intersection of two reference bearings called *etaks*. Lewis indicates that his intuitive translation of this strategy into one involving a bird's-eye map view puzzled his Micronesian informant Hipour:

On his telling me what they were, I drew a diagram to illustrate that Ngatik [Island] must necessarily be where these ETAK bearings intersected. ... Hipour could not grasp this idea at all. His concept is a wholly dynamic one of moving islands (Lewis, 1972, p. 142).

In a very interesting commentary on the miscommunication evident between Lewis and Hipour, Hutchins notes that the two men appear to have been using incompatible spatial frames, without realizing it:

The technique Lewis used ... contains some very powerful assumptions about the relation of the problem solver to the space in which the problem is being solved. First, it requires a global representation of the locations of the various pieces of land relative to each other. In addition, it requires a point of view on the space which we might call the "bird's eye view." ... This strategy then involves at least creating an abstract representation of a space and then assuming an imaginary point of view relative to the abstract representation. We can guess that Lewis did this because it is for him a natural framework in which to pose questions and solve problems having to do with the relative locations of objects in two-dimensional space (Hutchins, 1983, pp. 206-7).

Micronesian navigators, by contrast, appear to use mental maps that are dynamic, graded, and egocentric, what Hutchins terms "bearings which radiate out from the navigator" (Hutchins, 1983, pp. 206–207).[12] Despite the fact that humans universally possess the cognitive resources for both allocentric and egocentric representations of navigation, these spatial frames are clearly deployed in different ways in different cultural and historical settings.

4.9   Models Framing Social Orientation

Up to this point we have been looking at the role of allocentric and egocentric reference frames in "literal" spatial orientation. Numerous orientational cultural models governing *social orientation* also have allocentric and egocentric variants. Cultural models governing social orientation come in many formats and with varied content, but they all function to frame significant human interactions with the physical environment, with significant objects, with other people, and with ourselves. Because humans traverse complex plural environments—natural, constructed, social, emotional and psychic—they must develop competence in using a wide variety of different orientational models to locate and navigate these landscapes. Whether orientation is toward a physical space or a social space, humans can represent their relations with either egocentric or allocentric cultural models, depending on the circumstances.

To get some idea of the importance and diversity of orientational models, and their allocentric/egocentric variants, let's consider some of the orientational models that would be called upon during a "simple" walk across campus at my university.

Leaving the Anthropology building, I would have to access route maps of the corridors of our building and my knowledge of navigating my way to the exit door. Approaching the door, I would assess the weather outside (temperature, precipitation, quality of light, wind, temperature, and the appearance of the sky) and evaluate if I would need any additional gear (such as a coat or an umbrella) for walking outdoors.

If I am heading down a well-traveled route, I would access a model of my planned path, shaped by familiar landmarks (buildings, trees, roads, and signs) and path contours (directions, hills, turns). If I were going to an unfamiliar part of campus, I might need the aid of an overhead map of the campus, a complete set of verbal directions using either landmarks in an egocentric framing, or a verbal translation of the overhead map using allocentric coordinates such as north or east or "toward North Decatur Road."

Walking up the hill toward the main quadrangle, I would scan the oncoming people whom I might encounter, keeping to the right of most of the people passing in the opposite direction and avoiding any bicycles coming my way. Using my facial and postural recognition skills, I would continually evaluate the stream of oncoming people to recognize and prepare myself to appropriately greet anyone with whom I had a relationship. My evaluation would require modeling physical characteristics (relative age, ethnicity/race), gender markers, modes of dress (distinguishing students from faculty, from administrators, from janitors, from ground staff) and models of particular faces stored in long-term memory.

Seeing a familiar individual approaching, I would have to decide very quickly whether to stop and talk, nod, wave, or feign not having seen them and simply pass them by. If I stopped, I would have to access from long-term memory their name, and decide how formal my greeting should be. If the person I am addressing is the provost of the university, should I call him Earl, Mr. Lewis, or Provost Lewis? My decision might be in response to my interpretation of his reaction to seeing me. Is there a big smile? A look of nonrecognition? A wave? A curt nod of the head? Is he walking fast, as if late for a meeting, or is he sauntering slowly, signaling a readiness for conversation? What is the history of my relation with him? Do I show deference, casual familiarity, a lack of connection? Each sign would index a different kind and context of relationship, and the evaluation would draw on a host of orientational models assessing social status, mood, and the history of the relationship.

If we spoke, we would need to find some common ground for a conversation. I would have to know how much detail to go into in answering any questions he might ask me. And I would need to know what would be the appropriate range of questions or topics to address to him. We would need to use our models of polite conversation to negotiate how long our conversation should last and how to end it and move on. I would need to decide what sort of speech is appropriate to this situation.

Since we have a midsize campus, there might well be several such encounters with familiar consociates as I walked across campus, each one requiring the deployment of many of the same orientational models underwriting my recognition and performance of interaction rituals.

Throughout these encounters, I would be accessing a set of models of my significant social relationships on campus (e.g., a colleague, a staff member, a friend, a student, a former student whose name I know, a barely recognized former student who knows my name but whose name I have forgotten, etc.). I would be accessing sets of onomastic models of personal names, and submodels that convey different degrees of formality, intimacy, and deference in deciding whom I am speaking with.

I would also draw on my knowledge of the many models of speech registers that would govern my word choice, grammatical style, pronunciation style, intonation, and other variations that express the appropriate sort of relationship with which to converse with each person. I would need to summon my competence in reading postural, gestural, and intonational models that assess mood and relationship quality.

Finally, when I had moved on, I might use my knowledge of my internal emotional responses to assess my relationship with the person I had just spoken with. This might involve clear, consistent evaluations (e.g., "a good friend," "a person I don't like much," or "I think she's annoyed with me"). On the other hand, it might involve sorting through ambivalent responses and interpreting those to myself. The feelings that I experienced in my encounter would be powerfully modeled by my inner emotion system, which functions like an internal dialogue in somatized music as a relatively primitive way for me to assess my relationships to the world.

If I am heading to an unfamiliar place, I might have to get some navigational help to locate the place. Depending on the resources available to me, I might use a campus map that I had with me or that I found along my route. This would force me to translate my current position (understood egocentrically) to an allocentric position, and then, having located my current position and my goal, translate the route back into a point-to-point egocentric route. Alternately, I could ask someone for help, which could result in an egocentric or allocentric set of directions. Or, if I had a GPS system, I could have it locate my current position and map it or guide me verbally point-to-point to my destination.

All of this might reasonably occur in the course of just twenty minutes, but the sheer number and range of models that would be deployed is striking. Most of them are orientational models. Whether the content of the model is geography, climate, social intimacy or distance, emotional expression or feeling state, they all evoke the varieties of "landscape" through which people navigate in everyday life. In the following section, we turn to a familiar kind of social orientation model studied by anthropologists—

kin terms—to examine how allocentric and egocentric variants of kin terminologies differ in form and use.

4.10   Kin Terminologies: Egocentric and Allocentric Variants

Just as spatial navigation affords both egocentric and allocentric perspectives, many orientational cultural models turn out to have egocentric and allocentric variants. Egocentrically framed orientational models will typically define a continuous and dynamic gradient of terms, gestures, or speech variations that radiate outward from a perceptual center, an explicit "point of view." They map a gradient of social experience, often using some variation of the closer-further spatial metaphor.

The allocentrically framed variant of the model will translate this egocentric view into a conventional bird's-eye view model defined by discrete categories rather than by dynamic gradients. Allocentric orientational frames will be represented as perspectiveless or, more likely, through some approximation of perspective-neutrality such as the anthropologist's notions of "general ethnography," or "social structure" or "an experience-distant account."

An example of such homologous models familiar to anthropologists are kin terms, which are traditionally represented as either "terms of reference" (allocentric) or "terms of address" (egocentric). Referential kin terms are often represented as a table of kin-term categories defined by genealogical coordinates such as male, female, living, dead, married, lineal, collateral, generation (ascending/descending), and same-sex ties or opposite-sex ties.

While the referential terms will often be represented with an abstract ego at the center, the system does not in fact represent a particular ego's point of view of his or her relationships. Instead, it represents a summary bird's-eye view picture of all possible terminological kin relationships viewed by an observer outside of the system. In American English, terms of reference convey a cold, clinical vision of family relations: mother, father, great-grandmother, niece, nephew, great-granddaughter, etc. This is the model we would use in explaining our kin relationships to an outsider.

The equivalent egocentric "insider" model of kin terms is the set of "terms of address." Consistent with their egocentric and dynamic mapping of particular relationships, terms of address are often harder to pigeonhole (i.e., categorize) than terms of reference and are subject to considerable idiosyncratic variation. In other words, they are somewhat resistant to generalization. Nonetheless, members of a cultural group generally recognize a family resemblance in a group of idiosyncratic address terms. Grandmother may translate into Grandma, Granny Jones, Gram, Nana, etc. This is a hybrid system that has both generalized (i.e., allocentric) and idiosyncratic (i.e., egocentric) properties. Father may translate into Father, Dad, Daddy, Pops, Pop. Brother may translate into Brother, Bro, Henry, Hank, Hankie, etc. In this case, use of personal name rather than kinship term may represent a degree of intimacy defined partly by the system and partly by the individual user.[13] In egocentrically framed kin-term models, each variation marks a particular valence of relationships and context defined partly by properties of the name and partly by context of use. Thus, terms of address do not define simple genealogical relationships defined by global coordinates. They define particular contingent ongoing relationships, combining the closed general properties of genealogical categories with an open set of quasi-idiosyncratic designators that index particular and personal qualities of closeness and distance in relationships.

4.11   Speech Registers Mapping Social Distance

All languages possess powerful resources for modeling social distance and intimacy. Social distance can be modeled at any number of levels of language coding: word choice (e.g., "dining" vs. "chowing down"), grammatical forms (e.g., "Would you care to join us?" vs. "Come on over!"), forms of address (e.g., "Mr. President" vs. "Barack"), pronunciation shifts (e.g., hyperprecise pronunciation vs. deliberately slurred articulation), etc.

In use as active discourse (egocentric framing), these politeness forms tend to be deployed with fine gradations of variation, and with great dexterity as linguistic distance or closeness is shaped in response to interaction cues. The complex model that a native speaker has of these forms is generally not conscious (since it is too complex and rapidly changing to be subject to conscious control), and it is not accurately explained in terms of fixed categories of speech. Egocentric models of speech forms related to social distance are better described as constantly modulated and updated "flows" of language operating simultaneously at several levels of structure, operating in feedback loops with many kinds of cues from conversational partners. Such egocentric models of language styles govern implicit forms of language praxis rather than explicit representations of language structure (Bourdieu, 1999).[14]

Allocentric models of speech styles related to distance and intimacy are common for many languages, and generally specify explicit categorical distinctions in politeness, deference, social status, solidarity, and other aspects of social distance. Examples would include categorical models of speech dialects (e.g., Queen's English vs. Cockney English, or High German vs. a local dialect), or pronouns that mark intimacy vs. formality (e.g., French *tu* vs. *vous*). These allocentric models are often explicit parts of grammars used in teaching a language to foreigners. They tend to be framed as binary distinctions between alternative forms of language that are available as intersubjective reference-models *about* language. But they do not map easily onto the gradients of speech used by native speakers in linguistically navigating complex social spaces.

Samoan has a number of explicit, allocentrically framed models of social distance. At the phonological level, Samoans distinguish between what they call "good speaking" (*tautala lelei*) and "bad speaking" (*tautala leaga*). Good speaking (which appears to be a historically older form of Samoan pronunciation that antedates the arrival of the missionaries) uses *t* and *n* in both pronunciation and writing. Bad speaking (which is the form in use for most colloquial conversation outside of church contexts, formal government and school communication, and speaking to foreigners) shifts articulation to the back of the oral cavity so that *t* becomes *k* and *n* becomes *ng*. These shifts are only represented in speech and never (except humorously) in written Samoan.

The author's early research on these pronunciation registers revealed that these categories were highly cognized by Samoan speakers, and that most adults could provide a general account of the social contexts in which these distinctions would be used. They form an explicit part of Samoan language training for the Peace Corps. The use of the two forms is explicitly associated with categorically different contexts. Good speaking is to be used with foreigners who speak Samoan, in church, by pastors, by teachers and students in school, in written Samoan, and on public broadcasts in Western-based contexts (television, radio, parliamentary speeches).

Bad speaking is associated with intimate conversations among Samoans, the family, and formal oratory in traditional Samoan political settings.[15] The author was also able to get from informants other kinds of associations between pronunciation styles and moods or feelings. Bad speaking was associated with impulse-expression, anger, or extreme intimacy. Good speaking was associated with carefully controlled speech, politeness, and deference.[16]

In learning to use these pronunciation forms, it became clear that despite the apparently simple either-or character of the good speaking/bad speaking dichotomy and its public associations, Samoans navigate social relationships and contexts by fine-grained modulations of pronunciation (in addition to other linguistic markers of social distance). Rather than simply realizing the allocentric model in its duality, Samoans experienced it as a set of gradations of pronunciation, from "heavy" bad speaking articulated from the back areas of the oral cavity to the most delicate pronunciation using dental articulation of the *t*. The author was able to mimic this range of styles and get informants to identify the many kinds of contexts and the changes in social distance evoked in the mimicked speech. While the allocentric model is framed as categorical and dualistic, the egocentric version was actually tracking an analogous movement from back-to-front that paralleled closely the negotiation of village and domestic space.

4.12   Shifting Perspective: Interactions between Allocentric and Egocentric Cultural Models

Despite the clear distinctions drawn above between egocentric and allocentric models, our goal is to understand some of the ways in which the two perspectival frames interact. The interdependence of both egocentric and allocentric framing in navigational models used by animals and humans in negotiating physical environments has been noted; we have also seen important instances of hybrid models that employ alternations or combinations of egocentric and allocentric models.

Egocentric models of space are primary to navigation since they rely on relatively direct analog processing of sensory data from the environment. Moreover, egocentric navigation presumes a perceptual agent and a point of "view," both of which are basic to a navigating organism. Allocentric models of the environment introduce additional complexity and computation to navigation since they require an imaginary agent and an abstract, objectified representation of space and spatial coordinates that can be simulated, but never directly mapped from experience. For creatures who do not naturally fly, the ability to imagine and represent a bird's-eye view of anything requires a level of conceptual abstraction and integration beyond the immediate perceptual stream of direct experience.

In this sense, allocentric perspectives are not "natural" for human beings. They are achieved through analog-to-digital translations of particular experience into general categories, coordinates, and propositions. Allocentric navigation in the world has also been enhanced by the human invention of cognitive artifacts, such as navigation, communication, and computation tools, which have extended human perception and the range of coordination well beyond the zone of immediate experience (Hutchins, 1995).

Perhaps more than any other cognitive resource, the possession of language, with its great powers of paraphrase and its shift from analog to digital representation, gave humans the ability to reframe egocentric experience with its multiple analog sensory channels into generalized allocentric structures. What such allocentric structures lack in the way of phenomenal richness and particularity, they gain in their referential generalizability and their ability to coordinate the conceptual outlook of a theoretically unlimited number of individuals well beyond the horizon of common perception.

On the other hand, if egocentric framing of cultural models were completely nonfunctional or dysfunctional for humans, having no distinctive advantages over allocentric models, it would likely have given way to an exclusively allocentric framing strategy for humans. Both framing perspectives have survived as alternate navigation strategies because each has its adaptive advantages.

Allocentric framing has clear advantages for humans in its capacity to coordinate the orientations of an unlimited number of people. People can share a general orientational framework without any need for them to share the same physical location and an identical point of view. Aided by the powers of abstract language, allocentric models facilitate the intersubjective frameworks for mutual reference that are essential for a group that shares a common "worldview" (i.e., a cultural community). Egocentric models allow individuals within cultural groups to negotiate their ever-changing physical and social landscapes with models that use features from shared allocentric models, but in such a way as to allow the flexibility and the ongoing customization necessary to make possible individual journeys through shared landscapes.

Humans must reconcile their highly developed cognitive capacities for autonomous awareness with their equally highly developed social and cultural interdependence. Cultural communities vary significantly in their relative collective or individualistic orientations (Dumont, 1992; Markus & Kitayama, 1991; Kitayama, Markus, Matsumoto, & Norasakkunit, 1997). Yet whatever the cultural and institutional emphases of the community, their members must find a way to navigate their lives both as self-aware individuals and as socially coordinated members of a collectivity. The complex interactions between allocentric and egocentric framings of cultural models are centrally implicated in reconciling this double perspective that is a key aspect of the human condition.

We have seen this reconciliation at work in the coordination evident between egocentric and allocentric navigation strategies. In turning from navigating literal space to social space, we have noted some interesting examples of how kinship terminologies mediate between the demands of collective general orientation and particular personal orientation. General referential terminologies attempt to represent simultaneously all possible relationships (an allocentric frame), but cannot evade the postulating of an abstract and generalized "ego." The explicitly allocentric representation masks an embedded egocentric frame. Conversely, American terms of address reflect their purely local and egocentric function in their insistence on creative individual variations in address. But they also retain a strong echo of their allocentric common reference coordinates in the fact that the creatively manipulated kin terms (e.g., "momikins") or personal names (i.e., "Suzie-Q") are usually recognizable personal variants of conventional reference terms or names derived from allocentric models. Each framing turns out to be embedded in its opposite.

Another example of such perspective-crossing comes from the Kingdom of Tonga, a neighboring island group to Samoa. In what is perhaps the most thorough and detailed study ever done of the cultural cognition of space, Giovanni Bennardo traces in great detail what he terms "a founda-tional cultural model" governing the cultural framing of spatial relations in Tonga (Bennardo, 2009). He calls the model "radiality." Radiality is a mental model that is specifically spatial, and since it is shared within a community, i.e., Tongans, it is also cultural. Moreover, since it is repeated in other mental modules and domains therein, it becomes a foundational model. In this conception, radiality is a foundational cognitive process that is used to organize knowledge across mental modules. Its intrinsic nature is spatial, and as such it belongs to the spatial representations module. Tongans preferably adopt and use radiality in other domains of knowl-edge—exchanges, political action, social networks, religion, kinship, and social relationships (Bennardo, 2009, p. 13). As a foundational model, radiality is understood to structure through analogical transfer a large number of domains of experience beyond spatial relations. These include a wide variety of traditional knowledge domains including kin terminolo-gies, concepts of power (*mana*), linguistically framed concepts of posses-sion, and models of time.

For the very similar Samoan case, Shore (following Lévi-Strauss) juxta-posed this conception of a radial model (which Lévi-Strauss called "con-centric dualism") to its opposite (Lévi-Strauss' "diametric dualism"), which he identified as an essentially binary model, bisecting space into categorically opposed zones (Shore, 1996, chapter 11). Shore argued that the former model, was an egocentric, graded representation of cultural space, while the latter was an allocentric model framed by categories that afforded an intersubjective framework of collective reference.

Bennardo has proposed for the Tongan case a somewhat different and very compelling reconciliation of egocentric and allocentric framing. Through a series of careful experiments testing Tongan spatial cognition, Bennardo has discovered that for Tongans, the implied center of the radial structure is not the observing ego, but rather a displaced ego: the highest ranking chief in the community.

What is radiality? In its most abstract form, radiality is a structural organization in which a number of vectors share a common origin that may also function as an ending point for all these vectors. In spatial relationships, radiality is the relationship between two points where one of them functions as the origin or goal of the vectors that signals the relationship. This origin/goal remains constant over a number of relationships with any number of points. The origin/goal can be ego, i.e., cognizer, viewer, and/or speaker, or a point in the field of ego. *I decided to label radiality this latter specific case; that is, a point in the field of ego, i.e., other-than-ego, is chosen to function as the source/goal of a number of relationships with other points in the same field, including ego. Essentially, this type of radiality stands for the foregrounding of other-than-ego while at the same time ego is relegated to the background* (Bennardo, 2009, p. 173, emphasis added).

The model is structurally egocentric, and the graded perception defined by its vectors is clearly derived from the ordinary experience of egocentric navigation. But because the point of view is displaced from any ego to a ranking chief, this framework manages to create a representational paradox *whereby each ego is simultaneously himself and someone else.* The model requires that one see the world from the perspective of a superior other, embedding his own sense of self in the imagined persona of a ranking chief. An explicitly egocentric master model is culturally manipulated to produce a point of view that is simultaneously subjectified and objectified, egocentric and allocentric.

Like Samoa, Tonga is a chiefdom: a society dominated by social rank and pervasive hierarchy. While individuals have a sense of private self, it is embedded in an enveloping institutional framework that reframes the self not simply as collective, but as literally encompassed by ranking chiefs and nobles. This self is thus displaced through identification with encompassing relations to chiefs and villages. This kind of displacement is not uncommon in Samoan discourse. Here is how one individual described his relation to Tofilau Eti, a ranking chief who was to become Prime Minister of Western Samoa. The names referenced are Tofilau's chiefly title, each stemming from a different villages:

Tofilau Eti is a cousin of mine. One of his sides is the Tofilau [title] but another of his sides [the Va'elua title] is me. After the meeting in the village, the other chief said to Tofilau—for his is in one of his parts, my true body—that someone had to report to me what the council had decided (Shore, 1982, p. 137).

Through a remarkable crisscrossing of egocentric and allocentric spatial models, Samoans and Tongans are able to create a social space that at one and the same time foregrounds the subjective perspective of the perceiving individual and replaces it with a collective object of intersubjective identification.

A structural compromise between egocentric and allocentric representation, the Tongan radial model uses an inherently egocentric framing of space, but shifts the point of view from ego to other-than-ego, in this case a local chief. This egocentric-allocentric perspective shift is a form of *objectification*. It provides an intersubjective frame of social reference that produces a collective point of identification for a community, but retains the concrete immediacy and intimacy of its egocentric framing.

The converse of this inside-out cognitive play is also possible, and it represents a form of *subjectification*. Such an allocentric-egocentric perspective shift takes an allocentrically framed model and reframes it egocentrically. This allocentric-to-egocentric reframing can be seen in many rituals, where the aim is to enhance the memorability and meaningfulness of an important collective representation by creating for individuals a purely personal analog to a shared conventional cultural representation. Below, a few examples of this allocentric-egocentric shift are summarized; Shore has analyzed these examples elsewhere in much greater detail.[17]

In an intriguing study of fifteenth-century Italian painting, Michael Baxendall suggests that the religious paintings of the time were conceived to present easily identified and highly conventionalized pictures of key figures and scenes from biblical history in such a way as to allow individual viewers to imaginatively impose their own personal vision of faces of loved ones onto the painting.

The painter could not compete with the particularity of the private representation. Beholders might approach his painting with preconceived interior pictures of such detail, each person's different, the painter did not try as a rule to give detailed characterizations of the people and places. It would have been an interference with the individual's private visualization if he had (Baxandall, 1988).

This same process of imaginative reframing of a conventional model to personalize it and reconcile collective and private representations is evident in a theory of personal religious devotion that was the subject of a number of fifteenth-century religious tracts in Italy. Here is an example from such a tract, the *Zardino de Oration (The Garden of Prayer)*, a primer on religious devotion for young girls written in 1454 by Nicholas of Osmino and later printed in Venice. It takes the form of a journey using the power of guided imagery. Baxendall quotes at length a translation from this handbook.

The better to impress the story of the Passions on your mind, and to memorize each action of it more easily, it is helpful and necessary to fix the places and people in your mind: a city, for example, which will be the city of Jerusalem—taking for this purpose a city that is well known to you. In the city, find all the principal places in which all the episodes of the Passion would have taken place—for instance, a palace with a supper room where Christ has the Last Supper with the Disciples, and the house of Anne, and that of Caiaphus, with the place where Jesus was taken in the night, and the room where He was brought before Caiaphus and mocked and beaten. And the residence of Pilate where he spoke with the Jews, and in it the room where Jesus was bound to the Column. Also the site of Mount Calvary, where He was put on the cross; and other places like that. . . .

And then you must shape in your mind some people, people well known to you, to represent for you the people involved in the Passion—the person of Jesus Himself, of the Virgin, St. Peter, St. John the Evangelist, St. Mary Magdalen, Anne, Caiaphus, Pilate, Judas and others, every one of whom you will fashion in your mind.

When you have done all this, putting all your imagination into it, then go into your chamber. Alone and solitary, excluding every external thought from your mind, start thinking of the beginning of the Passion, starting with how Jesus entered Jerusalem on the ass. Moving slowly from episode to episode, meditate on each one, dwelling on each single stage and step of the story. And if at any point you feel a sensation of piety, stop; do not pass on as long as that sweet and devout sentiment lasts." (Baxendall 1988, pp. 46–47)

The meditation ritual described here is framed as a journey. The aim is to get young girls to reenact a mental journey using well-known biblical landmarks familiar from stories and pictures as cues for constructing a mental map of Christ's journey to Calvary. At first the story is to be imagined through a kind of third-person narrative—an allocentric model. Then the primer recommends that the devotee begin to recreate the story, episode by episode, replacing the faces of the biblical figures with faces of "people well known to you." The narrative is systematically transformed into a journey framed egocentrically, with the devotee identifying with Christ as he moves on the journey to Mount Calvary. As the frame of reference shifts, Christ's journey becomes her journey. By internalizing a conventional religious narrative using an egocentric framing, the devotee learns to identify with Christ's passion. A powerful cultural model is at once socialized and personalized through the reconciliation of two very different points of view.

4.13   Conclusion: Cultural Models in Perspective

Spatial cognition has a very special place for humans. Spatial orientation is important in its own right as the most basic form of orientation an individual has to the world. But spatial cognition is also important because so many cultural models governing orientation are derived by metaphoric extension from the navigation of physical space. This study has reviewed the basic cognitive structures that support human spatial navigation in order to understand these projections from spatial orientation to some of the key models organizing people's journeys though their complex social environments.

We have taken special note of the importance of the notion of frames of spatial reference, and examined in some detail the differences between egocentric and allocentric frames. These basic perspectival distinctions in models of spatial navigation turn out to have their analogs in many orientational cultural models. In consequence, these cultural models come in pairs, one model having an allocentric variant and the other an egocentric one. The functional basis of this duplication of orientational models appears to be the representation of two equally important kinds of coordination in human experience. Individuals need to coordinate themselves both personally and socially to their worlds. Social coordination results in intersubjectivity, and is scaffolded by allocentric models that provide the basis for joint cognition. Personal coordination results in the effective orientation of individuals to their changing environments, and to their ability to gain a basic sense of internal coherence and personal meaning in their lives.

One of the most interesting aspects of the role of egocentric and allocentric modeling in human life is that these two perspectives are so commonly combined or interchanged in human life. This is true for spatial navigation, and it is equally the case in more attenuated forms of social navigation. We have examined several cases where there seem to be egocentric-to-allocentric and allocentric-to-egocentric shifts in perspective. These have been explained using the notions of objectification and subjectification. It is notable that such perspective shifts are so frequently found in human ritual practices. This suggests that a drive to reconcile potentially problematic differences between conventional and personal aspects of human experience is at the heart of culturally mediated perspective shifts. It may well turn out that such fundamental reframing of experience has a deep, if poorly understood, relation to the place of ritual in human life.

Notes

1. These and similar objections to the notion of cultural models are taken up and discussed in considerable detail by Strauss and Quinn (1997). See especially chapter 2, "Anthropological Resistance."

2. For a fascinating exploration of such a foundational schema of spatial orientation in the South Pacific, see Bennardo (2009).

3. The fact that in recent decades football, a decidedly martial sport, has come to eclipse the more pastoral game of baseball in popularity and cultural salience is hardly insignificant.

4. The use of the notions of mental or social "frames" for conceptualizing distinct cognitive and social perspectives and contexts was first used by the sociologist Erving Goffman (1974). The concept of frame is quite common in sociology, anthropology, cognitive linguistics, psychology, and media studies. See Goffman, (1974); Gardner (1993); Tversky and Kahneman (1981); Lakoff (2002); Kuypers (2009). See also Fauconnier (1994) and Fauconnier and Sweetser (1996) for a conception of "mental spaces" in cognitive linguistics that is very similar to the notion of frames.

5. It is possible that the special appeal of still photography has to do with the freezing of action, and the translation of what is usually transient and ineffable visual stimuli into frozen forms subject to conscious recall, reflection, and analysis. The brain has an analogous capacity to the still camera. Under stress, the brain appears to have the capacity to "slow down" perceived egocentric motion so that it is subject to long-term memory and recall. In the case of traumatic events, this slowing down of perception results in what psychologists call "flashbulb memories" (Brown & Kulik, 1977; Conway, 1995; McCloskey, Wible, & Cohen, 1988). Alfred Hitchcock was a master at inducing in his film audience the flashbulb effect, so that certain scenes—always filmed from the victim's perspective (most notoriously, the shower scene in *Psycho*)—become literally unforgettable.

6. Until the advent of air and space travel and, more recently, the GPS system, direct human experience with a bird's-eye view was limited to viewing terrain from high natural or man-made elevations. While the view of the earth from space, reproduced by the program Google Earth in its opening image, along with the capacity to manually rotate the earth in any direction, create the allocentric impression of seeing

"everything at once," it is important to recognize that this is simply an approximation of perspectiveless vision by distance, a vision that for every particular locus
inevitably requires an egocentric perspective.

7. As a strategy in "social engineering," allocentric perspectives are related to the
idea of mass social control. One of the early political conceptualizations of allocentric perspective was implicit in the term *panopticon*, first used by Jeremy Bentham
in reference to a plan for a prison in which total control of prisoners would be
exercised by the all-seeing but invisible eyes of the prison guard looking down on
the entire prison from a height (1995). The idea anticipated the modern security
landscape, dominated traditionally by a web of covert spies, and more recently by
omnipresent video surveillance. Bentham borrowed this notion from his brother
Samuel, who used it as a way to produce efficient control of a large number of people
in his plans for a military school in Paris. Michel Foucault adopted Bentham's
concept of the panopticon to refer to the emergence of modern disciplinary societies, with their orientation toward mass control over populations by various forms
of surveillance and standardization of behavior. In these political contexts, extreme
allocentrism is related to the perspectiveless objectification of human life and to a
denial of epistemological reciprocity. Those subject to disciplinary institutions are
controlled by an all-seeing eye that itself cannot be seen (Foucault, 1995).

8. The languages studied are Arrernte (central desert of Australia); Jaminjung (Northern Territory of Australia); Kaurna Warra (northwest Australia); Yélî Dnye (Rossel
Island, Papua New Guinea); Kilivila (Milne Bay, Papua New Guinea); Tzeltal (Mayan
language, Chiapas, Mexico); Yukatek Maya, Tiriyó (Carib language, Brazil-Surinam
border); Ewe (Ghana-Togo); Tamil (Dravidian language, South India); Japanese; and
Dutch.

9. Frames of reference are defined by Levinson and Wilkens as "coordinate systems
whose function it is to designate angles or directions in which a figure can be found
in respect to a ground" (Levinson & Wilkens, 2006, p. 541).

10. Interestingly, if the reference were to a specific map located in the range of sight
of speaker and hearer, "left" and "right" would be acceptable.

11. Since language is a key source of cultural models (i.e., linguistically formatted
models), it should be considered an important subset of cultural models.

12. For a more extensive discussion of this cultural conflict between spatial representational frames, see Shore (1996, pp. 278–283).

13. This example suggests that while personal names have egocentric and allocentric representations as onomastic systems of their own, they are also sometimes parts
of kin address terminologies. Thus, in American English, personal names are commonly reciprocals of kin terms where kin terms address senior generations, and
personal names are used to address junior generations. This is a good example of a
hybrid egocentric-allocentric system, in which a widely shared structural constraint

interacts with a more idiosyncratic choice of naming variants. Thus the common exchange: "Bobby??" "Yes, Mom, I'm here." Similarly, grandparents tend to call their grandchildren by a familiar variant of their personal name, while the grandchild address the grandmother with a personal variant of their kin term, or kin term + name.

14. In Saussurian terms, they are aspects of *parole* (speech) rather than *langue* (language) (Saussure, 1959).

15. The interpenetration of the traditional Samoan chiefly system with colonially derived government institutions means that there is some variation with different speakers on whether they will use the *t* or *k* in political oratory.

16. For a more thorough analysis of these pronunciation styles, see Shore (1982, pp. 262–283), and Duranti (1994, pp. 43–46). Samoan also has registers of chiefly and commoner language forms with differences in word use and syntactic forms. Like the pronunciation registers, these are represented in allocentric models as binary alternatives. But in use via egocentric models, there is a sliding scale of modulated distance and closeness that can be represented by various degrees and combinations of these forms, a scale that has no formal representation in Samoan public models of their own culture.

17. Shore (1998).

References

Aginsky, V., Harris, C., Rensink, R., & Beusmans, J. (1997). Two strategies for learning a route in a driving simulator. *Journal of Environmental Psychology, 17*, 317–331.

Avons, S. C. (2007). Spatial span under translation: A study of reference frames. *Memory & Cognition, 35*(3), 402–417.

Bateson, G. (1972). Morale and national character. In *Steps to an ecology of mind* (pp. 88–106). Chicago: University of Chicago Press.

Baxandall, M. (1988). *Painting and experience in fifteenth-century Italy: A primer in the social history of pictorial style*. Oxford: Oxford University Press.

Benedict, R. (1932). Configurations of culture in North America. *American Anthropologist, 34*, 1–27.

Benedict, R. (1934). *Patterns of culture*. Boston: Houghton Mifflin.

Bennardo, G. (2009). *Language, space and social relationships: A foundational cultural model in Polynesia*. Cambridge: Cambridge University Press.

Bentham, J. (1995). Panopticon (preface). In M. Bozovic (Ed.), *The panopticon writings* (pp. 29–95). London: Verso.

Bourdieu, P. (1999). *Language and symbolic power*. Cambridge, MA: Harvard University Press.

Brese, C. R., Hampson, R., & Deadwyler, S. (1989). Hippocampal place cells: Stereotypy and plasticity. *Journal of Neuroscience, 9*, 1097–1111.

Brotons, J., O'Mara, S., & Sanchez-Vives, M. (2005). Neural processing of spatial information: What we know about place cells and what they can tell us about presence. *Presence, 15*(5), 59–67.

Brown, R., & Kulik, J. (1977). Flashbulb memories. *Cognition, 5*(1), 73–99.

Campbell, J. (1994). *Past, space, and self.* Cambridge, MA: MIT Press.

Carroll, J. (Ed.). (1956). *Language, thought, and reality: Selected writings of Benjamin Lee Whorf.* Cambridge, MA: MIT Press.

Conway, M. A. (1995). *Flashbulb memories.* Hove: Lawrence Erlbaum.

D'Andrade, R. (1989). Cultural cognition. In M. I. Posner (Ed.), *Foundations of cognitive science* (pp. 795–830). Cambridge, MA: MIT Press.

D'Andrade, R. (1995). *The development of cognitive anthropology.* New York: Cambridge University Press.

Dumont, L. (1992). *Essays on individualism: Modern ideology in anthropological perspective.* Chicago: The University of Chicago Press.

Duranti, A. (1994). *From grammar to politics: Linguistic anthropology in a Western Samoan village.* Berkeley, CA: University of California Press.

Fauconnier, G. (1994). *Mental spaces: Aspects of meaning construction in natural language.* Cambridge: Cambridge University Press.

Fauconnier, G., & Sweetser, E. (Eds.). (1996). *Spaces, worlds, and grammar.* Chicago: University of Chicago Press.

Finney, B. (1994). *A voyage of discovery: A cultural odyssey through Polynesia.* Berkeley, CA: University of California Press.

Foucault, M. (1995). *Discipline and punish: The birth of the prison.* New York: Vintage Books.

Friedman, A. (2005). Examining egocentric and allocentric frames of reference in virtual space systems. *Sprouts: Working papers on information systems* 5(16). Delft University of Technology, Delft, The Netherlands. Retrieved from http://sprouts.aisnet.org/5-16.

Gardner, H. (1993). *Frames of mind: The theory of multiple intelligences* (10th ed.). New York: Basic Books.

Geertz, C. (1973). *The interpretation of cultures*. New York: Basic Books.

Gentner, D. (1983). Structure-mapping: A theoretic framework for analogy. *Cognitive Science, 7*, 155–170.

Gentner, D. (1989). The mechanisms of analogical learning. In S. Vosniadou & A. Ortony (Eds.), *Similarity and analogical reasoning* (pp. 199–241). Cambridge: Cambridge University Press.

Gentner, D., & Stevens, A. (Eds.). (1983). *Mental models*. Hillsdale, NJ: Lawrence Erlbaum.

Gladwin, T. (1972). *East in a big bird*. Cambridge, MA: Harvard University Press.

Goffman, E. (1974). *Frame analysis: An essay on the organization of experience*. New York: Harper and Row.

Gorer, G. (1948). *The American people: A study in national character*. New York: W.W. Norton.

Gorer, G. (1950). *The people of Great Russia: A psychological study*. New York: Chanticleer Press.

Gorer, G. (1955). *Exploring English character*. New York: Criterion Books.

Grush, R. (2000). Self, world and space: The meaning and mechanisms of egocentric and allocentric spatial representation. *Brain and Mind, 1*(1), 59–92.

Hartley, T., Macguire, E., Spiers, H., & Burgess, N. (2003). The well-worn route and the path less traveled: Distinct neural bases of route following and way finding in humans. *Neuron, 37*, 877–888.

Haun, D., & Rapold, C. (2009). Variation in memory for body movements across cultures. *Current Biology, 19*(23), 1068–1069.

Holland, D., & Quinn, N. (1987). *Cultural models in language and thought*. New York: Cambridge University Press.

Howes, D. (*2003*). *Sensual relations: Engaging the senses in culture and social theory*. Ann Arbor, MI: University of Michigan Press.

Hutchins, E. (1983). Understanding Micronesian navigation. In D. Gentner, & A. Stevens, (Eds.) *Mental models* (pp. 191–226). Hillsdale, NJ: Lawrence Erlbaum.

Hutchins, E. (1989). The technology of team navigation. In J. Galegher, R. Kraut, & C. Egedo (Eds.), *Intellectual teamwork: The social and technical bases of cooperative work*. Hillsdale, NJ: Lawrence Erlbaum.

Hutchins, E. (1995). *Cognition in the wild*. Cambridge, MA: MIT Press.

Jackendoff, R. (1983). *Semantics and cognition*. Cambridge, MA: MIT Press.

Jackendoff, R. (1999). The architecture of the linguistic-spatial inference. In P. Bloom, M. A. Peterson, L. Nadel, & M. F. Garrett (Eds.), *Language and space* (pp. 1–30). Cambridge, MA: MIT Press.

Johnson-Laird, P. N. (1983). *Mental models: Towards a cognitive science of language, inference, and consciousness.* Cambridge, MA: Harvard University Press.

Johnston, J. R., & Slobin, D. (1979). The development of locative expressions in English, Italian, Serbo-Croatian and Turkish. *Journal of Child Language, 6,* 529–545.

Jordan, K., Schadow, J., Wuestenberg, T., Heinze, H., & Jancke, L. (2004). Different cortical activations for subjects using allocentric or egocentric strategies in a virtual navigation task. *Neuroreport, 15,* 135–140.

Kelly, J., & McNamara, T. (2008). Spatial memory and spatial orientation. In C. Freksa, N. S. Newcombe, P. Gärdenfors, & S. Wölfl (Eds.), *Spatial cognition VI: Learning, reasoning, and talking about space,* LNAI 5248 (pp. 22–38). Berlin: Springer-Verlag.

Kimmel, M. (2002). *Metaphor, imagery, and culture: Spatialized ontologies, mental tools, and multimedia in the making* (Unpublished doctoral dissertation). Department of Anthropology, University of Vienna, Vienna, Austria.

Kitayama, S., Markus, H., Matsumoto, H., & Norasakkunit, V. (1997). Individual and collective processes in the construction of the self: Self enhancement in the United States and self-criticism in Japan. *Journal of Personality and Social Psychology, 72,* 1245–1267.

Klatzky, R. (1998). Allocentric and egocentric representations: Definitions, distinctions, and interconnections. In C. L. Freska, C. Hable, & K. F. Wender (Eds.) *Spatial cognition: An interdisciplinary approach to representing and processing spatial knowledge* (Vol. 1, pp. 1–17). Berlin, Germany: Springer-Verlag.

Kuypers, J. (2009). *Rhetorical criticism: Perspectives in action.* Lanham, MD: Lexington Books.

Lakoff, G. (2002). *Moral politics: How liberals and conservatives think.* Chicago: University of Chicago Press.

Lakoff, G., & Johnson, M. (1980). *Metaphors we live by.* Chicago: University of Chicago Press.

Lakoff, G., & Johnson, M. (1999). *Philosophy in the flesh: The embodied mind and its challenge to Western thought.* New York: Basic Books.

Lakoff, G., & Kovecses, Z. (1987). The cognitive model of anger inherent in American English. In D. Holland & N. Quinn (Eds.), *Cultural models of language and thought* (pp. 195–221). Cambridge: Cambridge University Press.

Lakoff, G. & Nunñez, R. (2001). *Where mathematics comes from: How the embodied mind brings mathematics into being.* New York: Basic Books.

Lakoff, G., & Turner, M. (1989). *More than cool reason: A field guide to poetic metaphor.* Chicago: University of Chicago Press.

Lévi-Strauss, C. (1967). Do dual organizations exist? In *Structural Anthropology.* New York: Anchor Books.

Levi-Strauss, C. (1976). The meaning and use of the notion of model. In *Structural Anthropology* (Vol. Two, pp. 72–82). Chicago: University of Chicago Press.

Levinson, S. (1996). Frames of reference and Molyneux's question: Cross-linguistic evidence. In P. Bloom, M. A. Peterson, L. Nadel, & M. F. Garrett (Eds.), *Language and space* (pp. 109–169). Cambridge, MA: MIT Press.

Levinson, S. (2003). *Space in language and cognition.* Cambridge: Cambridge University Press.

Levinson, S., & Wilkins, D. (Eds.). (2006). *Grammars of space: Explorations in cognitive diversity.* Cambridge, MA: Cambridge University Press.

Lewis, D. (1972). *We the Navigators.* Honolulu: University of Hawaii Press.

Majid, A., Bowerman, M., Kita, S., Haun, D., & Levinson, S. (2004). Can language restructure cognition? The case for space. *Trends in Cognitive Sciences, 8,* 108–114.

Markus, H., & Kitayama, S. (1991). Culture and the self: Implications for cognition, emotion, and motivation. *Psychological Review, 98,* 224–253.

McCloskey, M., Wible, C., & Cohen, N. (1988). Is there a special flashbulb-memory mechanism? *Journal of Experimental Psychology. General, 117,* 171–181.

McNeill, D. (2005). *Gesture and thought.* Chicago: University of Chicago Press.

Mead, M. (1971). *Sex and temperament in three primitive societies.* New York: Harper Perennial. (Original work published 1935)

Mou, W., McNamara, T., Valiquette, C., & Rump, B. (2004). Allocentric and egocentric updating of spatial memories. *Journal of Experimental Psychology: Learning, Memory, and Cognition, 30*(1), 142–157.

O'Keefe, J., & Dostrovsky, J. (1971). The hippocampus as a spatial map: Preliminary evidence from unit activity in a freely moving rat. *Brain Research, 34,* 171–175.

Piaget, J., & Inhelder, B. (1956). *The child's conception of space.* London: Routledge and Kegan Paul.

Poucet, B., Lencke-Santini, P., Paz-Villagran, V., & Save, E. (2003). Place cells, neocortex and spatial navigation: A short review. *Journal of Physiology, Paris, 97,* 537–546.

Quinn, N. (1991). The cultural basis of metaphor. In J. Fernandez (Ed.), *Beyond metaphor: Trope theory in anthropology* (pp. 56–93). Palo Alto, CA: Stanford University Press.

Saussure, F. (1959). *A course in general linguistics.* C. Bally & A. Sechehaye (Eds.), W. Baskin (Trans.). New York: Philosophical Library.

Shore, B. (1977). *A Samoan theory of action: Social control and social order in a polynesian paradox.* Dissertation, Department of Anthropology, The University of Chicago.

Shore, B. (1982). *Sala'ilua: A Samoan mystery.* New York: Columbia University Press.

Shore, B. (1991). Twice-born, once conceived: Meaning construction and cultural cognition. *American Anthropologist, 93*, 9–27.

Shore, B. (1996). *Culture and mind: Cognition, culture and the problem of meaning.* New York: Oxford University Press.

Shore, B. (1998). The double life of models: Personal and cultural meaning. In *How culture means: The Heinz Werner lectures.* Worcester, MA: Clark University Press.

Sperber, D. (1985). Anthropology and psychology: Towards an epidemiology of representations. *Man, 20*, 73–87.

Sperber, D. (1996). *Explaining culture: A naturalistic approach.* Oxford: Blackwell.

Strauss, C., & Quinn, N. (1997). *A cognitive theory of cultural meaning.* New York: Cambridge University Press.

Suthana, N., Ekstrom, A., Moshivaziri, S., Knowlton, B., & Bookheimer, S. (2009). Human hippocampal ca1 involvement during allocentric encoding of spatial information. *Journal of Neuroscience, 29*(24), 10512–10519.

Tanz, C. (1982). An experimental investigation of children's comprehension of the locutionary verb ask. In C. E. Johnson & C. L. Thew (Eds.), *Proceedings of the Second International Congress for the Study of Child Language* (Vol. 1, pp. 357–371). Washington, D.C.: University Press of America.

Tolman, E. (1948). Cognitive maps in rats and men. *Psychological Review, 55*(4), 189–208.

Tversky, A., & Kahneman, D. (1981). The framing of decisions and the psychology of choice. *Science, 211*, 453–458.

Wexler, M. (2003). Voluntary head movement and allocentric perception of space. *Psychological Science, 14*, 340–346.

Whishaw, I. Q., Hines, D. J., & Wallace, D. G. (2001). Dead reckoning (path integration) requires the hippocampal formation: evidence from spontaneous exploration and spatial learning tasks in light (allothetic) and dark (idiothetic) tests. *Behavioral Brain Research*, 127, 49–69.

Wolbers, T., Weiller, C., & Buchel, C. (2004). Neural foundations of emerging route knowledge in complex spatial environments. *Brain Research. Cognitive Brain Research*, *21*, 401–411.

III POLITICS

5 Emotional Underpinnings of Political Behavior

Stanley Feldman, Leonie Huddy, and Erin Cassese

5.1 Introduction

Research on emotion has afforded new insight into the fields of public opinion and political behavior. Political scientists have long noted politicians' efforts to engage voters' emotions in political campaigns (Lazarsfeld, Berelson, & Gaudet, 1944) and understood that public opinion was not simply a cold calculation based on one's self-interest (Kinder, 1998; Sears & Funk, 1991). Over the last several decades, public opinion researchers have identified a variety of factors that influence political thinking, including powerful symbols linked to values, political identities, and salient social groups (Sears, Lau, Tyler, & Allen, 1980; Sears, Hensler, & Speer, 1979). Within Sears and colleagues' influential symbolic politics model, for example, public opinion is driven by Americans' positive and negative feelings for political parties, liberals, conservatives, the nation, and various sociodemographic, racial, and religious groups. Conover and Feldman (1981) extended this research to demonstrate that positive and negative feelings toward various social groups explicitly shape Americans' ideological self-placements as liberal or conservative. Taking this research as a whole, it is impossible to escape the conclusion that public opinion is shaped by powerful emotional forces.

Emotion became further integrated with cognition in the 1980s with the study of political information processing, helping to explain why strong partisans and committed ideologues seek out like-minded information in a biased form of information processing referred to as motivated reasoning (Kunda, 1990; Lord, Ross, & Lepper, 1979; Westen, 1985). Motivated reasoning maximizes the positive feeling aroused by having one's beliefs confirmed by new information and leads to the rejection of arguments and information that emanate from one's political opponents (Westen, 1985; Zaller, 1992). Lodge and Taber (2005) further developed this research into a "hot cognition" model of information processing in which political concepts are affectively tagged with a positive or negative emotional charge. Positive feelings arise in response to new information that confirms or is consistent with one's beliefs, whereas contrary information is associated with negative feelings. In this way, affective information is incorporated into a running tally or overall impression of policies and political candidates. Emotion thus shapes public opinion and is central to the maintenance of political beliefs by leading to a biased processing of new political information.

Political behavior research has implicitly acknowledged the powerful influence of affect on political attitudes. But until recently, emotion was not studied as a central feature of political behavior, and research remained disconnected from more general theories of emotion. In past political behavior research, emotions were crudely categorized as positive or negative, belying the richness of human emotional experience reflected in differentiated emotions such as anger, anxiety, sadness, hope, joy, and pride. Past research also offered little insight into the conditions under which people experience different emotions, how such differentiated feelings arise, the origins of variation in emotional intensity, or the conditions under which people break old emotional associations and form new ones.

In the last fifteen years, there has been an explosion of social science research on emotions that has had a growing influence on the study of political behavior. George Marcus, Michael MacKuen, and colleagues advanced the study of emotion in politics with the development of affective intelligence theory (Marcus, 2003; Marcus & MacKuen, 1993; Marcus, Neuman, & MacKuen, 2000). The theory emphasizes the distinct effects of two weakly related emotions—anxiety and enthusiasm—on political thinking and behavior. According to affective intelligence theory, anxiety arises in response to novelty or uncertainty and produces interest in and attention to politics, while decreasing reliance on long-standing predispositions (such as partisanship). Anxious citizens seek out more information than the non-anxious and ultimately learn more about the candidates. In contrast, enthusiasm elicited by candidates during a political campaign enhances participation, heightens reliance on long-standing predispositions, and motivates political action in support of one's preferred party or candidate. In this respect, affective intelligence theory goes beyond validating the political import of emotion to differentiate specific emotions and underscore the interconnectedness of emotion, cognition, and behavior.

Affective intelligence theory's emphasis on two distinct dimensions of emotion is consistent with other political research (Abelson, Kinder, Peters,

& Fiske, 1982; Marcus, 1988; Brader, 2006). Two dimensions of emotion—one positive and one negative—commonly emerge in social psychological research as well (Marcus, 2003). The two dimensions are only weakly related and have been linked to the competing behavioral tendencies of approach and avoidance. In a political context, this means that a candidate, campaign, or issue can simultaneously arouse positive and negative emotions that produce a complex set of political attitudes and behaviors (Cacioppo, Gardner, & Berntson, 1999; Watson, Wiese, Vaidya, & Tellegen, 1999; Tellegen, Watson, & Clark, 1999).

This relatively neat two-dimensional model of positive and negative emotions fails, however, to account for the existence of diverse negative emotions such as fear, sadness, and anger that can have very different political consequences. For example, Marcus et al. (2000) and MacKuen, Wolak, Keele, and Marcus (2010) demonstrate that aversion—an emotional state characterized by anger and disgust—depresses learning, whereas anxiety increases the desire to learn about both sides of a political debate (see also Valentino, Hutchings, Banks, & Davis, 2008b; Feldman & Huddy, 2005). In addition, anger and anxiety have divergent effects on an array of political attitudes, including the perceived risk of future terrorism (Lerner, Gonzalez, Small, & Fischhoff, 2003); political tolerance after the terrorist attacks of September 11, 2001 (Skitka, Bauman, & Mullen, 2004); attributions of criminal behavior (Petersen, 2010); presidential approval (Conover & Feldman, 1986; Lambert, Scherer, Schott, Olson et al., 2010); attitudes toward risk (Druckman & McDermott, 2008; Lerner & Keltner, 2000, 2001); and support for war (Huddy, Feldman, & Cassese, 2007). Other research distinguishes between the political consequences of anger and sadness (Small & Lerner, 2008) and anger and hatred (Halperin, Canetti-Nisim, & Hirsch-Hoefler, 2009)—suggesting a richness to negative emotional experience that is obscured by viewing emotions as simply falling on a positive or a negative dimension.

Overall, political emotions research suggests the need to "split" rather than "lump" emotional experience. To best understand the political implications of emotion, one must distinguish among emotions of the same valence and simultaneously contrast them in order to parse out their distinct (and sometimes opposing) effects. This chapter begins with a brief overview of research evidence on the distinct origins of anger and anxiety, and their divergent effects on risk assessment, risky behavior, political action, and depth of cognitive processing. This is followed by an overview of the research program developed by Feldman, Huddy, and colleagues on the empirical effects of anger and anxiety on American support for military action in Iraq in late 2002 and early 2003 (Huddy, Feldman, Taber, & Lahav, 2005; Huddy et al., 2007). As part of this overview, clear evidence is provided of the integral connection between emotion and cognition in the development of political attitudes toward foreign policy. The research findings discussed in this chapter highlight the utility of looking beyond positive and negative emotional valence to the distinct effects of anger and anxiety on opinion toward the Iraq war, and public opinion more generally.

5.2   From One to Two Dimensions of Emotion

The most basic conceptualization of human emotion suggests that all feelings can be arrayed on a single, bipolar dimension running from strongly negative to strongly positive. This notion of bipolarity is reflected in research on implicit feelings and attitudes in both psychology and political science (Lodge & Taber, 2005; Murphy & Zajonc, 1993; Bargh, 1994, 1997). For instance, Milton Lodge and Charles Taber's formulation of hot cognition (2005) posits that all social and political concepts a person encounters and evaluates, however fleetingly, are affectively charged in stored memory. The positive or negative affective "tag" for each object is activated automatically upon exposure to the object, without conscious deliberation. Similarly, Green and colleagues suggest that most variation in emotional response can be explained by a single positive-negative affect dimension (Green, Goldman, & Salovey, 1993). These unidimensional models of emotion suggest that positive and negative feelings will not co-occur, and that an increase in positive feeling necessitates a concomitant decrease in negative feeling.

Much research, however, suggests that positive and negative emotions are actually situated along two moderately correlated dimensions. When conceptualized as a single, bipolar dimension, extreme values on one end of the dimension would correspond to very negative emotions, while extreme values at the other end are characterized as very positive. In the two-dimensional model, extreme values on the negative-emotion dimension would correspond to very low levels of negative emotion, on one hand, and very intensive feelings of negative emotion on the other. The same is true for the positive-emotion dimension; it ranges from extremely low amounts of positive feelings to extremely high amounts of positive feelings. Unlike the unidimensional model, the two-dimensional model allows for people to feel simultaneously positively and negatively about the same event, person, or experience (Cacioppo, Gardner, & Berntson, 1999; Marcus, 2003). Watson and colleagues (1999), for example, find a correlation of -0.45 between self-reported positive and negative emotions. The correlation increases under some conditions, such as the presence of strongly aversive stimuli.

Watson and colleagues take this one step further. They present a two-dimensional model of human emotion (Watson & Clark, 1992; Watson et al., 1999; Tellegen et al., 1999), but argue that affect is best understood in terms of three hierarchical levels. The highest level is a bipolar positive/negative dimension, as embodied in the unidimensional valence model discussed earlier. The second, intermediate, level consists of the two separable dimensions of positive and negative affect. The third level distinguishes among discrete emotions, such as hope, pride, anger, and anxiety. The intermediate level of dissociated positive and negative affect is the cornerstone of Watson's conception of emotional structure. As a result, his hierarchical model of human emotion is considered a two-dimensional valence model.

Researchers employing the two-dimensional valence model have linked positive and negative emotions to other important psychological constructs, such as personality traits. For example, Watson and colleagues (1999) find a close link between positive emotion and extraversion, and negative emotion and neuroticism. The positive and negative dimensions of emotion also map cleanly onto two motivational systems that developed through the course of human evolution—the approach and avoidance systems (Barkow, Cosmides, & Tooby, 1992).[1]

The two-dimensional affective-motivational system is functional in nature. Positive and negative emotions activate different regions of the brain. Positive emotions tend to activate the left hemisphere of the brain—associated with approach behavior—and negative emotions tend to activate the right hemisphere, resulting in avoidance behavior (Davidson, 1995). Emotions cause changes in the brain, followed by physiological changes in the body, which ready it for a particular course of action (Damasio, 1994; LeDoux, 1996. This could involve goal-seeking behavior, or a motivation to pursue opportunities for reward, as in the case of positive emotions. Alternatively, negative emotions are experienced under conditions of threat and serve to produce behaviors associated with avoidance of damage, loss, or harm. In this respect, they serve an adaptive function, readying the individual for an appropriate response to cues in their environment.

5.3   Beyond Two Dimensions: The Distinct Effects of Anxiety and Anger

The two-dimensional valence model is parsimonious and conceptually tidy, mapping cleanly onto approach-avoidance mechanisms as well as important aspects of personality and political ideology. Despite its appeal, however, the model is unable to explain or accommodate observed differences among various negative (e.g., anger and anxiety) and positive (e.g., hope and pride) emotions. The two-dimensional valence model assumes uniform consequences for emotions of a similar valence. For instance, the model predicts that both anger and anxiety will increase vigilance for novel and threatening information and lead to the avoidance of harm. In this respect, it fails to account for consequential differences between the two emotional states. A growing body of work in psychology and political science has documented the distinct effects of anger on sociopolitical attitudes and behaviors, when compared to other negative emotions. In fact, anger looks more like a positive than negative emotion (Lerner & Tiedens, 2005; Carver, 2004; Berkowitz & Harmon-Jones, 2004). Typical negative emotions such as anxiety increase attention to threat and lead to avoidant behavior. In contrast, anger generates approach behavior and lowers risk estimates, responses typically associated with positive emotions (Lerner & Keltner, 2000, 2001; Litvak, Lerner, Tiedens, & Shonk, 2010; Marcus, 2003; Rydell, Mackie, Maitner, Claypool et al., 2008).

5.3.1   Divergent Origins of Anger and Anxiety

Anger's psychological effects are not only distinct from those of other negative emotions, they are also grounded in different experiences. Consider anger and anxiety. According to popular appraisal theory explanations of emotion, anxiety arises in response to an external threat, in situations characterized by high levels of uncertainty when perceptions of personal agency or control are low (Bower, 1988; Eysenck, 1992; Smith & Ellsworth, 1985). Anger, by contrast, is elicited by a negative event that frustrates a personally relevant or desired goal in a situation characterized by a sense of personal control over events. Anger can be further aroused by viewing the source of obstruction as illegitimate (Carver, 2004; Clore & Centerbar, 2004; Lazarus, 1991; Stein, Trabasso, & Liwag, 2000; Winterich et al., 2010) or as a threat to group goals and security (Cosmides & Tooby, 2006). Anger and anxiety thus differ in the degree to which a precipitating event has known origins and is controllable, revealing their distinct adaptive functions. According to evolutionary psychologists, anxiety is an adaptive response to an external hazard, whereas anger is an adaptation to social threat—social rule and norm violations that threaten one's group (Cosmides & Tooby, 2006 Petersen, 2010; see also Haidt, 2003).

The link between anger and social threat is developed further in intergroup emotions theory (IET), in which anger is seen as a collective response to threat from a weaker out-group that is felt most intensely by strong group identifiers (Mackie, Devos, & Smith, 2000). IET combines a basic understanding of emotions drawn from appraisal theory with insight into intergroup dynamics derived from social identity theory (Mackie et al., 2000; Maitner et al., 2006; Smith, Seger, & Mackie, 2007). From this perspective, threat is most likely to produce anger among group members who see their group as strong relative to a threatening out-group, whereas it is more likely to produce anxiety among those who see their group as relatively weak and lack effective collective protection. Perceived group strength is not simply a reflection of reality, but also increases with level of group identification and a group ideology that enhances perceived group efficacy (van Zomeren et al., 2008). This explains why someone who belongs to an objectively weak group can view it as strong and able to defeat a more powerful enemy. IET thus predicts that individuals who identify most strongly with their nation or group are most likely to overestimate their group or country's might (Mackie et al., 2000; Rydell et al., 2008). When applied to national threat, IET suggests that citizens who strongly identify with their country will feel the greatest anger when attacked by an apparently weaker opponent (e.g., as many Americans viewed Al Qaeda in the wake of 9/11). In essence, IET views anger as a luxury afforded to members of stronger groups who are more certain of the success of retaliatory action against a weaker opponent. In the context of politics, IET views anger as a response to a collective threat (see also Huddy & Feldman, 2011).

In contrast, anxiety is best characterized as an atomized and isolated response to a personal rather than collective threat. This connection between personal threat and anxiety has been amply demonstrated in research on political violence. Arian and Gordon (1993) found that Israelis who believed Saddam Hussein was out to get them personally during the 1991 Gulf War experienced higher levels of anxiety four weeks after the war than did those who believed Hussein posed a collective threat to the Jewish people. Living near the World Trade Center site significantly increased anxiety after the 9/11 terrorist attacks (Galea, Ahern, Resnick, Kilpatrick et al., 2002; Huddy et al., 2005; Skitka et al., 2004). In general, direct personal experience with terrorism has an especially powerful effect on the development of anxiety and related psychological symptoms

(Gordon & Arian, 2001; Piotrkowski & Brannen, 2002; Schlenger, Caddell, Ebert, Jordan et al., 2002; Schuster, Stein, Jaycox, Collins, et al., 2001; Silver, Holman, McIntosh, Poulin et al., 2002). Social or psychological proximity in the form of knowing someone who was victimized by the 9/11 terrorist attack heightened a sense of personal vulnerability and exacerbated feelings of anxiety (Huddy et al., 2005; Silver et al., 2002).

The differing origins of anger and anxiety pose a challenge to two-dimensional valence models of emotion. If negative emotions are interchangeable, why does collective threat more commonly lead to anger, while a sense of personal vulnerability generates anxiety? Such differences require scholars to look beyond valence to evaluate the social and political consequences of discrete emotional states. The next section examines the divergent effects of anger and anxiety on risk assessment, behavior, and cognition, each broadly construed. The distinction drawn between anger and anxiety helps to explain important facets of public opinion. In a context dominated by feelings of anxiety, citizens are likely to clamor for strong personal protections that could erode civil liberties, for example. In contrast, pervasive anger is likely to generate support for retaliation and a muscular response to threat that involves the use of force. Researchers who combine anger and anxiety as a single negative reaction to political threat run the risk of combining two opposing psychological forces that cancel each other's influence, obscuring the true dynamics of public opinion.

5.3.2 Risk Assessment and Action

There are robust differences in the effects of anger and anxiety on risk assessment and risk-seeking behavior. Experimental work consistently demonstrates that anxiety heightens perceived risk and promotes risk-averse preferences and behaviors, whereas anger lowers perceived risk and elevates risky behavior (Butler & Mathews, 1987; Eysenck, 1992; Lerner et al., 2003; Lerner & Keltner, 2000, 2001; Rydell et al., 2008). For instance, Druckman and McDermott (2008) experimentally manipulated emotion and compared the responses of angry and anxious respondents to classic problems employed in prospect theory, such as the Asian disease problem (which is framed in terms of lives either gained or lost; Tversky & Kahneman, 1981). They found anger heightened enhanced risk seeking, whereas anxiety depressed support for risky alternatives. In this research, anger and the positive emotion of enthusiasm have comparable effects on risk propensity. Both contribute to greater optimism or reduced concern about risks, a finding the authors attribute to greater preference confidence. Anxiety, alternatively, is associated with higher levels of uncertainty.

The link between emotion and risk-tolerant behavior and judgment is also apparent in attitudes toward government policy. In experimental research on reactions to 9/11, anger dampened and anxiety increased the perceived likelihood of a future attack (Lambert et al., 2010; Lerner et al., 2003; Sadler, Lineberger, Correll, & Park, 2005; Skitka, Bauman, Aramovich, & Morgan, 2006). Such findings underscore the importance of distinct negative emotions and the need to contrast their political effects.

5.3.3 Political Involvement

Anger and anxiety also have differing effects on political participation. In Marcus and MacKuen's model of affective intelligence, enthusiasm bolsters electoral activity (1993; Marcus et al., 2000). Later work on this topic shows, however, that anger motivates political participation much like enthusiasm. Valentino and colleagues (2008b, 2011) find that anger bolsters political activity in their research, especially when it is coupled with feelings of political efficacy. Anger is especially likely to mobilize action among those who feel politically efficacious and experience a threat to a preferred public policy. Using American National Election Studies panel data, the authors demonstrate that feelings of anger and efficacy operate in a recursive loop over time, serving to establish and strengthen the habit of political participation. Miller and colleagues (2009) find a similar result in their study of collective action problems. In their study, anger and perceptions of group efficacy contribute to collective action, whereas fear inhibits such action.

To some extent, the divergent effects of anger and anxiety on political involvement reflect this simple association, as noted by Berenbaum and colleagues (1995). Anger results in less thought and more action, whereas anxiety results in less action and more thought. While anger bolsters political action, it seems to reduce political interest and learning. In both the lab and naturalistic settings, anxiety heightens attention to, and interest in, novel and potentially threatening information (Eysenck, 1992; LeDoux, 1996; MacLeod & Mathews, 1988; Mathews & MacLeod, 1986; Mogg, Mathews, Bird, & MacGregor-Morris, 1990; Öhman, 2000; Yiend & Mathews, 2001; Williams, Watts, MacLeod, & Mathews, 1997). Marcus and colleagues have shown that anxious responses to one or more political candidates contributes to higher levels of campaign interest, even when controlling for general levels of political interest (1993, 2000; MacKuen et al., 2010).

Anxious citizens also learn more about the candidates' characteristics and issue positions. While anxiety typically promotes heightened attention and more thorough, effortful information search, anger has the opposite effect, depressing information search—particularly information involving the opposition viewpoint (Valentino, Banks, Hutchings, & Davis, 2009; MacKuen et al., 2010). In these studies, anger was also associated with lower levels of expressed interest in the topic and lower levels of conciliatory attitudes toward policy conflict. Political interest and learning is a second area in which anger operates more like a positive emotion than a negative emotion. Work on mood maintenance suggests people in a positive emotional state will selectively attend to information in such a way that maintains the positive mood (Isen, 1984).

5.3.4 Depth of Cognitive Processing and Learning

Differences in political attention, interest, and learning are probably attributable to the different cognitive styles associated with anger and anxiety. Emotions associated with high levels of certainty, such as anger, have been linked to a decreased depth of information processing (Tiedens & Linton, 2001; Rydell et al., 2008), while anxious individuals take a more thorough, central processing approach (Berenbaum et al., 1995; Bless, 2001; Bless, Mackie, & Schwarz, 1992; Clore, Wyer, Dienes, Gasper et al., 2001; Schwarz, 1990). For example, Bodenhausen and colleagues (1994) found angry people engaged in more stereotyping and were more easily convinced by superficial aspects of persuasive speech. Similarly, Tiedens (2001) associated anger with character inferences based on chronically accessible scripts, which indicated a more superficial level of processing. And Moons and Mackie (2007) find anger increases reliance on heuristic processing.

The positive effects of anxiety on political interest suggest that it is a plus within the context of political deliberation and public opinion. The general consensus among psychologists, however, is that anxiety undermines the efficiency and quality of cognitive deliberation in two ways (Eysenck, 1997). First, it clutters working memory with anxious thoughts, making it difficult for individuals to perform complex cognitive tasks (Calvo & Carreiras, 1992; Eysenck, 1992; Eysenck & Calvo, 1992). Second, anxiety impairs the control of attention so that anxious individuals have difficulty ignoring distractions, especially threatening stimuli (Eysenck, Derakshan, Santos, & Calvo, 2007). Heightened attention to political information may improve anxious individuals' knowledge of politics, as suggested by Marcus and colleagues (2000; see also Valentino et al., 2008b) but it is also likely to impede the acquisition of needed information under personally threatening circumstances. Huddy and colleagues (2005) found that Americans made anxious by the 9/11 terrorist attacks consumed more news about events related to the attacks than non-anxious individuals, but nonetheless knew less about Afghanistan, Islam, and bin Laden in the months after 9/11 compared to the non-anxious.

5.4   Data and Key Measures

The differing origins and political consequences of anger and anxiety are examined in this section by extending the research program developed by Feldman, Huddy, and colleagues on American support for military action in Iraq in the lead-up to the war (Huddy et al., 2005; Huddy et al., 2007). The analysis reported here builds on past work, providing further evidence of the link between emotion and political cognition. In doing so, it contributes to the growing body of work on emotion and public reactions to 9/11 (e.g., Lerner et al., 2003; Skitka et al., 2006). It also compliments work on the political correlates of emotion in other domains, such as political learning (Marcus et al., 2000; MacKuen et al., 2010), policy attitudes (Valentino, Brader, & Suhay, 2008a; Petersen, 2010) and political participation (Valentino et al., 2011).

5.4.1   Sample

The data are drawn from the Threat and National Security Survey (TNSS), a three-wave national panel study. The analyses presented in this chapter are based largely on wave 2 of the study, which focuses on political reactions to the Iraq war. Wave 1 of the survey was conducted via telephone with a national sample of 1,549 adults aged 18 or older between early October, 2001, and early March, 2002, focusing on psychological reactions to 9/11 and support for government anti-terrorism policy. The initial sample was drawn as a weekly rolling cross-section with roughly 100 individuals interviewed each week throughout this period. The first month of data was collected by Shulman, Ronca, and Bukuvalis; the remainder of the data (including waves 2 and 3) was collected by the Stony Brook University Center for Survey Research. The cooperation rate for the survey was 52% (Association for Public Opinion Research cooperation rate 3, or COOP3).[2]

Wave 2 of data collection occurred in October of 2002, after congressional debate on the Iraq war had ended. Of the original interviewees, 858 were re-interviewed between 7 and 12 months later, for a re-interview rate of 55%. An additional 221 respondents were added to the panel from a fresh random-digit dialing sample drawn to the same specifications as the original. This new component was designed to serve as a check on panel effects, attrition, and composition. The cooperation rate for this new component was 56%. Wave 3, a more complex data collection, occurred in 2003. Half the sample was recontacted during the Iraq war, starting on the day after the war's onset (3/20/03) and continuing until 4/10/03, roughly at the official end of the war as announced by then President George Bush. The other half was interviewed some time after the war had ended, from 5/20/03 until 6/18/03. 612 individuals from the original panel and 117 of those introduced in wave 2 were re-interviewed, for a re-interview rate of 68% between waves 2 and 3. Interviews during all three waves of the survey were roughly 20 minutes in length. The second and third waves focused on reactions to terrorism and support for the Iraq war.

5.4.2 Measuring Emotion

Waves 2 and 3 of the TNSS included a battery of questions focused on emotional responses to a number of targets: anti-American terrorists, Saddam Hussein, a war with Iraq, and antiwar protesters (in wave 3 only). Anxiety was measured using three items. Respondents indicated the degree to which each target made them feel nervous, scared, and afraid on a four-point scale. Responses to these items were combined to form an anxiety scale. Anger was measured in a similar fashion. Respondents indicated the degree to which each target made them feel angry, hostile, and disgusted on four-point scales, which were combined into a single measure of anger. For the Iraq war and antiwar protesters positive emotions were also assessed using three items: enthusiastic, proud, and hopeful. Respondents were asked to report how much a given target had made them feel each emotion (very, somewhat, not very, and not at all). The items for anger, anxiety, and positive emotions were intermixed for each target. Overall, respondents reported a total of nine (anxiety, anger, and positive) emotions toward the Iraq war and antiwar protesters, and six emotions (anxiety, anger) toward Saddam Hussein and terrorists. Of these feelings, only anxiety toward terrorists was measured in the first wave of the study. Respondents were asked: "As you think about the terrorist attacks and the U.S. response, how often have you felt: anxious, scared, worried, frightened?"

Nine emotions toward President Bush were also assessed in the second and third waves of the survey. They are not used here because emotional reactions to the president formed a single, bipolar dimension, making it possible only to observe positive or negative evaluations of him. This empirical structure matches Watson, Tellegen, and colleagues' first-level bipolar dimension (Tellegen et al., 1999). This raises an important question about when different emotions can be distinguished. In this case, a combination of high levels of negative emotion and post-decisional consistency may explain the strong bipolar reactions to President Bush.

Many Americans had made a decision about Bush in the 2000 presidential election. More importantly, Bush was widely disliked at the time by Democrats, and strong Democratic negativity toward him may have fueled equally strong defensive support for him among Republicans, resulting in the single bipolar positive-negative reaction observed in our data. There has been too little research to fully understand the conditions under which anger and anxiety are separable (as opposed to forming a single negative dimension). It is clear from our data that complex, negative objects such as war and terrorism elicit diverse negative reactions.

5.5 Analysis

5.5.1 The Origins of Political Emotions

Our analysis begins by examining the factors that lead to anxiety and anger. Past research suggests that anxiety emerges out of personal feelings of uncertainty and threat. In the context of the 9/11 terrorist attacks, this generates the prediction that personal terrorist risk—the belief that the individual or a close friend or relative could be killed or hurt in an attack—should primarily drive feelings of anxiety. In contrast, perceived terrorist threat to the nation should have less influence on anxiety but greater effect on anger, which is driven by a sense of collective threat. Strong American patriots should be especially likely to react with anger but not anxiety to the sense that the United States is subject to terrorist threat.

To test these hypotheses, measures of anxiety and anger[3] in wave 2 of the TNSS were regressed on three key predictor variables: measures of personal[4] and national terrorist threat,[5] and a measure of patriotism.[6] The patriotism measure was used as an indicator of the strength of attachment to the United States and its symbols. In addition, a number of other background variables were added as controls to this and the remaining analytic models: age, education (in years), gender, race/ethnicity (black, Hispanic, and other, with white as the excluded category), party identification, ideology, and authoritarianism.[7] The estimates of the two regression equations for anger and anxiety are shown in table 5.1.

Table 5.1
Determinants of anxiety and anger

	Anxiety	Anger
Personal threat	.39 (.04)	.14 (.04)
National threat	.15 (.04)	.06 (.04)
Patriotism	.14 (.05)	.34 (.04)
Age	−.003 (.006)	.010 (.006)
Education	.009 (.004)	.000 (.004)
Gender	.13 (.02)	−.01 (.02)
Race/ethnicity		
Black	−.05 (.03)	−.03 (.03)
Hispanic	−.07 (.04)	.03 (.04)
Other	−.04 (.04)	−.02 (.04)
Party ID	−.04 (.03)	.07 (.03)
Ideology	−.02 (.03)	−.03 (.03)
Authoritarianism	−.04 (.03)	−.04 (.03)
R^2	.35	.17

Note: Entries are regression coefficients with standard errors in parentheses. The dependent variables are coded to range from 0 to 1. Age is coded in 10-year intervals and education is coded in years. White is the excluded category for the race/ethnicity dummy variables. All of the other predictors range from 0 to 1.

As expected, perceived personal threat has the largest substantive effect on reported anxiety (see the first column of table 5.1). Perceptions of national threat and patriotism also have significant effects, but their estimated coefficients are less than half the size of the coefficient for personal threat. Thus, the substantive effect of personal threat is quite large. Personal threat increases anxiety by almost 40% of its range when one compares anxiety levels for those at the lowest and highest levels of threat, holding all the other variables constant. Of the other variables, only gender has a substantively large effect, with women reporting greater levels of anxiety than men.

The results are very different for reported feelings of anger. The largest coefficient in column 2 is for patriotism. The stronger someone's attachment to American symbols such as the flag, the greater their anger in response to the 9/11 terrorist attacks. This effect is also large, moving anger by a third of its range as patriotism varies from its lowest to highest values. Personal threat has a significant but much weaker effect on anger than on anxiety, and the coefficient for national threat does not significantly increase anger. None of the other variables in this equation have statistically significant effects on anger.

5.5.2 Perceived Risks

Anger and anxiety have divergent origins in the research program examined here. They also have differing effects on cognition, especially the assessment of risk and threat. In the fall of 2002, some months before the Iraq war, anger and anxiety were expected to have opposite effects on risk assessment. Anxiety was expected to elevate, and anger minimize, the perceived risks of the war. To assess the war's perceived risks, respondents were asked about their concerns that a war in Iraq would "hurt the U.S. economy"; "increase the threat of terrorism against the United States"; "make the situation in the Middle East less stable"; or "decrease help for U.S. allies in the war on terrorism." These four questions were combined into a measure of risk assessment that ranges from 0 to 1 to reflect the number of potential risks perceived by the respondent. The risk Saddam Hussein posed to the United States and its allies was included as a second measure of perceived risk. The threat posed by Saddam was assessed by combining the perceived risk among participants that he would attack the United States with weapons of mass destruction (WMD) or use WMD against neighbors in the Middle-East with the likelihood that he actively supported anti-U.S. terrorist groups. This measure was also coded to range from 0 to 1. The regression estimates assessing the determinants of war risk and Saddam threat are shown in table 5.2.

Consider first the perceived risk of going to war. Increased anxiety about Saddam and terrorists is associated with a heightened perception that the war in Iraq involved substantial risks. The coefficient for anxiety is significant and large. In contrast, anger has the opposite effect. In line with much past research, anger substantially lowers the perceived risk of deploying military troops in Iraq. It is also important to note that if anger and anxiety are combined into a single measure of negative emotion, this combined measure has no significant effect on perceptions of risk. It is therefore necessary to distinguish among negative emotions to see their powerful political effects, which are otherwise obscured in analyses that simply assess the effects of negative emotion on risk and threat.

Anxious people not only view the war as risky, they are also more likely than non-anxious people to view Saddam as a threat. In the case of Saddam, angry people share the same view. Anger increases the perception that Saddam posed a risk, perhaps because this perception boosted support for the war and military action. Thus, the effects of anxiety and anger go together, but for different reasons. Anxiety promoted the perception that the war was risky *and* that Saddam posed a threat to the United

Table 5.2
Effects of anxiety and anger on risk and threat perceptions

	Risk	Saddam Threat
Anxiety	.19 (.04)	.07 (.03)
Anger	−.29 (.04)	.26 (.03)
Age	.010 (.006)	−.009 (.004)
Education	.012 (.004)	.004 (.003)
Gender	.04 (.02)	.04 (.01)
Race/ethnicity		
Black	.04 (.04)	−.10 (.03)
Hispanic	.02 (.04)	−.08 (.03)
Other	.06 (.04)	−.10 (.03)
Party ID	−.15 (.03)	.06 (.02)
Ideology	−.06 (.03)	.10 (.02)
Authoritarianism	−.05 (.03)	.06 (.02)
R^2	.18	.25

Note: Entries are regression coefficients with standard errors in parentheses. The dependent variables are coded to range from 0 to 1. Age is coded in 10-year intervals and education is coded in years. White is the excluded category for the race/ethnicity dummy variables. All of the other predictors range from 0 to 1.

States and his neighbors, consistent with the general effects of anxiety on heightened risk perceptions. In contrast, anger diminished the perceived risk inherent in a war with Iraq but boosted the threat posed by Saddam; both perceptions are consistent with support for military action against Iraq.

5.5.3 Political Motivation and Media Attention

Emotions not only affect risk assessment, they have additional significant consequences for cognition through their influence on attention and depth of information processing. The effects of anxiety and anger on attention to the Iraq war and media news consumption were assessed by four dependent variables. Respondents' self-reported amount of thought regarding the war was assessed with a scale formed from six questions. The first asked, "Over the last week or two, how much thought have you given to a possible war with Iraq?" The other five questions asked, "How much thought have you given to the effects of U.S. military action in Iraq on:" (a) "the U.S. economy," (b) "the threat of terrorism to the United States," (c) "the situation in the Middle-East," (d) "the help we would get from U.S. allies in the war on terrorism," and (e) "on Saddam Hussein's willingness

Table 5.3

Effects of anxiety and anger on cognitive effort and attention

	Thinking	Talking	Television	Newspapers
Anxiety	.21 (.03)	.16 (.04)	.09 (.05)	.06 (.05)
Anger	.12 (.03)	.06 (.05)	.12 (.05)	.10 (.06)
Age	.025 (.004)	.012 (.007)	.085 (.008)	.077 (.008)
Education	.005 (.003)	−.001 (.004)	.008 (.005)	.023 (.005)
Gender	−.08 (.02)	−.08 (.02)	−.11 (.02)	−.17 (.03)
Race/ethnicity				
Black	.02 (.03)	.10 (.04)	.16 (.05)	−.02 (.05)
Hispanic	.06 (.03)	.08 (.05)	.06 (.05)	−.02 (.06)
Other	−.04 (.03)	−.06 (.05)	.03 (.05)	−.00 (.06)
Party ID	.03 (.02)	.02 (.03)	.04 (.04)	.02 (.04)
Ideology	−.02 (.02)	−.04 (.04)	.04 (.04)	−.02 (.04)
Authoritarianism	−.05 (.02)	−.09 (.03)	−.07 (.04)	−.14 (.04)
R^2	.18	.05	.15	.14

Note: Entries are regression coefficients with standard errors in parentheses. The dependent variables are coded to range from 0 to 1. Age is coded in 10-year intervals and education is coded in years. White is the excluded category for the race/ethnicity dummy variables. All of the other predictors range from 0 to 1.

to use weapons of mass destruction against U.S. troops?" The scale is constructed to range from 0 to 1. The second measure of attention is based on a single question that asked respondents how often they had talked to friends, family, co-workers, or neighbors about a possible war with Iraq. Responses ranged from very often, somewhat often, not very often, or not at all, and were recoded to range from 0 (not at all) to 1 (very often). The final two measures of attention to the war are based on the number of days in the past week respondents reported watching national TV news and reading about national events in a newspaper. These measures were also rescaled to range from 0 to 1. The regression results examining the determinants of these four measures of attention to the impending Iraq war are shown in table 5.3.

In past research on the political consequences of emotion, anxiety typically predicts political interest and attention. This relationship is replicated in these data. Individuals who felt anxious about Saddam and terrorists thought and talked more than others about the war. The effects of anxiety on attention are substantively large. Anxious people were not, however, significantly more likely than non-anxious individuals to watch TV or read a daily newspaper. Findings are in the correct direction, with anxiety increasing news consumption, but do not quite reach statistical significance.

Past research provides less guidance on how anger influences political attention. Some studies have found that general negative affect causes heightened vigilance to threatening stimuli, a finding that may extend to feelings of anger. There is some support for this prediction in table 5.3. Angry individuals were more likely than non-angry individuals to think about the war and consume news—particularly TV news, although they were no more likely to talk to others about the war. Thus, increasing levels of anxiety and anger are both related to higher levels of political involvement and news attention.

Overall, our results are consistent with theoretical models of the effects of negative affect on attention. Both negative emotions considered here— anxiety and anger—have broadly similar effects on attention to politics. As both emotions increase, respondents are more likely to report thinking about the Iraq war, talking about it, and, to a more limited extent, attending to national news, TV, and newspapers.

Finally, anger and anxiety are expected to have differing effects on political learning. Although anxiety promotes greater attention to politics, there is evidence, discussed earlier, that anxiety can also impede the learning of new information. Data from wave 2 of the TNSS are supplemented with results from wave 1 to test this hypothesis. Learning is assessed in wave 1 by a knowledge scale based on answers to four factual questions about Afghanistan and Al Qaeda.[8] In wave 2, the knowledge scale is made up of answers to five factual questions about Iraq.[9] Both knowledge scales were coded to range from 0 (no correct answers) to 1 (all questions answered correctly). The results of regression analyses predicting knowledge in both waves are shown in table 5.4.

The estimates in columns 1 and 2 of table 5.4 show that anxiety is negatively related to knowledge: as anxiety increases, the number of correct answers to the knowledge questions decreases. Anger was not measured in wave 1, but the results in column 2 show that anger is not a strong predictor of knowledge, contrary to some research discussed earlier. There appears to be a weak positive effect of anger on knowledge, but the coefficient does not reach conventional levels of significance. Other significant effects in these models are consistent with prior research. Knowledge increases with additional years of education; women score lower on foreign policy knowledge than men; and those high on authoritarianism are less knowledgeable than low authoritarians.

Table 5.4

Effects of anxiety and anger on knowledge

	Wave 1	Wave 2	Wave 2
Anxiety	−.14 (.03)	−.12 (.03)	−.15 (.03)
Anger		.05 (.03)	.02 (.03)
Age	.020 (.005)	.008 (.005)	−.003 (.005)
Education	.039 (.003)	.026 (.003)	.023 (.003)
Gender	−.13 (.02)	−.21 (.02)	−.18 (.02)
Race/ethnicity			
Black	−.05 (.03)	−.09 (.03)	−.09 (.03)
Hispanic	−.04 (.03)	−.07 (.03)	−.08 (.03)
Other	−.04 (.03)	−.02 (.03)	−.01 (.04)
Party ID	.08 (.02)	.08 (.02)	.07 (.02)
Ideology	−.04 (.03)	−.04 (.03)	−.03 (.03)
Authoritarianism	−.20 (.02)	−.17 (.02)	−.14 (.02)
Thinking			.19 (.04)
Talking			.01 (.03)
Television			−.02 (.02)
Newspapers			.10 (.02)
R^2	.27	.31	.35

Note: Entries are regression coefficients with standard errors in parentheses. The dependent variables are coded to range from 0 to 1. Age is coded in 10-year intervals and education is coded in years. White is the excluded category for the race/ethnicity dummy variables. All of the other predictors range from 0 to 1.

The effect of anxiety on knowledge was more closely examined by adding measures of attention to the war (analyzed in table 5.3) to the regression models shown in table 5.4 for wave 2. The regression estimates are shown in the third column of table 5.4. Two of the new variables have significant effects on knowledge: thinking about war with Iraq and reading about politics in newspapers increase knowledge about Iraq. With attention included in the model, the effect of anxiety on knowledge is increased by 25%. Anxiety thus appears to have two different effects on knowledge acquisition. First, it indirectly increases knowledge by increasing attention to politics and heightening cognitive activity (thinking). Second, it has a negative direct effect. Leaving aside its ability to increase attention, anxiety seriously hinders the acquisition of knowledge. The significant negative effect of anxiety in column 2 of table 5.4 shows that the direct negative effect of anxiety on learning substantially overwhelms its beneficial effects

on attention. Overall, it is clear that anxiety interferes with political learning, whereas anger has little or no effect on the ability to acquire political
information.

5.6   General Discussion

Anger and anxiety have distinct effects on attitudes toward war, as well as
attention to and retention of conflict-related information. When political
scientists fail to account for the differences between these emotions and
instead treat them as indicators of the same underlying dimension of negative affect, they may be faced with the misleading conclusion that negative
emotion has no effect on political attitudes and behavior. In combination,
the cognitive biases associated with anger and anxiety often cancel each
other out, producing a null result. Emotion then appears to offer no insight
whatsoever into citizen reactions to political events, in this case the 9/11
terror attacks and impending Iraq war. The reality could not be more
different.

Findings reported in this chapter on the distinct origins and divergent
consequences of anger and anxiety on attitudes toward the Iraq war pose
a clear challenge to the two-dimension valence model in which all negative
emotions are viewed as a single entity. Anger and anxiety are related, but
they are also distinct. Consider their origins. Anger toward terrorism is
more prevalent among strong patriots, whereas anxiety is more common
among those who feel personally threatened by terrorism. Anxiety is
grounded in personal vulnerability and insecurity, whereas anger is a
collective response to threat that is heightened among the strongest
group identifiers. This result is consistent with evolutionary arguments
regarding the distinctive origins and functions served by anger and anxiety
(Cosmides & Tooby, 2006; Petersen, 2010)

Anger and anxiety have distinct effects on political cognition, and serve
to bias cognitive processing in directly opposing ways. We found that the
relationship between anger, anxiety, and risk assessment uncovered in
classic behavioral decision theory applications, such as the Asian disease
problem (Lerner & Keltner, 2001; Druckman & McDermott, 2008), also
hold in response to real-world threats. Specifically, anger leads to a reduced
perception of the war's risks, whereas anxiety heightens risk perception. In
other work, it has been shown how this relationship between emotion and
risk perception translates into war support (Huddy et al., 2007). Anxiety
influenced risk assessments and risk-averse preferences and depressed
support for military intervention. Alternatively, anger deflated perceptions

of risk associated with military intervention and heightened support for intervention. This pattern of results likely generalizes to citizen responses to a broad class of threats, along with proposed government actions to address these threats.

Both anger and anxiety heightened interest in the Iraq war, but anxiety reduced the ability to learn factual information about it. Anxiety increased self-reported thought about the war and heightened conversations with others about it. This suggests greater attention to the war. But anxious individuals also knew less about Iraq, suggesting they had greater difficulty learning despite their concerted efforts to obtain information. These findings on the interplay of anger, anxiety, and cognition lead to the inescapable conclusions that emotion and cognition are thoroughly interconnected, and that emotion increased biased processing of information about the Iraq war, although anxiety had a more deleterious effect than anger on learning.

In conclusion, our findings point to the political importance of distinct negative emotions, underscoring the societal relevance of recent developments in the study of emotion. The replacement of two-dimensional models of emotion by more differentiated multi-dimensional models represents an important breakthrough in understanding the diverse ways in which emotions color political reasoning. The two-dimensional models do not explain why negative emotions such as anxiety and anger have competing and diametrically opposed effects on political attitudes. A functional neuroscience approach in which anger is linked to approach and anxiety to avoidance, or cognitive appraisal theories that posit a link between anger and greater action and anxiety and caution, are a much better fit to the findings reported in this chapter than a simple two-dimensional valence model (Carver, 2004; Lerner & Tiedens, 2005).

Ultimately, more research is needed to verify the collective nature of anger and the personal origins of anxiety, and to explain why anger and anxiety are so closely associated in self-report data yet have such distinct consequences. But with or without further research, it is imperative for scholars of political behavior to consider distinct negative emotions and the differing ways in which they influence information processing to fully understand political reactions to threat. Failure to do so will lead researchers to overlook the offsetting effects of emotions like anxiety and anger, leading to the mistaken conclusion that emotion is of little consequence to the study of public opinion and political behavior. The reality is that emotions have a complex structure and influence political thinking and learning in complex and interesting ways.

Acknowledgments

This research was supported by grants SES-0201650, SES-9975063, and SES-0241282 from the National Science Foundation. We gratefully acknowledge the assistance of Gallya Lahav, George E. Marcus, Michael MacKuen, and Charles Taber in data collection.

Notes

1. While Cacioppo and colleagues refer to these systems as approach and avoidance, George Marcus and colleagues (1993, 2000) refer to them as the disposition and surveillance systems. Carver and White (1994), alternatively, relate the experience of positive and negative emotion to the behavioral activation and behavioral inhibition systems.

2. There was no difference in response rate between the two survey organizations.

3. The anxiety and anger measures used in this chapter are based on reported feelings toward "anti-American terrorists" and Saddam Hussein. The only exception is in wave 1 of the TNSS, in which anxiety is assessed toward "terrorists."

4. Personal threat in wave 1 was measured with a scale composed of two items: "How concerned are you personally about you yourself, a friend, or a relative being the victim of a future terrorist attack in the United States?" and "How much, if any, have the terrorist attacks shaken your own sense of personal safety and security?"

5. National threat was measured with a scale composed of two items: "How concerned are you that there will be another terrorist attack on U.S. soil in the near future?" and "How concerned are you that terrorists will attack the United States with biological or chemical weapons?"

6. Patriotism was measured with three questions: "How angry does it make you feel, if at all, when you hear someone criticizing the United States?" "How proud are you to be an American?" and "How good does it make you feel when you see the American flag flying?"

7. Age was coded in 10-year intervals; education coded in years; gender coded 0 if male and 1 if female; race/ethnicity composed of three dummy variables for black, Hispanic, and other (with white as the excluded category); party identification and ideology were standard seven-category measures recoded to range from 0 to 1, with 1 indicating Republican and conservative identifications and 0 indicating Democratic and liberal/progressive identifications; and authoritarianism was a three-item scale ranging from 0 to 1 (see Feldman & Stenner, 1997). Both anxiety and anger are coded to range from 0 to 1.

8. Knowledge was assessed with four questions: "Can you tell me the name of one country that shares a border with Afghanistan?" "Is Afghanistan an Arab country?" "Can you name the country that Osama Bin Laden is originally from?" and "Can you tell me the name of the Muslim holy book?"

9. Learning was measured with four questions: "Can you tell me the name of one country that shares a border with Iraq?" "What is the capital city of Iraq?" "What is the name of the middle-eastern TV network that has broadcast statements by Osama bin Laden and Al Qaeda?" "What is the name of the ruling political party in Iraq?" and "What is the name of the major ethnic group that lives in the north of Iraq?"

References

Abelson, R. P., Kinder, D. R., Peters, M. D., & Fiske, S. T. (1982). Affective and semantic components in political personal perception. *Journal of Personality and Social Psychology*, *42*(4), 619–630.

Arian, A., & Gordon, C. (1993). The political and psychological impact of the Gulf War on the Israeli public. In S. A. Renshon (Ed.), *The political psychology of the Gulf War: Leaders, publics and the process of conflict* (pp. 227–250). Pittsburgh: University of Pittsburgh Press.

Bargh, J. A. (1994). The four horseman of automaticity: Awareness, intention, efficiency, and control in social cognition. In R. S. Wyer & T. K. Srull (Eds.), *The handbook of social cognition: Basic processes* (pp. 1–40). Hillsdale, NJ: Lawrence Erlbaum.

Bargh, J. A. (1997). The automaticity of everyday life. In R. S. Wyer (Ed.), *The automaticity of everyday life* (pp. 1–62). Mahwah, NJ: Lawrence Erlbaum.

Barkow, J., Cosmides, L., & Tooby, J. (1992). *The adapted mind.* Oxford: Oxford University Press.

Berenbaum, H., Fujita, F., & Pfenning, J. (1995). Consistency, specificity, and correlates of negative emotions. *Journal of Personality and Social Psychology*, *68*(2), 342–352.

Berkowitz, L., & Harmon-Jones, E. (2004). Towards an understanding of the determinants of anger. *Emotion*, *4*, 107–130.

Bless, H. (2001). Mood and the use of general knowledge structures. In L. L. Martin & G. L. Clore (Eds.), *Theories of mood and cognition: A user's handbook* (pp. 9–28). Mahwah, NJ: Lawrence Erlbaum.

Bless, H., Mackie, D. M., & Schwarz, N. (1992). Mood effects on attitude judgments: Independent effects of mood before and after message elaboration. *Journal of Personality and Social Psychology*, *63*(4), 585–595.

Bodenhausen, G., Sheppard, L. A., & Kramer, G. P. (1994). Negative affect and social judgment: The differential impact of anger and sadness. *European Journal of Social Psychology, 24,* 45–62.

Bower, D. H. (1988). *Anxiety and its disorders: The nature and treatment of anxiety and panic.* New York: Guilford Press.

Brader, T. (2006). *Campaigning for hearts and minds: How political ads use emotion to sway the electorate.* Chicago: Chicago University Press.

Butler, G., & Mathews, A. (1987). Anticipatory anxiety and risk perception. *Cognitive Therapy and Research, 11,* 551–555.

Cacioppo, J. T., Gardner, W. L., & Berntson, G. G. (1999). The affect system has parallel and integrative processing components: Form follows function. *Journal of Personality and Social Psychology, 76*(5), 839–855.

Calvo, M. G., & Carreiras, M. (1992). Selective influence of test anxiety on reading processes. *British Journal of Psychology, 84*(3), 375–388.

Carver, C. S. (2004). Negative affects deriving from the behavioral approach system. *Emotion, 4*(1), 3–22.

Carver, C. S., & White, T. L. (1994). Behavioral inhibition, behavioral activation, and affective responses to impending reward and punishment: The BIS/BAS scales. *Journal of Personality and Social Psychology, 67*(2), 319–333.

Clore, G., & Centerbar, D. (2004). Analyzing anger: How to make people mad. *Emotion, 4,* 139–144.

Clore, G. L., & Wyer, R. S., Dienes, B., Gasper, K., Gohm, C., & Isabel L. (2001). Affective feelings as feedback: Some cognitive consequences. In L. L. Martin & G. L. Clore (Eds.), *Theories of mood and cognition: A user's handbook.* Mahwah, NJ: Lawrence Erlbaum.

Conover, P. J., & Feldman, S. (1981). The origins and meaning of liberal/conservative identification. *American Journal of Political Science, 25*(4), 617–645.

Conover, P. J., & Feldman, S. (1986). Emotional reactions to the economy: I'm mad as hell and I'm not going to take it anymore. *American Journal of Political Science, 30*(1), 50–78.

Cosmides, L., & Tooby, J. (2006). Evolutionary psychology, moral heuristics, and the law. In G. Gigerenzer & C. Engel (Eds.), *Heuristics and the law* (pp. 175–205). Cambridge, MA: MIT Press.

Damasio, A. R. (1994). *Descartes' error.* New York: Putnam.

Davidson, R. J. (1995). Cerebral asymmetry, emotion, and affective style. In R. J. Davidson & K. Hugdahl (Eds.), *Brain asymmetry* (pp. 361–387). Cambridge, MA: MIT Press.

Druckman, J. N., & McDermott, R. (2008). Emotion and the framing of risky choices. *Political Behavior, 30,* 297–321.

Eysenck, M. W. (1992). *Anxiety: The cognitive perspective.* Hove, England: Lawrence Erlbaum.

Eysenck, M. W. (1997). *Anxiety and cognition: A unified theory.* Hove, England: Psychology Press.

Eysenck, M. W., & Calvo, M. G. (1992). Anxiety and performance: The processing efficiency theory. *Cognition and Emotion, 6*(6), 409–434.

Eysenck, M. W., Derakshan, N., Santos, R., & Calvo, M. G. (2007). Anxiety and cognitive performance: Attentional control theory. *Emotion, 7,* 336–353.

Feldman, S., & Huddy, L. (2005). The paradoxical effects of anxiety on political learning. Unpublished manuscript.

Feldman, S., & Stenner, K. (1997). Perceived threat and authoritarianism. *Political Psychology, 18,* 741–770.

Galea, S., Ahern, J., Resnick, H., Kilpatrick, D., Bucuvalas, M., Gold, J., et al. (2002). Psychological sequelae of the September 11 terrorist attacks on New York City. *New England Journal of Medicine, 346*(13), 982–987.

Gordon, C., & Arian, A. (2001). Threat and decision making. *Journal of Conflict Resolution, 45*(2), 197–215.

Green, D. P., Goldman, S. L., & Salovey, P. (1993). Measurement error masks bipolarity in affect ratings. *Journal of Personality and Social Psychology, 64*(6), 1029–1041.

Haidt, J. (2003). The moral emotions. In R. Davidson, K. Scherer, & H. H. Goldsmith (Eds.), *Handbook of affective sciences* (pp. 852–870). Oxford: Oxford University Press.

Halperin, E., Canetti-Nisim, D., & Hirsch-Hoefler, S. (2009). The central role of group-based hatred as an emotional antecedent of political intolerance: Evidence from Israel. *Political Psychology, 30*(1), 93–123.

Huddy, L., & Feldman, S. (2011). Americans respond politically to 9/11: Diverse psychological reactions to threat. Unpublished manuscript.

Huddy, L., Feldman, S., & Cassese, E. (2007). On the distinct political effects of anger and anxiety. In R. Neumann, G. Marcus, A. Crigler, & M. MacKuen (Eds.), *The affect effect* (pp. 202–230). Cambridge: Cambridge University Press.

Huddy, L., Feldman, S., Taber, C., & Lahav, G. (2005). Threat, anxiety, and support of anti-terrorism policies. *American Journal of Political Science, 49*(3), 610–625.

Isen, A. M. (1984). *Toward understanding the role of affect in cognition.* In R.S. Wyer & T. K. Srull (Eds.). *Handbook of social cognition* (Vol. 1, pp. 179–236). Hillsdale, NJ: Lawrence Erlbaum.

Kinder, D. R. (1998). Opinion and action in the realm of politics. In D. Gilbert, S. Fiske, & G. Lindsey (Eds.), *Handbook of social psychology* (pp. 778–867). Boston: McGraw-Hill.

Kunda, Z. (1990). The case for motivated reasoning. *Psychological Bulletin, 108,* 480–498.

Lambert, A., Scherer, L. D., Schott, J. P., Olson, K. R., Andrews, R. K., O'Brien, T. C., et al. (2010). Rally effects, threat, and attitude change: An integrative approach to understanding the role of emotion. *Journal of Personality and Social Psychology, 98*(6), 886–903.

Lazarsfeld, P., Berelson, B., & Gaudet, H. (1944). *The people's choice.* New York: Duell, Sloan and Pearce.

Lazarus, R. S. (1991). *Emotion and adaptation.* New York: Oxford University Press.

LeDoux, J. (1996). *The emotional brain.* New York: Simon and Schuster.

Lerner, J. S., Gonzalez, R. S., Small, D. S., & Fischhoff, B. (2003). Effects of fear and anger on perceived risks of terrorism: A national field experiment. *Psychological Science, 14,* 144–150.

Lerner, J. S., & Keltner, D. (2000). Beyond valence: Toward a model of emotion-specific influences on judgment and choice. *Cognition and Emotion, 14*(4), 473–493.

Lerner, J. S., & Keltner, D. (2001). Fear, anger, and risk. *Journal of Personality and Social Psychology, 81*(1), 146–159.

Lerner, J. S., & Tiedens, L. Z. (2005). Portrait of the angry decision maker: How appraisal tendencies shape anger's influence on cognition. *Journal of Behavioral Decision Making, 19,* 115–137.

Litvak, P. M., & Lerner, J. S., Tiedens, L. Z., & Shonk, K. (2010). Fuel in the fire: How anger impacts judgment and decision making. In M. Potegal, G. Stemmler, C. Spielberger (Eds.), *International handbook of anger* (pp. 287–310). New York: Springer.

Lodge, M., & Taber, C. S. (2005). The automaticity of affect for political leaders, groups, and issues: An experimental test of the hot cognition hypothesis. *Political Psychology, 26*(3), 455–482.

Lord, C. G., Ross, L., & Lepper, M. R. (1979). Biased assimilation and attitude polarization: The effects of prior theories on subsequently considered evidence. *Journal of Personality and Social Psychology, 37,* 2098–2109.

Mackie, D. M., Devos, T., & Smith, E. R. (2000). Intergroup emotions: Explaining offensive action tendencies in an intergroup context. *Journal of Personality and Social Psychology, 79,* 602–616.

MacLeod, C., & Mathews, A. (1988). Anxiety and the allocation of attention to threat. *Quarterly Journal of Experimental Psychology, 38,* 659–670.

Maitner, A. T., Mackie, D. M., & Smith, E. R. (2006). Evidence for the regulatory function of intergroup emotion: Emotional consequences of implemented or impeded intergroup action tendencies. *Journal of Experimental Social Psychology, 42*(6), 720–728.

Marcus, G. E. (1988). The structure of emotional response: 1984 presidential candidates. *American Political Science Review, 82*(3), 737–761.

Marcus, G. E. (2003). The psychology of emotion and politics. In D. O. Sears, L. Huddy, & R. Jervis (Eds.) *Handbook of political psychology* (pp. 182–221). Oxford: Oxford University Press.

Marcus, G. E., & MacKuen, M. B. (1993). Anxiety, enthusiasm, and the vote: The emotional underpinnings of learning and involvement during presidential campaigns. *American Political Science Review, 87*(3), 672–685.

Marcus, G. E., Neuman, W. R., & MacKuen, M. B. (2000). *Affective intelligence and political judgment.* Chicago: University of Chicago Press.

MacKuen, M., Wolak, J., Keele, L., & Marcus, G. E. (2010). Civic engagements: Resolute partisanship or reflective deliberation? *American Journal of Political Science, 54*(2), 440–458.

Mathews, A., & MacLeod, C. (1986). Discrimination of threat cues without awareness in anxiety states. *Journal of Abnormal Psychology, 95,* 131–138.

Miller, D. A., Cronin, T., Garcia, A. L., & Branscombe, N. R. (2009). The relative impact of anger and efficacy on collective action is affected by feelings of fear. *Group Processes & Intergroup Relations, 12*(4), 445–462.

Mogg, K., Mathews, A., Bird, C., & Macgregor-Morris, R. (1990). Effects of stress and anxiety on the processing of threat stimuli. *Journal of Personality and Social Psychology, 59,* 1230–1237.

Moons, W. G., & Mackie, D. M. (2007). Thinking straight while seeing red: The influence of anger on information processing. *Personality and Social Psychology Bulletin, 33,* 706–720.

Murphy, S. T., & Zajonc, R. B. (1993). Affect, cognition, and awareness: Affective priming with optimal and suboptimal stimulus exposures. *Journal of Personality and Social Psychology, 64*(5), 723–739.

Öhman, A. (2000). Fear and anxiety: Evolutionary, cognitive, and clinical perspectives. In M. Lewis & J. M. Haviland-Jones (Eds.) *Handbook of emotions* (pp. 573–593). New York: Guilford Press.

Petersen, M. B. (2010). Distinct emotions, distinct domains: Anger, anxiety, and perceptions of intentionality. *Journal of Politics, 72*(2), 357–365.

Piotrkowski, C. S., & Brannen, S. J. (2002). Exposure, threat appraisal, and lost confidence as predictors of PTSD symptoms following September 11, 2001. *American Journal of Orthopsychiatry, 72*(4), 476–485.

Rydell, R. J., Mackie, D. M., Maitner, A. T., Claypool, H. M., Ryan, M. J., & Smith, E. R. (2008). Arousal, processing, and risk-taking: Consequences of intergroup anger. *Personality and Social Psychology Bulletin, 34*(8), 1141–1152.

Sadler, M. S., Lineberger, M., Correll, J., & Park, B. (2005). Emotions, attributions, and policy endorsements in response to the September 11th terrorist attacks. *Basic and Applied Social Psychology, 27*, 249–258.

Schlenger, W. E., Caddell, J. M., Ebert, L., Jordan, B. K., Batts, K. R., Wilson, D., et al. (2002). Psychological reactions to terrorist attacks: Findings from the National Study of Americans' Reactions to September 11. *Journal of the American Medical Association, 288*(5), 581–588.

Schuster, M. A., Stein, B. D., Jaycox, L. H., Collins, R. L., Marshall, G. N., Elliott, M. N., et al. (2001). A national survey of stress reactions after the September 11, 2001, terrorist attacks. *New England Journal of Medicine, 345*(20), 1507–1512.

Schwarz, N. (1990). Feelings as information: Informational and motivational functions of affective states. In E. T. Higgins & R. M. Sorrentino (Eds.), *Handbook of motivation and cognition* (pp. 527–561). New York: Guilford Press.

Sears, D. O., & Funk, C. L. (1991). The role of self-interest in social and political attitudes. In M.P. Zanna (Ed.), *Advances in experimental psychology* (Vol. 24., pp. 1–91). San Diego, CA: Academic Press.

Sears, D. O., Hensler, C. P., & Speer, L. K. (1979). Whites' opposition to "busing": Self-interest or symbolic politics? *American Political Science Review, 73*(2), 369–384.

Sears, D. O., Lau, R. R., Tyler, T. R., & Allen, H. M. (1980). Self-interest vs. symbolic politics in policy attitudes and presidential voting. *American Political Science Review, 74*(3), 670–684.

Silver, R. C., Holman, A. E., McIntosh, D. N., Poulin, M., & Gil-Rivas, V. (2002). Nationwide longitudinal study of psychological responses to September 11. *Journal of the American Medical Association, 288*(10), 1235–1244.

Skitka, L. J., Bauman, C. W., Aramovich, N. P., & Morgan, G. S. (2006). Confrontational and preventative policy responses to terrorism: Anger wants a fight and fear wants "them" to go away. *Basic and Applied Social Psychology, 28*(4), 375–384.

Skitka, L. J., Bauman, C. W., & Mullen, E. (2004). Political tolerance and coming to psychological closure following the September 11, 2001, terrorist attacks: A model comparison approach. *Personality and Social Psychology Bulletin, 30*, 743–756.

Small, D., & Lerner, J. S. (2008). Emotional policy: Personal sadness and anger shape judgments about a welfare case. *Political Psychology, 29*(2), 149–168.

Smith, C. A., & Ellsworth, P. C. (1985). Patterns of cognitive appraisal in emotion. *Journal of Personality and Social Psychology, 48*, 813–838.

Smith, E., Seger, C., & Mackie, D. (2007). Can emotions be truly group level? Evidence regarding four conceptual criteria. *Journal of Personality and Social Psychology, 93*(3), 431–446.

Stein, N. L., Trabasso, T., & Liwag, M. D. (2000). A goal appraisal theory of emotional understanding: Implications for development and learning. In M. Lewis & J. M. Haviland-Jones (Eds.), *Handbook of emotions* (pp. 436–457). New York: Guilford Press.

Tiedens, L. Z., & Linton, S. (2001). Judgment under emotional certainty and uncertainty: The effects of specific emotions on information processing. *Journal of Personality and Social Psychology, 81*, 973–988.

Tellegen, A., Watson, D., & Clark, L. (1999). On the dimensional and hierarchical structure of affect. *Psychological Science, 10*(4), 297–303.

Tversky, A., & Kahneman, D. (1981). The framing of decisions and the psychology of choice. *Science, 211*, 453–458.

Valentino, N. A., Banks, A. J., Hutchings, V. L., & Davis, A. K. (2009). Selective exposure in the Internet age: The interaction between anxiety and information utility. *Political Psychology, 30*(4), 591–613.

Valentino, N. A., Brader, T., Groenendyk, E. W., Gregorowicz, K., & Hutchings, V. L. (2011). Election night's alright for fighting: The role of emotions in political participation. *Journal of Politics, 73*(1), 156–170.

Valentino, N. A., Brader, T., & Suhay, E. (2008a). What triggers public opposition to immigration? Anxiety, group cues, and immigration threat. *American Journal of Political Science, 52*(4), 959–978.

Valentino, N. A., Hutchings, A. L., Banks, A. J., & Davis, A. K. (2008b). Is a worried citizen a good citizen? *Political Psychology, 29*(2), 247–273.

Van Zomeren, M., Postmes, T., & Spears, R. (2008). Toward an integrative social identity model of collective action: Quantitative research synthesis of three socio-psychological perspectives. *Psychological Bulletin, 134*(4), 504–535.

Watson, D., & Clark, L. A. (1992). Affects separable and inseparable: On the hierarchical arrangement of the negative affects. *Journal of Personality and Social Psychology, 62*(3), 489–505.

Watson, D., Wiese, D., Vaidya, J., & Tellegen, A. (1999). The two general activation systems of affect: Structural findings, evolutionary considerations, and psychobiological evidence. *Journal of Personality and Social Psychology, 76*(5), 820–838.

Westen, D. (1985). *Self and society: Narcissism, collectivism, and the development of morals.* New York: Cambridge University Press.

Williams, J. M. G., Watts, F. N., MacLeod, C., & Mathews, A. (1997). *Cognitive psychology and emotional disorders* (2nd ed.). Chichester: Wiley.

Winterich, K. P., Han, S., & Lerner, J. S. (2010). Now that I'm sad it's hard to be mad: The role of cognitive appraisals in emotion blunting. *Personality and Social Psychology Bulletin, 36*(11), 1467–1483.

Yiend, J., & Mathews, A. (2001). Anxiety and attention to threatening pictures. *Quarterly Journal of Experimental Psychology, 54,* 665–681.

Zaller, J. R. (1992). *The nature and origins of mass opinion.* Cambridge: Cambridge University Press.

6 Theory and Practice in Political Discourse Research

Peter Bull and Ofer Feldman

6.1 Introduction

"Political activity does not exist without the use of language. It is true that other behaviors are involved: for instance, physical coercion. But the doing of politics is predominantly constituted in language" (Chilton & Schäffner, 2002, p. 3). From this perspective, to study political language is to study one of the prime means through which politics is conducted. The term *discourse* is used to refer to both text and talk (Chilton & Schäffner, 2002), to the analysis not only of written documents, but also of different forms of spoken communication (e.g., speeches, broadcast interviews, parliamentary debates). The term *political discourse analysis* can have a pronounced ideological slant; for example, critical discourse analysis (e.g., Fairclough, 2001) focuses on the ways social and political domination are reproduced by text and talk. However, the focus of this chapter might be characterized rather as ethnographic (Chilton & Schäffner, 2002), because it focuses on the extent to which we can enhance our understanding of the political process through the analysis of political discourse.

Recent decades have seen a dramatic and growing interest in political discourse research, which has been conducted in a variety of different contexts: for example, parliamentary debates (Ilie, 2010), broadcast interviews (Clayman & Heritage, 2002), broadcast political speeches (Atkinson, 1984), and general election campaigns (Bull & Feldman, 2011). This research has focused on many different features of language use, such as metaphor (Lakoff & Johnson, 1980), verb forms (Fetzer, 2008), questions (Sivenkova, 2008), pronouns (Bull & Fetzer, 2006), equivocation (Bull, 2008), rhetorical devices (Atkinson, 1984), interruptions (Beattie, 1982), and narrative stories (Fetzer, 2010).

Notably, much of this research has been based on theories that were in the first instance not devised for the analysis of political discourse, but have proved readily applicable to this field of academic endeavor. This chapter is focused on four such theoretical perspectives—namely, the social skills model of interaction (Argyle & Kendon, 1967); theories of face and facework (Brown & Levinson, 1978, 1987; Goffman, 1955); equivocation theory (Bavelas, Black, Chovil, & Mullett, 1990; Bull, 2008); and metaphor theory (Lakoff & Johnson, 1980). The next four sections review each theory and consider its application to political discourse. The final section comprises a general discussion, in which an overview is presented of the four main theories, together with a consideration of broader implications.

6.2 The Social Skills Model

In "The experimental analysis of social performance," Argyle and Kendon (1967) proposed that social behavior can be understood as a form of skill. Social behavior, they argued, involves processes comparable to those involved in motor skills, such as driving a car or playing a game of tennis. Given that we already know a great deal about motor skill processes, they proposed that this knowledge could be used to advance our understanding of social interaction. Six processes were considered to be common to motor skills and social performance: distinctive goals, selective perception of cues, central translation processes, motor responses, feedback and corrective action, and the timing of responses.

In recent years, Argyle and Kendon's (1967) model has been significantly revised and updated by Hargie (e.g., Hargie & Marshall, 1986; Hargie, 1997, 2006a, 2006b). Although neither version of the social skills model was intended to encompass political behavior, Bull (2011) has argued that the model has significant implications for our understanding of what makes a successful politician. These are detailed below, taking into account Hargie's revisions of the original Argyle and Kendon model.

1. *Distinctive goals* The proposal that social behavior is goal directed is particularly relevant to politics. For example, in a general election campaign, a political party needs a coherent set of policies to bring to the electorate. Indeed, politicians may be criticized for lacking clear vision or purpose. Furthermore, politicians must consider not only their own goals, but also the goals of others, most notably those of opposing politicians.

2. *Person perception* Undoubtedly, it is important for politicians to read people and situations well, since this will affect how they behave toward others. But not only do politicians need to be good at perceiving others, as public figures they need to be aware of how others perceive them. Thus,

in a study based on the 2001 British general election, ratings of political leaders were shown to be one of the best predictors of how people voted. Factor analysis of these ratings showed two distinct but interrelated dimensions, labeled *competence* and *responsiveness*. Ratings of "keeps promises," "decisive," and "principled" loaded on competence; ratings of "caring," "listens to reason," and "not arrogant" loaded on responsiveness (Clarke, Sanders, Stewart, & Whiteley, 2004). Accordingly, Clarke and his colleagues (2004) proposed that competence and responsiveness may be regarded as two enduring dimensions of how British political leaders are perceived. From this perspective, politicians must endeavor to be seen as both competent and responsive, since failing on either dimension may lose them electoral support (Bull & Fetzer, 2010).

3. *Mediating factors*   In the original Argyle and Kendon (1967) model, *central translation processes* referred to the planning aspect of behavior, the rules by which a particular signal is interpreted to prompt a particular action. However, this term was widely regarded as too restrictive, and was replaced by Hargie (e.g., 2006a) with the broader term *mediating factors*. A good example of a mediating factor is the concept of face and facework. *Face* is defined by Goffman as "the positive social value a person effectively claims for himself by the line others assume he has taken during a particular contact" (Goffman, 1955, p. 5), and *facework* the actions people take to protect threats to the face of both themselves and others. There is now an extensive research literature attesting to the importance of face and facework in political discourse (e.g., Bull & Fetzer, 2010), which is discussed more fully in section 6.3 below.

4. *Motor responses*   These refer to the performance of actual behavior. It is not enough for a politician to be a skilled perceiver, or to be able to translate perceptions into appropriate behavioral strategies; the behavior itself has to be performed in a convincing and effective manner. Motor responses can refer to both speech and nonverbal behavior, as well as their respective integration and synchronization.

5. *Feedback and corrective action*   An individual may modify his behavior in light of feedback from others. There are many different forms of feedback available to politicians. Political activity receives intense coverage through television, the Internet, and newspapers. Politicians themselves continuously monitor each other's activities, evaluating and criticizing each other's performance. The electorate can also give feedback through opinion polls, focus groups, writing to their member of parliament (MP), and of course through elections themselves. In fact, so much feedback is available to politicians that their real skill may lie in knowing how to

respond appropriately while avoiding the twin dangers of overreaction and underreaction.

6. *Good timing and rhythm*   Synchronization of behavior and good timing are important aspects of social skills. This may involve both synchronizing one's own behavior and coordinating the behavior of others. For example, making effective points in a political debate may require skilled timing, such that a point is made neither too early nor too late. Jumping in too soon to interrupt an opponent before s/he has finished a point may be perceived as rude and aggressive, but attempting to make a counterargument to a previous point once the discussion has moved on to a different topic may seem pedestrian and inappropriate.

7. *Person-situation context*   Social skills also need to be understood in terms of what Hargie (e.g., 1997) has termed the person-situation context. With regard to political behavior, contextual factors may be considered from a whole variety of perspectives. Notoriously, the discourse politicians employ during an election campaign may differ substantively from that which they employ after winning office. A speech made by a politician after an election victory will differ substantively from one made in the context of a political scandal. Furthermore, the communication skills required for a politician to perform well in one context may differ from those required in another. In section 6.3 of this chapter, four specific contexts (broadcast interviews; prime minister's questions; parliamentary deliberations and committee meetings; and political speeches) are discussed with particular respect to the concepts of face and facework.

Not only has the social skills model contributed to our understanding of social interaction and interpersonal communication (Bull, 2002), it also has significant practical applications. If social interaction is a skill, then it should be possible for people to learn to interact more effectively, just as it is possible to improve performance in any other skill (Argyle & Kendon, 1967). This proposal was formalized in what has been termed social skills training. Currently, it is better known as communication skills training (CST), and has been used extensively in a wide variety of social contexts (see Hargie, 2006c). There is now a substantive research literature on CST (e.g., Hargie, 2006c), although there are no published studies of formal CST with politicians; nevertheless, there are plenty of anecdotal examples. For example, the recent British general election of 2010 saw the introduction for the first time of televised prime ministerial debates between the leaders of the three main political parties (Labour, Conservative, and Liberal Democrat). Each party conducted its own rehearsals for the debates, with other well-known political figures playing the role of each of the party leader's opponents. Such rehearsals can readily be understood as a form of CST for what is a novel genre of political communication in the UK (Bull, in press).

6.3   Theories of Face and Facework

In the sphere of politics, an important mediating factor is what Johansson (2008) has termed the presentation of the political self. This concept is based on Goffman's seminal book *The Presentation of Self in Everyday Life* (1959), in which Goffman analyzed social interaction as if it were part of a theatrical performance, arguing that people in everyday life are like actors on a stage, managing settings, clothing, words, and nonverbal behavior to give a particular impression. Similarly, democratically elected politicians must strive to create a favorable impression on their voters, through controlling or managing their impressions and perceptions. This process can be usefully analyzed through the related concepts of face and facework (Bull & Fetzer, 2010).

Face is important in all cultures; it can be lost, maintained, or enhanced, according to Brown and Levinson's highly influential theory of politeness (1978, 1987). Thus, face preservation is a primary constraint on the achievement of goals in social interaction. "Some acts are intrinsically threatening to face and thus require 'softening'" (Brown & Levinson, 1978, p. 24). Communicative actions such as commands or complaints may be performed in such a way as to minimize the threat to positive and negative face, where *positive face* is defined as "the want of every ... [person] that his wants be desirable to at least some others," and *negative face* as "the want of every 'competent adult ... [person]' that his actions be unimpeded by others" (Brown & Levinson, 1987, p. 62). So, for example, a request to do something may threaten someone's negative face (by restricting their freedom of action), whereas disagreements may threaten positive face (by showing a lack of approval).

Although Brown and Levinson's (1978, 1987) conceptualization of face is known as politeness theory, the terms face and politeness are not synonymous. Politeness is a form of facework, but not the only one. Indeed, Goffman (1955) specified three kinds of facework: an avoidance process (avoiding potentially face-threatening acts), a corrective process (performing a variety of redressive acts), and also what he called making points (the aggressive use of facework). Regarding this last type, Goffman's extended 1967 essay, "Where the Action Is," discussed incidents in which adversaries deliberately antagonize one another; the focus is on who will back down in such situations, and on what counts as backing down. Interestingly, although Brown and Levinson (1978, 1987) were indebted to Goffman's (1955) analysis of facework, they overlooked such instances of deliberate face aggravation. Within the framework of politeness theory, rudeness is envisaged simply as a deviation from or violation of rules of cooperative/ polite communication. In fact, rudeness can be deliberate and motivated, if not calculated and strategic (Kienpointner, 1997). Culpeper (1996) has argued that in some contexts (e.g., army training and literary drama) impoliteness is not a marginal activity, but central to the interaction that takes place.

Another such context is arguably that of adversarial political discourse (Bull & Fetzer, 2010). Insults are one very characteristic form of face aggravation, and were analyzed by Ilie (2001, 2004) in the context of parliamentary debates in the UK and Sweden. Other forms of face aggravation were analyzed by Harris (2001) in the context of prime minister's questions (PMQ), the weekly sessions in the British House of Commons in which the prime minister is open to questions from any MP. Harris argued that much of PMQ discourse comprises intentional and explicitly face-threatening acts, and provides a number of illustrative examples of how strategic impoliteness is central to the interaction that takes place.

Maintaining positive and negative face is also of central importance in political discourse. According to Jucker (1986, p. 71), "It is clear that what is primarily at issue in news interviews is the interviewee's positive face." Upholding positive face, Jucker proposed, is of particular importance for democratically elected politicians in the context of political interviews. This is because their political survival ultimately depends on the approval of a majority of people in their own constituency. Conversely, Jucker argued that negative face is of little importance in news interviews, because the politician by consenting to be interviewed has already consented to his or her freedom of action being limited in this way.

In fact, positive face is of fundamental importance for politicians not only in interviews, but also in other situations, such as making a speech, debating with another politician, or responding to a question from a member of the public. A politician who suffers serious loss of positive face may come to be regarded as a liability by his or her political party. A government minister or an opposition front-bench spokesperson may come under pressure to resign; an MP may be defeated at the next general election or, if deselected, may not even be allowed to stand for election as the party's parliamentary candidate.

But negative face is also important. Even in news interviews, where Jucker (1986) downplayed its significance, politicians may suffer serious potential face damage through responses to questions which circumscribe future freedom of action. If, for example, the leader of the British Conservative or Labour Party categorically asserted that the party would never to go into coalition with the Liberal Democrats, s/he might suffer serious face loss in the event of a hung Parliament, if s/he formed such a coalition because it was the only way of securing a parliamentary majority. To protect face against even the possibility of threat, people avoid performing actions which, although acceptable in the present, may reflect badly upon them in the future (Goffman, 1955). Hence, a politician will be careful to avoid making statements that may hamper or constrain his or her future freedom of action. This point is effectively summed up in the old political saw: "Never say never." Thus, issues of positive and negative face may both be addressed in political discourse, but should not be seen as alternatives: rather, their relative importance may vary according to situational context, as may the role of face aggravation.

Although the term "facework" was introduced by Goffman (1955) to refer to actions taken to counteract the threats to face by avoidance or corrective processes, it may also be performed in instances of face aggravation, as described above. Notably, facework can be seen as a form of communicative skill, and a highly important one for democratically elected politicians. However, the kinds of facework a politician needs to perform will vary according to communicative context. The following contexts are discussed below: broadcast interviews, PMQs, parliamentary deliberations and committee meetings, and political speeches.

Broadcast interviews, at least as now conducted in the US and the UK, are essentially adversarial encounters (Clayman & Heritage, 2002), in which interviewers pose questions which may be challenging or even confrontational. At the same time, interviewers are expected to be impartial. This acts as a restraint on open face aggravation, although interviewers may utilize a variety of implicit techniques to attack politicians, while still maintaining a stance of neutrality (Clayman & Heritage, 2002). Thus, as a consequence of adversarial questioning, politicians must seek to defend both their positive and negative face. They may also on occasion use interviews as an opportunity for open face aggravation by attacking either the interviewers or their political opponents.

In contrast, parliamentary questions (e.g., PMQs) are posed not by professional interviewers, but by other politicians, who can be as partial and as unashamedly partisan as they choose. In PMQs, all participants must orient to the expectation both that the dialogue should follow a question-answer pattern, and that they should refrain from unacceptable, unparliamentary language. Nevertheless, within these constraints, face aggravation is a prime feature of PMQ discourse (Harris, 2001), although both positive and negative face will also come into play. Thus, the positive face of the leader of the opposition party depends at least in part on an ability to undermine the face of the prime minister, while the positive face of the prime minister depends in part on an ability to defend him/herself against such attacks, while also protecting his or her own freedom of action (negative face).

In a different legislative context, Feldman (2004, pp. 49–53) analyzed face-saving techniques utilized by members of the Japanese National Diet. During deliberations and committee meetings, Diet members refrain from expressing definite or concrete opinions. They shy away from clear-cut positions and avoid passing judgment, criticizing, conveying personal emotions, or expressing annoyance in any way that might provoke anger, embarrass colleagues, or offend anyone. Toward this end, Diet members use a kind of code that defies lexicographical definition. They often say "respectfully listen" to mean that they will listen (but probably do nothing); "cautiously" when they intend to do nothing (e.g., when a situation is virtually hopeless, but they do not want to say so); "consider" when something is meant to stay tabled indefinitely; "deal with appropriately" to avoid saying anything concrete; or "look into" or "study" if they mean to kick around an idea without acting on it. They habitually avoid using vocabulary that makes a commitment to any position, hedging their comments with words like "probably," "perhaps," or "could be." They frequently use terms like "positively" or "constructively" to give a vague impression that they intend to move on an issue at some unspecified time in the future, "assiduous" or "energetic" to convey a sense of effort when prospects for accomplishment are poor, "adequately" or "thoroughly" when stalling for time, and "to endeavor" or "work hard" when they intend to take no personal responsibility.

Thus, speakers are able to appear to say something without revealing any personal opinion, and to phrase comments in ways that make it impossible for the listener to determine where the speaker stands on a particular issue, thereby preserving an image of neutrality regarding sensitive issues and protecting their positive face. One has to understand the hidden meaning of these terms in order to grasp the true intentions of Japanese politicians or government officials at a given time. This point was taken to an extreme expression recently (November 2010), when it was disclosed by Japan's justice minister—who had eventually to step down from his post over his comments—that when he faced questions from other politicians in parliament, he only needed to remember two responses: "I won't comment on individual cases," and "We are dealing with the matter appropriately based on law and evidence."

Broadcast interviews, PMQs, and parliamentary committee meetings and deliberations are all highly interactive encounters. So too are political speeches, although audience reactions are typically limited to gross displays of affiliation or disaffiliation. Notably, Atkinson (1984) identified rhetorical devices whereby speakers invite audience applause, thereby enhancing their own positive face. But facework in monologue is not just confined to applause invitations. Throughout a speech, self-references may be performed either explicitly or implicitly to celebrate a speaker's actions, intentions, thoughts, or feelings (Bull & Fetzer, 2010). The particular advantage of monologue (in contrast to responding to awkward questions in broadcast interviews or PMQs) is that it provides politicians with a platform whereby they can present themselves to best advantage, thereby enhancing positive face. However, they may also seek to protect negative face by avoiding awkward commitments, as well as take the opportunity for face aggravation by attacking their political opponents (Bull & Fetzer, 2010).

In summary, it is proposed that positive face, negative face, and face aggravation are all important aspects of political facework. That is to say, politicians will seek to present themselves in a favorable light, to defend their freedom of action, and—in an adversarial political system—to attack the face of their political opponents while defending their own. Through skilled facework, politicians may seek both to defend their own face and to undermine the face of their political opponents. However, the particular ways in which facework is performed will vary according to communicative context (Bull & Fetzer, 2010).

6.4 Equivocation Theory

According to Bavelas and her colleagues (1990), people typically equivocate when posed a question to which all of the possible replies have potentially negative consequences, but where nevertheless a reply is still expected (referred to in this chapter as the situational theory of communicative conflict [STCC]). Many everyday situations can be seen to create these kinds of conflicts. Perhaps the most common involves a choice between saying something false but kind and something true but hurtful. For example, a person receives a highly unsuitable gift from a well-liked friend, who then asks directly, "Did you like the gift?" In responding, the person has two negative choices: saying, falsely, that s/he likes the gift or saying, hurtfully, that s/he does not. According to equivocation theory, the person will, if possible, avoid both of these negative alternatives—especially when a hurtful truth serves no purpose. What s/he does instead is to equivocate. For example, someone might say "I appreciate your thoughtfulness," with no mention of what s/he liked or disliked about the actual gift, thereby not answering the question. Bavelas and her colleagues stressed that although it is individuals who equivocate, such responses must always be understood in the situational context in which they occur.

Bavelas and her colleagues (1990) further proposed that equivocation can be understood in terms of four dimensions—namely, *sender, content, receiver* and *context*. They state (1990, p. 34): "All messages that would (intuitively or otherwise) be called equivocal are ambiguous in at least one of these four elements." The *sender* dimension refers to the extent to which the response is the speaker's own opinion; a statement is considered more equivocal if the speaker fails to acknowledge it as his own opinion, or attributes it to another person. *Content* refers to comprehensibility, an unclear statement being considered more equivocal. *Receiver* refers to the extent to which the message is addressed to the other person in the situation, the less so the more equivocal the message. *Context* refers to the extent to which the response is a direct answer to the question— the less relevance, the more equivocal the message. Although these four dimensions are arguably common to non-equivocal as well as to equivocal communications, Bavelas and her colleagues use this approach to stress that equivocation should be understood as a multidimensional phenomenon.

Bavelas and her colleagues (1990) conducted a series of experiments in which a number of communicative conflicts were described. Participants were then asked to indicate how they would respond to these scenarios. Their responses were rated by observers along the four dimensions of sender, content, receiver, and context. In comparison to non-conflictual scenarios, conflictual situations were consistently found to receive significantly more equivocal responses.

However, this research was based primarily on laboratory experiments, in which participants were asked to describe how they would respond to hypothetical scenarios, which have no serious real-life implications for the participants. An alternative approach is to analyze discourse in a naturally occurring situation, characterized by a high degree of equivocation, with a high degree of risk for the participants. For this purpose, the analysis of broadcast political interviews seemed ideal, and presented an exciting opportunity to apply equivocation theory to a very different situation from that for which it was originally devised. Research on broadcast political interviews has provided impressive support for both the main propositions of equivocation theory discussed below: equivocation as a multidimensional construct, and the situational theory of communicative conflict.

6.4.1 Equivocation as a Multidimensional Construct

Feldman (2004, pp. 76–110) conducted a study of 67 one-on-one televised interviews with Japanese politicians. Their responses were rated by two Japanese postgraduate students on the four equivocation dimensions devised by Bavelas and her colleagues (1990), using 6-point Likert-type scales; any disagreements between the raters were resolved through discussion with Feldman, the project director. The higher the rated score, the more equivocal the message was judged. The mean equivocation ratings were context, 3.51; content, 2.79; sender, 3.39; and receiver, 1.68. Thus, messages were rated as most equivocal on the context dimension.

The main revision that Feldman (2004, pp. 95–96) had to make was to the receiver scale. The original scale was devised to rate informal dyadic conversations, where it is typically clear who the receiver might be. However, this is not the case in broadcast news interviews (Bull, 2003b). When a politician is posed a question, the receiver may be the interviewer, the viewing audience, some particular section of the viewing audience, another politician or group of politicians, or any of these. On the original receiver dimension, the rater is asked, "To what extent is the message addressed to the other person in the situation?" Feldman replaced this with two questions. Firstly, "To what extent is the message addressed to the person who asked the question?" Secondly, "To whom does the message seem to be addressed?" Six possible options for this second question were listed (e.g., "members of the speaker's own political party"), together with a seventh option, "not clear to whom the message is addressed." A message was considered to be equivocal only if the intended recipient was unclear.

Thus, although the original scales were devised by Bavelas and her colleagues (1990) on the basis of laboratory experiments in North America, they proved readily applicable to the widely different context of televised political interviews in Japan. The main modification required was to the receiver dimension, which had to be elaborated to take account of the issue of multiple addressees.

6.4.2 The Situational Theory of Communicative Conflict

Bull, Elliott, Palmer, and Walker (1996) analyzed 557 questions from 18 televised interviews in the 1992 British general election according to whether they created a communicative conflict. A reliability study, in which an independent observer coded questions from three of the interviews, showed a Cohen's (1960) kappa of 0.80 when compared with the three main scorers. Overall, results showed that 41% of questions were judged as conflictual, for which the modal response was to equivocate (64% of questions). The remaining 59% of questions were judged as non-conflictual, for which the modal response was to reply (60% of responses). These results provide impressive support for the STCC. Equivocation by politicians was closely associated with the high proportion of conflictual questions posed in broadcast interviews.

This proposition was further investigated in a second study, based on six interviews from the 2001 British general election (Bull, 2003a). This was the first UK general election in which both major television channels gave members of the general public the opportunity, alongside professional interviewers, to directly question the leaders of the three main political parties. It was hypothesized that, in comparison to professional political interviewers, members of the public might pose fewer conflictual questions. For example, whereas interviewers might use conflictual questions to highlight policy inconsistencies, members of the public might prefer non-conflictual questions intended to establish a party's stance on a particular policy issue. Indeed, results showed that political interviewers used a significantly higher proportion of conflictual questions (58%) than members of the public (19%). Furthermore, the politicians replied to significantly more questions from members of the public (73%) than from political interviewers (47%). Finally, a correlation of 0.76 between question type (conflictual/non-conflictual) and response type (reply/equivocate) showed a significant tendency for politicians to equivocate in response to conflictual questions from the political interviewers. The comparable correlation for questions from members of the audience just missed significance (0.70).

In summary, the results of these two studies provide impressive support for the STCC. Bull and his colleagues (1996) found that politicians equivocated in response to conflictual questions and replied to non-conflictual questions. Bull (2003a) found that professional political interviewers asked significantly more conflictual questions than members of the general public, and received significantly fewer replies.

6.4.3   Face and Face Management

Both the main propositions of the equivocation theory proposed by Bavelas and her colleagues (1990) are well supported by the research on televised political interviews reviewed above. Nevertheless, a major revision to the STCC was proposed by Bull and his colleagues (1996), who argued that communicative conflicts can be understood in terms of what are called *threats to face*. That is to say, questions may be formulated in such a way that politicians constantly run the risk of making *face-damaging responses* (responses which make themselves and/or their political allies look bad, or constrain their future freedom of action). Bull and his colleagues further argued that communicative conflicts may occur when *all* the principal ways of responding to a question are potentially face-damaging.

A typology of face-threatening questions in political interviews was devised by Bull et al. (1996) as part of the study discussed in section 6.4.2 of questioning techniques utilized in televised interviews broadcast during the 1992 British general election. The typology is organized in terms of the three faces which it is proposed politicians need to defend: personal, political party, and significant others. These three components were regarded as superordinate categories, and further subdivided into a total of 19 subcategories. For example, the question "Why do you think your party is doing so badly in the opinion polls?" would threaten the face of the party the politician represents. In contrast, the question "Do you not think the public is entitled to regard your own expenses' claims as unreasonable?" would threaten the face of the individual politician.

Bull and his colleagues (1996) categorized all 557 questions from 18 televised interviews in the 1992 British general election in terms of this typology, and found that all the identified communicative conflicts could be understood as a response to threats to face. For example, a yes-no (or polar) question projects three principal forms of response—to answer in the affirmative, to answer in the negative, or not to answer (through some form of equivocation). Each of these possible forms of response may or may not present a threat to face. When all the possible forms of response are potentially face-threatening, it was argued that this sets up the kind of communicative conflict described by Bavelas and her colleagues (1990).

Equivocation by politicians is often ascribed to their personalities—they are the sort of devious, slippery people who will never give a straight answer to a straight question. But from the perspective presented above, conflictual questions are anything but straight; they create strong pressures toward equivocation through what has been termed their *face-threatening structure* (Bull, 2008). Notably, Bull and his colleagues (1996) found that where all the principal forms of response presented a threat to face, politicians typically equivocated. Furthermore, this analysis can also be extended to instances when politicians *do* reply to questions. For example, the leader of the opposition in Britain might be asked by an interviewer in a general election campaign, "If you become prime minister, can you give me a specific vision of how our lives will be different?" Not answering or equivocating to such a question would be extremely face-damaging; it would make the leader of the opposition look totally incompetent. Hence, the face-threatening structure of this question is such that an answer can be predicted with confidence. Thus, a major theoretical merit of this face analysis is parsimony: whether a politician equivocates or answers can arguably be understood within the same underlying theoretical conceptualization (Bull, 2008).

6.5 Metaphor Theory

Focusing on how individuals deal with the complexities and ambiguities of the world that surrounds them, cognitive psychologists view *metaphorical reasoning* as one of the tools that people use to make sense of their world. When individuals need to understand aspects of their milieu that might be abstract or unknown to them, they tend to search their memories and compare them to things they have knowledge about that are more concrete. Once the comparison is drawn, individuals try to understand one thing in terms of another. In cognitive linguistics, *conceptual metaphor theory* (also called *cognitive metaphor theory*) focuses on this process. It was developed initially by Lakoff and Johnson (1980) and subsequently elaborated by Gibbs (1994), Kövecses (2005), and Evans and Green (2006). According to this theory, metaphors are a means of conceiving or understanding something in terms of something else by "mapping" one conceptual domain to another. In this section, the main tenets of the theory are examined, and then discussed in relation to the use of metaphors in politics.

6.5.1 Conceptual Metaphor Theory
The essential principle of conceptual metaphor theory is that metaphors link two conceptual domains. The first is the *vehicle* or *source domain*, which comprises a set of literal entities, characteristics, and processes, linked

semantically and apparently stored together in the mind; these are expressed in language through related words and expressions, which can be grouped in a way similar to those described by linguists as *lexical sets* or *lexical fields*. The second is the *topic, tenor,* or *target domain,* which tends to be abstract and takes its structure from the source domain through the metaphorical link, or *conceptual metaphor.* Target domains are therefore believed to have relationships between entities, traits, and processes that mirror those found in the source domain. Metaphors thus allow people to talk, think about, and understand one area or domain of experience (target domain) in terms of another (source domain). In other words, metaphors present abstract, complex, or unfamiliar target domains (e.g., love) in terms of ideas from source domains that are taken from more concrete, clear, and familiar spheres of life (e.g., a journey). The latter half of such a phrase as *love is a journey* (or *love is a war*) invokes certain assumptions about concrete experience and requires the listener or reader to relate them to the preceding abstract concept of love in order to understand the sentence in which the conceptual metaphor is used. Entities in the domain of *love* (e.g., the lovers, the love relationship) correspond systematically to entities in the domain of a *journey* (the travelers, destinations, etc.) or of a *war* (line of attack, approach, triumph, etc.).

Another central thrust of conceptual metaphor theory is that metaphors are first and foremost a conceptual construction, and that thought and reason have primacy over language. Metaphor operates at the level of thought, which in itself is metaphorical, as our understanding is based on analogies that ultimately originate from physical perception. What constitutes the metaphor is not any particular word or expression, but the way we conceptualize one mental domain in terms of another (hence the term conceptual metaphor). The general theory of metaphor is given by distinguishing such *cross-domain mappings* (such as from the source domain of *journeys* to the target domain of *love*). The mapping—that is, the systematic set of associations that exist between elements of the source and the target domain—tells us how *love* is being conceptualized as a *journey*. In the process, daily abstract notions such as time, circumstances, and goals also turn out to be metaphorical. Hence, metaphor is integral to the semantics of our everyday language, and each metaphor is characterized by many cross-domain mappings.

According to conceptual metaphor theory, everyday language is filled with metaphors that people may not always notice, and which shape their communication as well as the way they think and act. Arguably, it is not possible to converse about abstract concepts without the use of metaphor;

there is no direct way of perceiving them, and we can only understand them through directly experienced, concrete notions. To illustrate: the metaphor *argument is war* shapes the way we view argument as war or as a combat to be won; those involved in an argument "attack" each other's "positions" and "shield" against debaters' "force" and "assaults."

6.5.2 Metaphorical Politics

Thus, metaphor is not simply a device that is typical of poetic or highly rhetorical language, but an essential feature of all types of language. Besides embellishing descriptions of well-understood realities for rhetorical purposes, metaphors also contribute to the very construction and understanding of social reality. This notion has special significance in politics, where structuring one domain in terms of another can influence the way large numbers of people perceive sensitive and controversial aspects of their social environment.

A social environment is too complex and changeable to be known directly, as individuals are simply not equipped to deal with more than a certain amount of subtlety and variety, or with too many changes and associations. Thus, decision-makers, politicians, and the news media tend to transfer to political issues qualities from other spheres of life, such as family, nature, sports, health, or work (Beer & De Landtsheer, 2004, pp. 15–22). In order to communicate more efficiently, they utilize topics that are familiar to the audience, thereby making political issues easier to explain. Arguably, support for policies is less likely to be shaped by conventional understanding of ideology than by policy metaphors evoked by particular social issues or policy proposals (Schlesinger & Lau, 2000). Policy metaphors are cognitive frames that provide easy and accessible ways to understand and evaluate social and political problems, often by comparing complicated political issues with more familiar experiences. Metaphors are therefore an essential component of communication between elites and the general public through the translation of complex political events and situations into common terms more comprehensible for the non-politician. Metaphors play a significant role in providing information as well as perspectives through which the public perceives political activity and issues, and in maintaining and shifting political meanings and behavior (e.g., Mio, 1997).

However, the use of metaphors is affected by various factors such as the political, social, and economic situation at a given time. During recessions, health metaphors frequently appear, as the image of the doctor may fulfill the popular need for a strong political leader who knows what to do.

During periods of international crisis, politicians speak with "more 'pathos,' and as a consequence they use more metaphor power," according to De Landtsheer and De Vrij (2004, p. 166). They found that the use of metaphor by Dutch political leaders during the 1995 Dutch involvement at Srebrenica (in former Yugoslavia) was more conspicuous during periods of crisis, with a greater reliance on metaphor topics that were more original and powerful (e.g., death, disease, drama, sports, and games).

Members of different political groups tend to use different metaphors in their speeches or writings, and political speakers or writers may deliberately choose particular ways of expressing ideas metaphorically to convey an ideological or persuasive point. In the United States, Lakoff (1996) identified differences in metaphor usage between liberals and conservatives in the context of the relationship of the state to its citizens. Both groups perceive political authority through metaphors of the family, but whereas conservatives use the so-called strict father model, liberals use the nurturant parent model. The strict father model depicts a family structured around a dominant "father" (government), in which the "children" (citizens) need to be disciplined to become responsible "adults," after which the "father" (government) should not interfere in their lives. In contrast, the nurturant parent model depicts both "fathers" and "mothers" working to keep the "children" away from "corrupting influences" (e.g., pollution, social injustice, poverty). According to Lakoff, most citizens use a combination of both metaphors at different times, while political speakers utilize them to appeal to their audiences.

In the Ukraine, Taran (2000) argued from an analysis of parliamentary rhetoric that decision-makers on the margins of the political spectrum tended to make greater use of metaphor and elements of mythical thinking, whereas politicians at the center of the political spectrum favored more logical thinking. For political extremists, any current situation appeared to be critical, and they were the least satisfied with the status quo. According to Taran, extremists of both the left and the right try to destroy the semantic construction of the current political discourse, and metaphor appears to be the ideal means for them to do this.

Notably, the use of specific metaphors depends on culture, as different societies tend to use different metaphors. As an integral element of the social and political discourse of a given society, metaphors vary in content and intensity, or "strength" (Mooij, 1976, pp. 121). Since metaphors do not exist within a cultural vacuum, they cannot be treated as though they are independent of a social context. Their various forms and uses are related to specific sets of historical and cultural antecedents. A metaphor that can be frequently used, understood, and expected to reliably generate a certain reaction in one society cannot be used, or at least not as effectively, in another society. In this sense, both the general public and elites in a specific society share a common understanding of the broad meaning of the metaphors used in their society. Knowledge of a society's culture, especially of its language and history, thus facilitates understanding of how that society uses metaphors in politics, as well as their meaning and potential effect.

The significant role of culture can be well illustrated through the use of "war" as a political metaphor. In Japan, war has a particular meaning; it almost invariably refers to the Second World War. This war symbolizes a dark era in Japanese contemporary history, the outcome of a fascist and authoritarian regime that put Japan on the road to devastation and humiliation, followed by starvation and poverty. Most Japanese associate "war" with fascism, death, and the atom bomb. War is something that all citizens over a certain age have experienced personally, and which none of them want to experience again. As a consequence, it probably occurs less frequently as a metaphor in Japanese than Western political discourse in the context of policy issues, or measures that a government intends to take to solve important issues.

For example, several American presidents used war as a metaphor to delineate their stance on a domestic issue. These included Lyndon Johnson's use of *war on poverty*, Gerald Ford's *war on inflation*, and Ronald Reagan's and George Bush's *war on drugs* to emphasize their determination to seriously and aggressively address issues on the national agenda until those issues were resolved (Elwood, 1995; Barry, 1998). Japan has fewer problems related to drugs and poverty than Western societies, yet it confronts a growing number of other social, economic, and environmental problems. Many candidates mention these and other issues in their campaign speeches, as many politicians have done in National Diet speeches and deliberations. But one never hears a Japanese politician "declare war" on a specific social or economic problem. Candidates never use the term *war* in regard to environmental problems in their campaign slogans, the government does not declare "war" on unemployment, and the prime minister never announces "war" against immorality among youth, or against deteriorating values (Feldman, 2004, pp. 114–115).

Feldman has further detailed the extent to which Japan's culture determines the scope, nature, and effects of metaphors used by politicians, news media, and government officials, which in turn help to mold other aspects of the political culture (2004, pp. 111–151). Notably, many Japanese political metaphors are based on traditional customs and characters or roles from Japan's history. For example, a *portable shrine* is an object of religious veneration, which may be carried by bearers on two or four poles in a traditional religious Japanese festival. As a political metaphor, it is used to describe a prime minister who is no more than a figurehead, whereas it is the bearers who hold the real power. Metaphorically, these people may be termed *shoguns in the dark* (a shogun is a military leader, typically a military dictator), *chief magistrates* (who executed orders for a feudal lord in the Edo period, 1603–1868), *stagehands* (the running crew in *kabuki* theater), or simply *kingmakers*. All these metaphors depict a powerless and indecisive national leader without clear policy goals, reliant on powerbrokers who control his administration and political agenda. Other illustrative examples of Japanese political metaphor include *choosing a koma* (a Japanese chess piece), which in political jargon means handpicking a candidate; *entering the sumô ring*, which symbolizes entering a contest for political office; and *raising the gunbai*, which refers to a type of wooden fan used by samurai military officers to convey commands to their troops. In modern *sumô* wrestling, a referee "raises the *gunbai*" to indicate the winner of a match; in a political context, it means winning a political debate or winning support.

6.5.3 The Effects of Metaphors

The effects of metaphors used in political rhetoric—including their aesthetic and manipulative powers—have been widely discussed. Metaphors can stir emotions, justify action, or bridge the gap between logical (rational) and emotional (irrational) forms of persuasion. The power of metaphors derives mainly from the emotional responses that they evoke in a variety of political and social contexts. Gibbs, Leggitt, and Turner (2002, p. 139) argued that "metaphorical language reflects and conveys greater emotional intensity" than literal language, as metaphors create "a sense of intimacy" centered on the "sharing of emotions."

When applied consistently and systematically in a particular political context, metaphors can be powerful tools. First, metaphors can be vehicles for political change. Anderson (2004), for example, contrasted metaphors utilized by the authoritarian rulers of the Soviet Union with those used by politicians in the post-Soviet Russian political system. Soviet and post-Soviet politicians made very different use of metaphors that express relative size (e.g., big, large, and wide) and those that express personal superiority or subordination (e.g., metaphors that compared authoritarian rule to parenting, and those that compared political activity to an assignment set by an authority, such as a teacher). The use of such metaphors often constitutes and supports the status quo—the existing government with its current institutions, rules, and practices.

Second, metaphors can be a means for changing people's attitudes, diverting attention, strengthening legitimacy, and creating positive images of political groups and regimes, while portraying rival groups and their leaders in adverse ways. Thus, Lakoff (1992) showed how the 1991 Gulf War was justified by the American media in terms of a metaphorical "fairy tale scenario" in which President Saddam Hussein was the villain, Kuwait the victim, and the United States the hero. Similarly, Sandikcioglu (2000, p. 300) argued that the metaphors appearing at that time in magazines such as *Time* and *Newsweek* polarized the world into "the Orient vs. the West, Us vs. Them."

Thirdly, metaphors can be used to justify leaders' activities. Rosati and Campbell (2004) argued that the administration of President Jimmy Carter (1977–1981) used metaphor to justify policies in order to legitimize its actions and receive public support. Metaphors, they proposed, smooth communication of distant political realities both among decision makers, and between them and the public, thereby furthering understanding of such phenomena. Moreover, metaphors that mirror the beliefs and objectives of decision makers served as functional cognitive guides in foreign policy.

Finally, effective metaphors can facilitate relationships between leaders and the general public. Political leaders may use conventional metaphors effectively to present their own views of reality as "natural" and "commonsense," thereby reducing the chances that the public will challenge them. In the context of US politics, Shimko (2004) examined so-called metaphors of power, so-called because they provided a cognitive, intellectual, discursive foundation for the exercise of US power, which they explain, justify, and enhance. Shimko (2004, p. 211) distinguished between *dynamic metaphors* that include "power vacuums, falling dominoes, contagious diseases, and spreading fires" and *relational metaphors*, which indicate an association between objects' or actors' "families" and "neighbors." Such metaphors frame or structure international relations in ways that promote policies of expansion and intervention through an image of the world and the United States' role in it, thereby making it difficult to imagine any alternative policies.

In summary, according to conceptual metaphor theory, metaphors are not merely "ornaments" or "figures of speech," but vehicles that form and shape an individual's world outlook. It has been argued that metaphors play a significant role in the rhetoric utilized by politicians, news reporters, and government officials. They help citizens to make sense of political events and practices, as well as contribute to their structuring of social and political reality. They guide individuals' understanding of political issues, roles, processes, and relationships—thereby crucially affecting the decisions individuals make as participants in the overall political process.

6.6   General Discussion

In this chapter, four cognitive theories were discussed that have substantively influenced the authors' own research on political discourse: the social skills model of interaction; theories of face and facework; equivocation theory; and metaphor theory. These theories have significant implications not only for political discourse research, but also for the practice of politics. As Lewin (1951, p. 169) famously remarked, "There is nothing so practical as a good theory." The practical significance of these theories is considered below in relation to politicians, to political journalists, and to the electorate as a whole.

In contemporary politics, the mass media are of central importance, as is the ability of politicians to communicate through mass media. Contemporary politics is mediated politics: politicians communicate both with each other and with the electorate, especially through television. Hence, the communication skills of politicians now play a crucial role. The social skills model as applied to political behavior specifies the nature of those skills, and the means whereby politicians can improve through appropriate learning, training, and feedback. In particular, the concepts of face and facework highlight how politicians need appropriate communicative skills to present themselves and their party in a favorable light, to defend their freedom of action and, in an adversarial political system, to attack the face of their political opponents while defending their own. Political discourse research provides important insights into how politicians can improve their communication skills.

Such research also has significant implications for the practice of political journalism. In British broadcast interviews, adversarialism, characterized by tough, confrontational questioning, has become the norm. Arguably, adversarialism can be conceptualized in terms of equivocation theory; thus, political interviewers perceived as tough have been shown to pose a higher proportion of conflictual questions (Bull & Elliott, 1998). Furthermore, when compared to members of the general public, interviewers posed a significantly higher proportion of conflictual questions and received significantly fewer replies (Bull, 2003a). Media critics (e.g., Lloyd, 2004) have begun to question how well the public is served by adversarial interviewing, which creates confrontation rather than dialogue. From the perspective of equivocation theory, it is highly debatable whether posing unanswerable questions is the most effective way of creating political dialogue.

Stereotypically, politicians are often perceived as slippery and evasive, the sort of people who will never give a straight answer to a straight question. But from the perspective of equivocation theory, conflictual questions are anything but straight. Arguably, a greater awareness of the relationship between equivocation and forms of questioning is also of value for members of the general public, enabling them to make more sophisticated judgments about situational factors that affect political discourse. Again, a greater awareness of techniques of hidden persuasion is also of value. For example, from the perspective of metaphor theory, if politicians can persuade us to think in terms of their metaphors, they may also succeed in persuading us of their political viewpoint, without us even being aware that they have done so. Thus, a further practical application of political discourse research lies in enhancing the political awareness and sophistication of the electorate.

Although none of the four theories discussed in this chapter were developed specifically for the analysis of political discourse, their application, this chapter proposes, has led to significant advances in the understanding of political behavior. Not only do these theories have significant practical applications, they are also important in framing research aims, questions, and methods. As such, they arguably provide significant insights not only into how political communication can be better researched, but also into how it can be better understood and evaluated.

References

Anderson, R. D. (2004). The causal power of metaphor: Cueing democratic identities in Russia and beyond. In F. A. Beer & C. De Landtsheer (Eds.), *Metaphorical world politics* (pp. 91–108). East Lansing, MI: Michigan State University Press.

Argyle, M., & Kendon, A. (1967). The experimental analysis of social performance. *Advances in Experimental Social Psychology, 3,* 55–97.

Atkinson, J. M. (1984). *Our masters' voices.* London: Methuen.

Barry, H. (1998). Functions of recent US presidential slogans. In O. Feldman & C. De Landtsheer (Eds.), *Politically speaking* (pp. 161–169). Westport, CT: Praeger.

Bavelas, J. B., Black, A., Chovil, N., & Mullett, J. (1990). *Equivocal communication.* Newbury Park, CA: Sage.

Beattie, G. W. (1982). Turn-taking and interruption in political interviews: Margaret Thatcher and Jim Callaghan compared and contrasted. *Semiotica, 39,* 93–114.

Beer, F. A., & De Landtsheer, C. (2004). Metaphors, politics, and world politics. In F. A. Beer & C. De Landtsheer (Eds.), *Metaphorical world politics* (pp. 5–52). East Lansing, MI: Michigan State University Press.

Brown, P., & Levinson, S. C. (1978). Universals in language usage: Politeness phenomena. In E. Goody (Ed.), *Questions and politeness* (pp. 56–310). Cambridge: Cambridge University Press.

Brown, P., & Levinson, S. C. (1987). *Politeness.* Cambridge: Cambridge University Press.

Bull, P. (2002). *Communication under the microscope.* London: Psychology Press.

Bull, P. (2003a). *The microanalysis of political communication.* London: Routledge.

Bull, P. (2003b). The analysis of equivocation in political interviews. In G. Breakwell (Ed.), *Doing social psychology research* (pp. 205–228). Oxford: Blackwell.

Bull, P. (2008). Slipperiness, evasion and ambiguity: Equivocation and facework in non-committal political discourse. *Journal of Language and Social Psychology, 27,* 324–332.

Bull, P. (2011). What makes a successful politician? The social skills of politics. In A. Weinberg, (Ed.), *The psychology of politicians* (pp. 51–75). Cambridge: Cambridge University Press.

Bull, P. (in press). The microanalysis of political discourse. *Philologia Hispalensis.*

Bull, P., & Elliott, J. (1998). Level of threat: Means of assessing interviewer toughness and neutrality. *Journal of Language and Social Psychology, 17,* 220–244.

Bull, P., & Feldman, O. (2011). Invitations to affiliative audience responses in Japanese political speeches. *Journal of Language and Social Psychology, 30,* 156–176.

Bull, P., & Fetzer, A. (2006). Who are *we* and who are *you*? The strategic use of forms of address in political interviews. *Text and Talk, 26,* 1–35.

Bull, P., & Fetzer, A. (2010). Face, facework and political discourse. *International Review of Social Psychology, 23,* 155–185.

Bull, P. E., Elliott, J., Palmer, D., & Walker, L. (1996). Why politicians are three-faced: The face model of political interviews. *British Journal of Social Psychology, 35,* 267–284.

Chilton, P., & Schäffner, C. (Eds.). (2002). *Politics as text and talk*. Amsterdam, The Netherlands: John Benjamins Publishing.

Clarke, H. D., Sanders, D., Stewart, M. C., & Whiteley, P. F. (2004). *Political choice in Britain*. Oxford: Oxford University Press.

Clayman, S., & Heritage, J. (2002). *The news interview*. Cambridge: Cambridge University Press.

Cohen, J. (1960). A coefficient of agreement for nominal scales. *Educational and Psychological Measurement, 20*, 37–46.

Culpeper, J. (1996). Towards an anatomy of impoliteness. *Journal of Pragmatics, 25*, 349–367.

De Landtsheer, C., & De Vrij, I. (2004). Talking about Srebrenica: Dutch elites and Dutchbat: How metaphors change during crisis. In F. A. Beer & C. De Landtsheer (Eds.), *Metaphorical world politics* (pp. 163–189). East Lansing, MI: Michigan State University Press.

Elwood, W. N. (1995). Declaring war on the home front: Metaphor, presidents, and the war on drugs. *Metaphor and Symbolic Activity, 10*, 93–114.

Evans, V., & Green, M. (2006). *Cognitive linguistics: An introduction*. Mahwah, NJ: Lawrence Erlbaum.

Fairclough, N. (2001). *Language and power* (2nd ed.). London: Longman.

Feldman, O. (2004). *Talking politics in Japan today*. Eastbourne, England: Academic Press.

Fetzer, A. (2008). 'And I think that is a very straightforward way of dealing with it': The communicative function of cognitive verbs in political discourse. *Journal of Language and Social Psychology, 27*, 384–396.

Fetzer, A. (2010). Small stories in political discourse: The public self goes private. In C. R. Hoffman (Ed.), *Narrative revisited* (pp. 163–184). Amsterdam, The Netherlands: John Benjamins Publishing.

Gibbs, R. (1994). *The poetics of mind*. Cambridge: Cambridge University Press.

Gibbs, R. W., Jr., Leggitt, J. S., & Turner, E. A. (2002). What's special about figurative language in emotional communication? In S. R. Fussell (Ed.), *The verbal communication of emotions* (pp. 125–150). Mahwah, NJ: Lawrence Erlbaum.

Goffman, E. (1955). On face-work: An analysis of ritual elements in social interaction. *Psychiatry, 18*, 213–231.

Goffman, E. (1959). *The presentation of self in everyday life*. Penguin: Harmondsworth, England.

Goffman, E. (1967). Where the action is. In E. Goffman (Ed.), *Interaction ritual* (pp. 149–270). New York: Anchor.

Hargie, O. D. W. (1997). Interpersonal communication: A theoretical framework. In O. D. W. Hargie (Ed.), *The handbook of communication skills* (2nd ed., pp. 29–63). London: Routledge.

Hargie, O. D. W. (2006a). Skill in practice: An operational model of communicative performance. In O. D. W. Hargie (Ed.), *The handbook of communication skills* (3rd ed., pp. 37–70). London: Routledge.

Hargie, O. D. W. (2006b). Training in communication skills: Research, theory and practice. In O. D. W. Hargie (Ed.), *The handbook of communication skills* (3rd ed., pp. 553–565). London: Routledge.

Hargie, O. D. W. (Ed.). (2006c). *The handbook of communication skills* (3rd ed.). London: Routledge.

Hargie, O. D. W., & Marshall, P. (1986). Interpersonal communication: A theoretical framework. In O. D. W. Hargie (Ed.), *The handbook of communication skills* (1st ed., pp. 22–56). London: Croom Helm.

Harris, S. (2001). Being politically impolite: Extending politeness theory to adversarial political discourse. *Discourse & Society, 12*, 451–472.

Ilie, C. (2001). Unparliamentary language: Insults as cognitive forms of confrontation. In R. Dirven, R. Frank, & C. Ilie (Eds.), *Language and ideology: Vol. II. Descriptive cognitive approaches*, pp. 235–263). Amsterdam, The Netherlands: John Benjamins Publishing.

Ilie, C. (2004). Insulting as (un)parliamentary practice in the British and Swedish Parliaments: A rhetorical approach. In P. Bayley (Ed.), *Cross-cultural perspectives on parliamentary discourse* (pp. 45–86). Amsterdam, The Netherlands: John Benjamins Publishing.

Ilie, C. (2010). *European parliaments under scrutiny*. Amsterdam, The Netherlands: John Benjamins Publishing.

Johansson, M. (2008). Presentation of the political self: Commitment in electoral media dialogue. *Journal of Language and Social Psychology, 27*, 397–408.

Jucker, L. (1986). *News interviews*. Amsterdam, The Netherlands: John Benjamins Publishing.

Kienpointner, M. (1997). Varieties of rudeness: Types and functions of impolite utterances. *Functions of Language, 4*, 251–287.

Kövecses, Z. (2005). *Metaphor in culture*. Cambridge: Cambridge University Press.

Lakoff, G. (1992). Metaphor and war: The metaphor system used to justify war in the Gulf. In M. Putz (Ed.), *Thirty years of linguistic evolution* (pp. 463–481). Amsterdam, The Netherlands: John Benjamins Publishing.

Lakoff, G. (1996). *Moral politics*. Chicago: University of Chicago Press.

Lakoff, G., & Johnson, M. (1980). Metaphors we live by. Chicago: University of Chicago Press.

Lewin, K. (1951). D. Cartwright (Ed.), *Field theory in social science: Selected theoretical papers*. New York: Harper and Row.

Lloyd, J. (2004). *What the media are doing to our politics*. London: Constable & Robinson.

Mio, J. S. (1997). Metaphor and politics. *Metaphor and Symbol, 12*, 113–133.

Mooij, J. (1976). *A study of metaphor: On the nature of metaphorical expressions with special reference to their reference*. Amsterdam, The Netherlands: North Holland Publishing.

Rosati, J. A., & Campbell, A. J. (2004). Metaphors of U.S. global leadership: The psychological dynamics of metaphorical thinking during the Carter years. In F. A. Beer & C. De Landtsheer (Eds.), *Metaphorical world politics* (pp. 217–236). East Lansing, MI: Michigan State University Press.

Sandikcioglu, E. (2000). More metaphorical warfare in the Gulf: Orientalist frames and news coverage. In A. Barcelona (Ed.), *Metaphor and metonymy at the crossroads* (pp. 199–320). Berlin: De Gruyter Mouton.

Schlesinger, M., & Lau, R. R. (2000). The meaning and measure of policy metaphors. *American Political Science Review, 94*, 611–626.

Shimko, K. L. (2004). The power of metaphors and the metaphors of power: The United States in the Cold War and after. In F. A. Beer & C. De Landtsheer (Eds.), *Metaphorical world politics* (pp. 199–216). East Lansing, MI: Michigan State University Press.

Sivenkova, M. (2008). Expressing commitment when asking multiunit questions in parliamentary debates: English-Russian parallels. *Journal of Language and Social Psychology, 27*, 359–371.

Taran, S. (2000). Mythical thinking, Aristotelian logic, and metaphors in the parliament of Ukraine. In C. De Landtsheer & O. Feldman (Eds.), *Beyond public speech and symbols* (pp. 120–143). Westport, CT: Praeger.

7 Cognition and Moral Choice

Kristen Renwick Monroe

7.1 Introduction

The broad question posed by this volume is whether cognitive science can form a scientific foundation for social science, producing an integrated whole that helps us understand the traditional questions in social science from a cognitive-psychological perspective. Applying this general question to ethics and to moral choice in times of international political tension in particular, this chapter asks whether a theory of moral choice can be grounded in scientific knowledge, not in the more traditional first-principles approach so widespread in philosophy and religion.

This chapter thus approaches ethical theory—and moral choice more specifically—by attempting to wed science with ethics through an inquiry into the etiology of ethics, with a focus on moral psychology, conceptualized broadly to include interdisciplinary work drawing on the resources of philosophical ethics and the empirical resources of the human sciences concerned with the philosophy of mind.[1] This makes moral psychology closely related to political psychology, itself defined more narrowly to refer to the study of how the human mind thinks about politics and how these psychological processes then influence behavior (Monroe, Chiu, Martin, & Portman, 2009). This broader conceptualization makes the domain of moral psychology include, though it is not restricted to, research on topics such as moral decision making, choice, responsibility, character, luck, courage, imagination, disagreement, virtue ethics, and forgiveness, as well as work on psychological egoism and altruism and their behavioral manifestations at both the individual and group level.[2] Moral psychology as a field assumes the human articulation of moral ideals is appropriately constrained by knowledge of the basic architecture of the human mind, how the mind develops, what our core emotions are, our social psychology, and the limits on the human capacity for rational deliberation (Lakoff &

Johnson, 1999). Moral psychology thus owes much to cognitive psychology, but straddles two fields: normative ethics and empirical psychology, especially cognitive and social psychology.

The importance of the approach followed in this chapter is clear if we consider just one example as an illustration of work in this field. Traditional normative philosophical discussions of ethics tend to set standards of feeling, thinking, and conduct with little concern for whether or not people can or do measure up to these standards (Lakoff, 1996; Lakoff & Johnson, 1980, 1999). Traditional moral theories often presuppose a theory of the structure of agency and character, which then explains what motivates actors and how people's practical moral deliberations occur (Kant, 1797/1991; Bentham, 2002). This means we need to know what sensibilities should be developed and which must be suppressed. We might ask, in this regard, whether the capacities required for Kantian, contractarian, or utilitarian deliberation (for example) emerge naturally as a person matures, and as some developmental psychologists tend to assume (Piaget, 1928, 1932, 1952). Can this kind of moral reasoning process itself be taught? If so, how? For example, do certain emotional dispositions—e.g., trust or fellow feeling—have to be developed before moral reasoning can occur? These issues are directly relevant for social scientists, human rights activists, lawyers, and indeed any member of the general public concerned with public policy, insofar as we ask whether fellow feeling (an emotion) or democracy or equality (institutional norms and behavior) may be prerequisites for moral action, in both the public and the private realms.

The chapter is presented in four sections. It begins with a brief overview of the traditional approach to ethics (section 7.2). Section 7.3 then turns to a specific area of work on moral choice and suggests how a fresh approach, not anchored to the traditional paradigm, allows us to see things otherwise missed by analysts more rooted to the traditional way of thinking about ethics. This section also presents a specific research area for illustration of how these issues play out in practice. The focus on an important problem in international relations—genocide and ethnic cleansing—asks how an empirically grounded knowledge of cognitive psychology can better inform our understanding of human behavior during times of war and political upheaval. Section 7.4 summarizes a new theory of moral choice, one that fills a void in the existing literature, where Anglo-American ethics is dominated by Kantian and utilitarian ethics, both of which rely heavily on the assumption that human beings are conscious and rational calculators. In contrast, the theory presented here emphasizes cognitive perceptions and their influence on moral choice and political behavior. This approach, based on cognitive psychology, is better able to answer long-standing questions about the spontaneous aspect of moral choice and hence opens new avenues for research. The chapter concludes by posing a few questions to be considered in future work, laying out a research agenda for future scholars.

7.2 Traditional Theories of Moral Choice: Origins of Ethics

The first question usually asked when considering ethics is whether morality exists independent of human construction. Here we find both works that deny the existence of any abstract morality and works that affirm the existence of independent, abstract moral values. In doing so, we consider some of the most frequently mentioned origins of morality, such as natural law, religion, and reason. Important scholars who tend to deny the existence of abstract morality are philosophers, such as Friedrich Nietzsche and Martin Heidegger, or social scientists who are cultural relativists, such as anthropologist Clifford Geertz (1973), as well as postmodern philosophers such as Richard Rorty (1999). In contrast, there are many scholarly works that affirm the existence of independent, abstract moral values. Theorists of natural law and natural rights such as Thomas Aquinas offer some of the clearest articulations of the idea that morality does exist independently of human cultural construction (Dworkin, 1977a).[3] Natural rights theorists offer a closely related idea and one found throughout much of Anglo-American philosophy via social contract theory (e.g., John Locke), which takes the stance that people possess natural rights to life, property, and liberty and that these rights, which are not to be transgressed by others, are therefore part of a natural law that presupposes an abstract, universal morality. (See Montesquieu, 1748, or the United States Declaration of Independence for illustration in theory and in law.) This idea of natural law and natural rights is enshrined in the Geneva Convention. (See Telford Taylor, 1992, or Ronald Dworkin, 1996.)

Religion is an equally popular source for the origins of morality. Religious influences on morality tend to cluster around three ideas: (1) We need religion, or a god of some kind, to provide moral values in the first place. This deity creates the moral values; people do not. (2) People cannot be expected to do the right thing without religious motivation. (3) Religion is necessary to socialize and guide people in the search toward moral action. Examples here include the following scholars, who tried to reconcile traditional Greco-Roman philosophical traditions with their particular religion: St. Augustine (Christianity), Alfaro (Islam), and Maimonides (medieval Judaism).

Finally, we find reason is a frequently mentioned cause of morality. Reason is said to help us discover and agree on morality via a reasoning process—such as Kantian (1797/1991), contractarian (e.g., Locke, Hobbes, Rousseau) or utilitarian (Bentham)[4] deliberation—and reason is said to help us achieve morality by helping us as individuals dominate the baser passions (Adam Smith, 1759/1976). Authors who emphasize reason include the philosophers with whom students of ethics are most familiar: Plato, Immanuel Kant, Adam Smith, social contract theorists such as John Locke, and utilitarians such as Jeremy Bentham.[5] As we shall learn below, recent work in cognitive psychology provides support for theoretical approaches that do not rely so heavily on deliberation and reason. Two such theories are less popular, but nonetheless valuable, approaches to ethics: moral sense theory and virtue ethics.

Moral sense theory developed out of what is now a somewhat obscure seventeenth-century theory by philosophers Francis Hutcheson and A.A.C. Shaftesbury (Hutcheson, 1755/1968). Based on the work of these two thinkers, moral sense theory essentially argues that morality is born within us, man and beast, much as we all have the capacity for smell, taste, sight, and so on. Culture may shape the final product, but the initial propensity toward morality is inherent as part of our human nature. David Hume (1751/1998) rejects the views of these Scottish Enlightenment philosophers (especially Hutcheson, Hume's teacher), who argue that human beings have a distinct moral sense that allows them to perceive right and wrong. He argues instead that morality rests on sentiments or feelings. Hume does not, however, suggest that these have to be "fellow feelings," merely that the foundation of morality is sentiment or feeling, rather than religion, reason, etc. Hume and Kant together shifted the argument away from the idea of an innate moral sense. It was not until the late twentieth and early twenty-first century that non-philosophers, such as social scientists (Kagan, 1998; de Waal, 1996) and biological scientists (Damasio, 1994), helped revise this theory by arguing on the basis of scientific data that a moral sense is innate in humans, as in other sentient beings.

Virtue ethics originated in Aristotle's *Nichomachean Ethics* and was introduced in contemporary philosophy by G. E. M. Anscombe in the mid-twentieth century. A follower of Wittgenstein—Anscombe later held the same chair Wittgenstein had held—Anscombe published "Modern Moral Philosophy" in 1958, an article commonly taken to have led to the development of virtue ethics as an alternative to utilitarianism, Kantian ethics, and social contract theories. Anscombe argues that neither utilitarianism, nor Kantian ethics, nor social contract ethics is based on a solid foundation as a secular approach to moral theory. Hence, none of these popular theories of ethics can be used as a basis for ethics for those who do not rely on God or "mother nature." Anscombe admits that these theories do use concepts such as "morally ought," "morally right," "morally obligated," and so on, but she argues that these are legalistic concepts and thus require a source of moral authority. Systems that dispense with God as part of the foundation of their theory thus lack the proper foundation for the meaningful use of these concepts.

A strong indictment of the moral theories that dominated philosophical circles in the 1950s, Anscombe's work makes a powerful argument for the need to develop alternative theories of morality, especially those that do not rely on religion or a legislator to provide moral authority. This suggests the need to develop an alternative theory of ethics, one based on moral psychology, moral virtue and, insofar as is possible, the scientifically determined facts of human nature. Such an account would have to address what constitutes the good for humans based on this approach and would be an empirically grounded theory of ethics.

Space constraints preclude full discussion of the intricacies of the philosophical debate over virtue ethics. (See works by philosophers such as Williams, 1981; Nussbaum, 1986; Arnold, 1989; Hursthouse, 1999; Crisp & Slots, 1997; or Statman, 1997.) For our purposes, the salient point is that a rich and complex debate about virtue ethics seems to call for an empirically grounded theory of ethics, one that does not depend on religion or a legislator for its moral authority. Further, virtue ethics emphasizes developing a good character rather than a series of decision-making rules or ethical rules for conduct, as Kantian and utilitarian ethics tend to do.

How might we develop a theory that both responds to Anscombe's philosophical demands and incorporates recent findings in cognitive psychology? One route is to consider different theories of moral choice in light of recent empirical work relevant to ethics. Here, we draw on recent experimental work in cognitive science suggesting limitations to a reliance on reason as a source of ethics, work developed more formally in the body of literature known as moral psychology. We then turn to an empirical examination of one particular group of people most of us would agree are moral exemplars—rescuers of Jews during World War II—to ask what principles we can derive from understanding their laudable behavior.

7.3   Theory and Practice: Developing a Theory of Moral Choice

Contrasting actual human behavior with the behavior assumed in the dominant theories—which suggests reason and religion should explain how people respond to others during wars or genocides—reveals clear cognitive constraints on moral reasoning (see Lakoff & Johnson, 1999). Some of these constraints have been incorporated into moral psychology, a body of literature interdisciplinary in nature and initially quite narrowly defined to refer to the study of moral development with an emphasis on moral reasoning (Coles, 1967; Kohlberg, 1981; Piaget, 1928). In the last two decades of the twentieth century, however, this conceptualization expanded, as mainstream psychologists increasingly recognized the importance of emotions, intuitions, and innate predispositions for action.[6] Zajonc (1980) argued that cognitive and affective systems are largely independent, with affect being both more powerful than and preceding cognition. Zajonc's paper stimulated much interest in affect among psychologists and was critical in bringing the study of emotion and affective processes into the forefront of psychology in both the United States and Europe. (Also see Kahneman, Slovic, & Tversky, 1982, or Nisbett & Ross, 1980, for a review.) This work caused a critical shift in psychology after the 1980s, and we now find approaches—such as the social intuitionist approach— arguing that the initial organization of the mind is structured in advance of experience (Marcus, 2004). Reason still figures in accounts of moral psychology and plays a role in moral judgment, but reason now is frequently said to operate in a space that is prefigured by affect (Schnall, Haidt, Clore, & Jordan, 2008).

Research at the intersections of evolutionary biology, cognitive science, neuroscience, linguistics, philosophy of mind, and biological anthropology provides further insight into how the concept of a universal moral grammar might explain the nature and origin of moral knowledge by reference to the concepts and models utilized by Chomsky's linguistics (Hauser, 2006). The universal moral grammar effectively provides a toolkit for constructing specific moral systems. Researchers draw on experiments in child development to argue that there is a specific dedicated mechanism in the brain that gives us a moral grammar and that there appears to be a moral faculty, located in the area of the brain that specializes in recognizing certain kinds of problems as morally relevant. In doing so, scholars (Mikhail, 2007) distinguish the universal moral grammar from the dual-process model of moral judgment (Greene & Haidt, 2002; Margolis, 1984) and argue that moral cognition is linked to moral intuitions and emotions.

For example, Haidt (2001) argues that morality involves *post hoc* reasoning to justify what we have already decided what we want to do. Thus, Haidt suggests, moral reasoning is used primarily to find the arguments to justify our judgments arrived at via intuition and to encourage others to reach the same conclusions we have reached. Haidt further specifies morality as focusing on five principal concerns: (1) fairness/harm, (2) harm/care, (3) in-group loyalty, (4) authority/respect, and (5) purity/sanctity.

Building on Chomsky's work in linguistics and a virtual explosion of work in cognitive science illuminating how the mind works, researchers suggest ordinary language is susceptible to precise formal analysis (see Monroe, 2012, chapter 10 for an overview of this literature), and that the principles of grammar are rooted in the human bioprogram. A universal moral grammar holds out the prospect of doing something analogous for aspects of ordinary human moral cognition. This work has attracted the attention of legal scholars (Sunstein, 2005) and resonates with prior work by linguists (Lakoff & Johnson, 1980, 1999) who use linguistic and categorization theory to argue that humans employ metaphors of morality to parse the difficult ethical situations we face.[7]

Thus, an important shift in recent work in moral psychology is the move away from a Kantian emphasis on logic and reasoning. If the traditional route to moral action is said to involve perceptions, analysis, and the strength of will necessary to "do the right thing," perceptions may be the most important, since they involve the framing of the situation, establishing our relations to others involved, and so on. Beyond this, we now believe emotions work faster and more accurately than reason, with the mind resembling a pattern-matcher in which arguments and evidence work best when they tap into basic intuitions. This shift in scientific knowledge, and the experiments on which our new understanding of the brain is based, are reflected in empirical work designed to examine how this process works in real human beings in difficult ethical situations, such as the work done on moral exemplars during war. Prior empirical analysis (Oliner & Oliner, 1988; Monroe, 1996, 2004, 2012) of such moral exemplars suggests rescue behavior is not the product of a rational process—associated in philosophy and ethics with Plato, the utilitarians, the social contractarians, and Kant—whereby morality results from reason's dominance of the passions; nor does it emanate from the kind of moral reasoning described by developmental psychologists such as Kohlberg (1981) or Gilligan (1993). Instead, rescuers acted spontaneously, out of their sense of themselves in relation to others. Since rescue behavior at minimum intersects with other forms of ethical behavior, a fuller exploration of the psychological part of this process thus should yield insight into other forms of ethical political action. What is important to note at this point, however, is that there are scientific underpinnings suggesting the psychological mechanisms behind the empirical phenomena noted in many of these moral exemplars (Oliner & Oliner, 1988; Monroe, 1996, 2004). The general phenomenon is succinctly captured in a Czech rescuer's remark: "The hand of compassion was faster than the calculus of reason" (Monroe, 2004, p. 55).

All of this scientific work on ethics and morality resonates with the early philosophical work by moral sense theorists, who argued that moral psychology is the most powerful when and insofar as it taps into the basic intuitions or sentiments that make up a moral disposition.[8] While the original seventeenth- and eighteenth-century moral sense theory was eclipsed by Kant's magisterial work emphasizing the role of reason in moral choice and thus was not a major philosophical contender in twentieth-century discussions of moral choice,[9] it is time to rethink the value of this old theory. Certainly, much work in cognitive science and related fields justifies a reconsideration of an inborn moral sense.

Indeed, arguments that human beings have an inborn sense of morality, much as they have an instinct for survival, surface prominently in the contemporary literature of several disciplines. Anthropologists (Sapolsky, 2002) ask about human behavior in the ancestral environment to discern the role of culture in influencing moral behavior. Animal ethologists ask if the ethical nature of human beings is rooted in the biological nature we share with other species (Goodall, 1986). Developmental psychologists (Kagan, 1978) examine children in their earliest years, before culture and language have shaped what might be innate tendencies toward certain kinds of behavior. And, increasingly, as we have seen, moral psychologists and neuroscientists (Reimer et al., 2011) are making inroads into the biological substrates of moral behavior, not only in animals or infants but in adults and throughout the life cycle. This empirical research on an innate moral sense can be fragmentary and prelusive; it occasionally involves questions about the scientific reliability of certain findings. Nonetheless, this evidence is salient enough to justify a reconsideration of the existence of an innate moral sense and whether the assumption of such an innate sense, or at least its possibility, should be built into our ethical models. In doing so, it is important to note that utilizing the concept of an innate moral sense does not deny the important role of cognition. Instead of counterposing emotions and cognition, we should think of the two as working together. Probably all moral choice is cognitive, but some cognitions are fast, automatic, and more intuitive, while others originate in controlled reason and are relatively free of affect. Emotions influence cognition in ways that are still not fully understood, but which include framing, scripts, and schemas that originate in prior experience as well as biology in phenotypic fashion. We can think of schemas as "structured parcels of knowledge from memory situating the self in relation to others" (Reimer et al., 2011, p. 72). These schemas then "give rise to scripts, or conceptual representations of action sequences associated with particular social situations" (Reimer et al., 2011, p. 72). Even if the actor is not aware of these scripts or schema, their influence is still felt, thus possibly providing the route for behavior that appears spontaneous, as was the case for so much behavior during the Holocaust.

Moral psychology has benefited from work on appraisal theory, which addresses the relationship between emotion and decision making and suggests linkages between emotion and appraisal happen somewhere beyond our deliberative awareness. Much of the decision processes we attribute to "gut intuition" may be of this variety. (See Damasio, 1994; Haidt & Bjorklund, 2008; Scherer, 2003; Reimer et al., 2011.) Frames consist of a schema or a collection of stereotypes that people rely on to understand and respond to events; hence the terminology of *framing* events. The frame of reference frequently influences how an event is interpreted. Behavioral economists have conducted experiments on many aspects of framing. One is called loss aversion, the phenomenon wherein people will pay only $4 for a coffee mug, for example, but if given a mug will not relinquish it for less than $8; the experiment suggests people really don't like losing something once they have it. Similarly, framing a choice in terms of loss (95% chance of things going wrong) feels different to people than framing in terms of winning (5% chance of winning). (For an overview, see Kahneman, Knetsch, & Thaler, 1991.) Most individuals are unaware of the frames of reference they bring to an event, but these frames have been found to determine the choices made in critical ways. Framing has an influence over the individual's perception of the meaning attributed to words, phrases, or acts, defining the packaging of an event in a manner that encourages certain interpretations and discourages other interpretations.[10]

The scientific research on the human mind and its unconscious effects on decision making is developing quickly, and we can expect much interesting future research to focus on how emotions contribute to appraisals, cognition, and deliberative reflection about ethics. The hard-science understanding of spontaneous choice is not yet there, but it is being worked on by scholars like Reimer et al. (2011), whose recurrent, multilevel appraisal model of emotion's role in decision making is one intriguing route toward understanding those parts of decision making that occur outside the direct influence of deliberative reflection. Such work complements Haidt's social intuitionist model (2001), which effectively describes morality as built on intuitions that function as "evaluative feelings at the edge of consciousness" (Reimer et al., 2011). Haidt's work retains the concept of moral reasoning, but makes it the result of a wide range of intuitions that have been cataloged in our memory, thereby giving us emotional sensitivities that get recognized as decisions (Haidt & Bjorklund, 2008). The emotions aid in producing the kind of quick response appraisals that moral situations often require.

Moral psychology as a field seems poised to take advantage of the intellectual potential in recent work in the wide-ranging set of fields briefly described above. Adopting a broad conceptualization seems empirically justified, as the experiments cited above suggest. Such a conceptualization is the one most likely to yield scientific advances for theoretical work. Accordingly, it thus seems prudent to include interdisciplinary work, drawing on the conceptual resources of philosophical ethics and the empirical resources of the human sciences concerned with the philosophy of mind, to broadly conceptualize moral psychology.

7.4   An Identity-Based Theory of Moral Choice: How It Looks in Practice[11]

It is important to begin by grounding a theory of moral choice in empirical reality, accepting as a premise that there are central and universal tendencies in human behavior and that any scientific theories of political behavior should reflect these. Two tendencies are sufficiently documented to accept as givens, upon which we can construct empirically based theories:

1. Human nature is complex, but there is a basic human need to protect and nurture the self. The implication of this is that both self-interest and sociability are critical, but neither is necessarily or exclusively dominant. Our selfish desires are balanced by less self-centered, though often still individualized, yearnings for social respect, affection, memberships in groups, etc.[12] These needs provide limits to the selfishness that may accompany self-interest. (See Sun & Wilson, 2011, on essential human needs and motives.)

2. People need predictability and control. The vast literature on cognitive dissonance makes clear that a key source of an individual's psychic comport

and the maintenance of identity is the desire for cognitive consistency (Festinger, 1957; Aronson, 1997).

We next add our conceptualization of identity. Identity refers to an individual's sense of him or herself as an entity discrete and separate from others. In practice, identity is used interchangeably with concepts such as "the self" and "character," and we can adopt this practice here rather than parsing fine distinctions among the three different terms. An important part of this conceptualization of identity is its emphasis on the individual's own comprehension or self-concept. In philosophy, identity is often used interchangeably with the term "sameness" to define whatever it is that makes an entity recognizable and definable. Both of these conceptualizations can refer to either micro- or group-level identity, as can cultural identity or an individual's self-affiliation as a member of a cultural group. While individual identity can shift as the individual likes, an individual's cultural identity can perhaps more easily be designated by the categorization of others. This cultural identity then can feed into the individual identity, as it did for many secular Jews during World War II.

Analysts traditionally assume identity forms in some kind of nature-nurture way that will vary, and that identity formation refers to the process through which the distinct personality of an individual develops. It is worth noting that identity is a concept utilized in a wide variety of fields, from computer science to business accounting, and that definitions thus will vary immensely.

Finally, we add to the above assumptions and conceptualization of identity findings from the last twenty years of empirical research on moral exemplars.[13] This research suggests a further fundamental assumption: ethical acts emanate not so much from conscious choice, but rather from deep-seated instincts, predispositions, and habitual patterns of behavior that are related to our central identity. These spring from diverse forces, such as genetic predispositions, social roles, or culturally inculcated norms. Culture provides a range of self-images, but actors gravitate toward the image that strikes a chord with their genetic propensities, with a powerful filter coming from situational or contextual factors. The actor need not be consciously aware of this process; our moral sense is instinctual and powerful, often more powerful than any conscious calculus.

The corollary of the above suggests that by the time we reach adulthood, the main contours of our identity are set, and our basic values largely integrated into our underlying sense of self. We thus speak of adults as agents discovering rather than creating their identities, or choosing how those identities will let them do certain acts but not others. This leaves open the extent to which externally imposed identities fit into this picture, a subject worthy of future study. Consider just one of the many fascinating aspects of this topic, most of which cannot be addressed here. What if a person—call her Louisa—thinks of herself as non-Jewish, but the society in which Louisa lives designates her Jewish because she has a Jewish grandmother? This was the situation for many Jews in the Third Reich, individuals who were highly assimilated and did not think of themselves as of any particular faith. Does Louisa become Jewish once she is designated and treated as such by the society? Analogous situations can occur for many other categories: race, ethnicity, sexual preference, age, etc. The identity theory of moral choice described above is designed to alert us to the importance of the extent to which notions of individuality, the self, and relations to others are culturally inflected and constructed, as well as culturally variable.

Any scholar attempting to construct a moral theory must do so modestly and with an acute sense of the shortcomings that will exist in a first effort. Beyond this, scientific advances in our increased knowledge of how the brain works may soon make obsolete some of the findings on which the following identity theory of moral choice is based. For these reasons, the theoretical thoughts outlined below are initial ones. Other scholars can—and will—correct inaccuracies and fill in the details and omissions as our scientific knowledge progresses.

With these caveats, those interested in moral choice might profitably begin by looking at two critical components of any one individual: (1) *ethical framework* and (2) *ethical perspective*.

An ethical framework consists of an actor's underlying sense of self and worldview. Ethical perspective consists of the actor's cognitive classifications, relational self, and moral imagination. The ethical perspective is determined by both the actor's underlying ethical framework and the way in which events are framed for the actor by the external world. At the moment of action, it is the ethical perspective that produces both the menu of options from which the actor can choose, and the actor's moral sense regarding the nature of the action. (Is the act helpful? Morally neutral? Harmful?) These two factors—the menu of choice and the sense of felt imperative to act—are what result in ethical acts that appear spontaneous.

The psychological process through which identity leads to moral choice is complex, and it is not possible at this time to establish more than the general contours of how the influences operate. In general, however, we begin with an ethical framework. This underlying framework is innate to all individuals. It gets filled in through some as yet undetermined process in which innate predispositions are acted on by critical others and the general environment to create each person's individual character, self-image, and identity[14] or, more particularly, our sense of who we are in relation to others. Our self-images include our sense of ontological security, and key values are integrated into our basic sense of self. Beyond this self-image, each actor has a particular worldview. (Is it a harsh place or one filled with friends? What causes things to work as they do in the world? What does one expect to happen in any given situation? Are there familiar situations that usually play out the same way?)

This worldview is made up of our sense of agency, our canonical expectations, and our main idealized cognitive models. Agency is very important, and a critical factor in worldview is how the actor sees the prime movers of critical events. Is this a world in which individuals can make a difference? Are we puppets moved about by superior forces, such as economics or geopolitics or religion? If so, what are these forces? Canonical expectations are another critical component of worldview. What does an actor assume constitutes normal human behavior? What does the actor think is correct and proper behavior? For himself? For others? Under which circumstances and conditions? Finally, what are the actor's key idealized cognitive models? What does an actor think it means to be a human being? To be a good citizen or neighbor? To be happy and fulfilled? How does this relate to our sense of what comprises desirable behavior?

All of these influences will give specificity to an ethical framework. The nature of the specific content and how it is entered into the actor's ethical framework then produce the actor's general ethical perspective. We can locate an ethical perspective on a moral continuum, much as we locate a person on a moral continuum. (She's a good person. He's a cad.) Each individual has a kind of median position or ethical perspective, but individuals can fluctuate and relocate along this continuum according to stimuli from the external environment. Oskar Schindler provides a well-known example of this point. A gambler, a drunkard, a womanizer, a Nazi, and a war profiteer, Schindler was not someone who would ordinarily be judged morally commendable or expected to risk his life for other people. Yet something happened to give him a different way of viewing certain Jews, and he performed extraordinary deeds to keep them alive. The movie *Schindler's List* uses an interesting device to convey this image. Shot in black and white, the movie has two moments of color. The first is the moment when Schindler—riding a horse on the bluffs of Krakow with one of his mistresses—looks down to see the Nazis clearing the ghetto. His eyes focus on a little girl wearing a bright red dress. This color may be intended to signal the awakening of a tie Schindler feels, a moment when his sense of moral salience clicks in and he becomes someone who will help Jews, not simply benefit from their persecution as he had before. It was interesting that the color red was chosen, since that is the color visual cognitive psychologists use in a set of classic experiments on perception.[15]

An ethical framework works in conjunction with the external environment to produce an ethical perspective consisting of three factors that seem essential: cognitive classification or categorization, the relational self, and the moral imagination. The cognitive classification of others, which works through the framing of moral choice to set the boundaries of entitlement, effectively differentiates (a) those to whom we accord fellow feeling, (b) those to whom we deny fellow feeling, and (c) those toward whom we feel indifferent. This has a great influence on *moral salience*, the feeling that moves us beyond a generalized concern or sympathy in the face of another's plight of others to a felt imperative to act to alleviate another's suffering. Beyond this, the ethical perspective contains a relational self, or our sense of self in relation to others. Do we see ourselves linked to the other person in any way, for any reason? Do we see ourselves as linked to all others, through bonds of a common humanity? Or do we see ourselves as someone under threat from others? Each person will have a general, underlying sense of self in relation to others. But each of us will also have a shifting sense of self that depends on a wide variety of factors: the society around the actor, the immediate framing of the situation, the behavior of the other person, how that person is viewed by others, and so on. Finally, we need to consider the role played by the moral imagination. Are we able to conceptualize certain situations and mentally visualize certain possibilities? Can we distinguish between doing no harm and doing evil? Doing good? The possibilities we are able to imagine have a great influence on the next critical factor: the menu of choices.

This ethical perspective works in conjunction with events in the external environment—such as the framing of critical situations by others, like Nazis—to produce the final two critical influences determining action: a menu of choice options perceived as available, not just morally but cognitively; and a sense of moral salience, the feeling that the suffering of another is not just sad or deplorable but is something that demands action by oneself to ameliorate that suffering. The sense of moral salience provides the drive to act or the imperative to refrain from acting. The menu of choice determines the nature of that action. The result of the combination of the two will be a moral act, whether that act be one most of us would find morally commendable, one that is objectively speaking morally neutral, or morally evil.[16]

This theory proposes that our moral choices reflect our basic sense of who we are in relation to others. Identity constrains moral choice through setting the range of options we perceive as available, not just morally but cognitively. This distinction, for example, means we do not think of an option but reject it as ethically wrong; we do not think of it at all. Consider the following example to illustrate this point. Assume you are in a new town and are mugged while sightseeing. You have no money and no way to get back to your hotel. Few of us would even think of mugging someone to get money for a taxi, even though we must know in some objective sense that people get money by mugging others, since it has just happened to us. Why not? Because mugging is not something "people like us" do. It is not on our cognitive menu. This example illustrates identity's ability to exert its influence by filtering an actor's sense of available choices, while operating in conjunction with the actor's worldview. This ethical framework will work in conjunction with the external stimuli that frame the situation and how the actor sees himself and the world to produce an ethical perspective, which consists of the actor's cognitive classification or categorization system, relational self, and moral imagination. All of this will produce a menu of choice options along with a sense of moral salience, which creates a felt need to act in response to a situation. In our mugging example, we do not put ourselves in the class of people who mug others, and hence do not attempt to mug someone to get money for a taxi to our hotel.

This theory is designed to explain that part of moral behavior which emanates in a psychological process that appears spontaneous, reflecting intuitions and emotions that affect how we see ourselves in relation to others at the time of action. It makes moral behavior not merely the result of conscious deliberation, although such conscious deliberations may indeed enter the equation. But it also allows for spontaneous forces that lie outside the realm of conscious deliberation or reasoning and assumes that what we have chosen may reflect who we are as much as—perhaps even more than—any conscious calculus based on reasoning.

This theory makes identity central to moral choice by providing a framework through which ethical situations are perceived, analyzed, and acted upon. Predicting moral choice requires us to understand both the actor's underlying ethical framework and the ethical perspective of the actor at the moment action is taken. It is the ethical perspective that constitutes the link between the social and individual influences on behavior.

To understand moral choice, we need to understand how ethical framework and perspective in turn relate to the critical role played by identity perceptions in driving ethical behavior and moral choice, whether that choice takes the form of helping and peaceful cooperation or involves us in the stereotyping and prejudice that deteriorates into ill treatment based on any kind of difference: ethnic, religious, racial, gender or sexual preference, sectarian violence, etc., and which then results in behavior, ranging from rescuing or bystander behavior to ethnic cleansing and genocide.

This conceptualization of moral psychology makes our sense of who we are phenotypic. Our sense of self is composed of both genetic factors that constitute a predisposition toward a certain personality—shy, aggressive, risk-taking, etc.—plus cultural and environmental factors, such as socialization via parents and schools, that shape our initial genetic predispositions.[17] It assumes every person has an ethical framework through which we view the world and ourselves in it as we think about and relate to others. This ethical framework includes cognition as well as emotion, intuitions, predispositions, and sentiments.[18] Deciphering the critical parts of an individual's ethical framework will help us understand how that individual's particular ethical perspective at times of action influences how the individual will respond to situations that call for a moral choice.

7.5   Conclusion

This chapter considers the importance of cognitive psychology for ethics, using an example from international relations and politics: moral choice during wars and genocides. It reviews the literature on ethics, which is dominated by theories—such as Kant's or utilitarianism—that are based on reason and deliberation at their core. It then suggests traditional approaches need to respond to recent advances in our knowledge of how the human mind works. It next summarizes a body of literature known as moral psychology, designed to deal with cognitive psychology and its import for moral action. Finally, it considers behavior by rescuers of Jews during the Holocaust to develop an original theory of moral choice that is *not* grounded in religion, first principles, or reason but instead tries to use the archetype of the human mind to propose a theory of moral choice that allows for spontaneous choices.

As analysts try to build this general theory into more specific models of moral choice, we should begin by focusing first on the actor's underlying ethical framework as the entry point for detecting the importance of the actor's perceptions. The next challenge is to specify the ways in which the actor's ethical perspective works with external stimuli to contribute to the ways in which choice options are analyzed, again drawing heavily on the concept of identity and the subconscious forces that constitute identity. Finally, we need to specify the factors that lead to action, chief among them being an individual's established categorization system and sense of moral salience, both of which will influence action. Our scientific knowledge of identity and how the brain processes and operates is at an exciting stage, and much of our current knowledge will be surpassed and corrected in the years ahead. The above theory is offered as a starting point for social and political science work in what will be an exciting new frontier in understanding the importance of political psychology for social science and ethics. Others are encouraged to correct and build upon this work.

Notes

1. No clear scholarly consensus exists on the distinction between ethics and morality. In general, morality may be more personal, while ethics refers to a system of principles by which one lives one's life within a certain social structure. (Hence we speak of personal morality or personal ethics but also talk of business ethics, situational ethics, etc.) The terms *ethics* and the term *morality* tend to be used interchangeably in the field of moral psychology. There are differences that have occurred at different points in time and according to discipline, but these differences are not consistent.

2. Representatives of psychologists in this field include those initially designated moral psychologists using the older conceptualization of term, one that focuses on moral reasoning, such as Jean Piaget and Lawrence Kohlberg. But it also includes psychologists such as Bibb Latané and John Darley (1970), C. Daniel Batson (1991), Jonathan Haidt (2001; Haidt & Joseph, 2007), and Marc Hauser (2006). Philosophers are represented by Stephen Stich, John Doris, Joshua Knobe, Shau Nichols, Martha Nussbaum, Thomas Nagel, Bernard Williams, and R. Jay Wallace (for illustrations, see Batson, 1993; Monroe, 1996; Nussbaum, 1986, 2001; Toulmin, 1958, *inter alia*; Williams, 1981; Wilson, 1993).

3. See Hart (1977) and Dworkin (1977a), both in Dworkin's edited volume *The Philosophy of Law* (1977b).

4. For a more extended treatment, see Lyons (1965), Smart and Williams (1973), or Rosen (2003).

5. Kant disagrees with Hume, taking issue both with Hume's locating morality in sentiment and feeling and with Hume's emphasis on the consequences of an act as the test of morality. Kant argues instead that it is reason which provides the foundation for moral duty and which helps us discover what morality is.

6. See Zajonc's 1980 paper, "Feeling and thinking: Preferences need no inferences," presented as his lecture for the 1979 Distinguished Scientific Contribution Award from the American Psychological Association.

7. Full discussion of this important work deserves a volume of its own. See work by Damasio (1994) and Haidt (2001) or, for an easily accessible summary, the sequence of "Brain Series" episodes featured on the television program *Charlie Rose* (2010).

8. As discussed in section 2 of this chapter, moral sense theory holds that we are able to distinguish between right and wrong through a distinctive moral sense. See Monroe (2012) for a more systematic discussion.

9. Despite the philosophical and political theoretical lack of attention to the idea of an inborn moral sense, the idea nonetheless perseveres in other disciplines.

10. See Kahneman, Slovic, and Tversky's (1982) work on heuristics, the mental shortcuts people use to make decisions.

11. Much of this section comes from Monroe (2012). I am grateful to Princeton University Press for allowing me to reproduce the argument here.

12. There has been a virtual explosion of work in recent years suggesting happiness relates closely to our interactions with others, and thus calling into question the traditional distinction between behavior designed to further our self-interest versus behavior designed to help others. See Vaillant (2002) and Taylor (2002).

13. The rest of this part of the chapter is excerpted from my book, *Ethics in a Time of Terror and Genocide: Identity and Moral Choice*, published in 2012 by Princeton University Press. I am grateful to Princeton University Press for allowing me to reprint sections of this book here and encourage readers to consult the full volume for empirical data supporting this theory.

14. As noted previously, the terms are used interchangeably depending on the discipline and speaker.

15. See Rosch, Varela, and Thompson (1991). One illustration would be experiments designed to show how our eyes focus on a bright color in a beige room. The room is beige. The walls, the furniture, the rug, everything is beige. Except a pot of red flowers. The eye automatically goes to the flowers. So—I hypothesize—it was with Schindler's ethical perspective; something clicked so that Schindler felt that the Jews' suffering was relevant for him and demanded his help.

16. I use moral choice not to refer only to acts that are considered morally commendable, but to refer to all acts that touch on ethics.

17. Dennis, Easton, and Easton (1969); Greenstein (1965). Studies involving twins offer evidence for the genetic basis of certain personality characteristics, including the "big five" factors of neuroticism, extraversion, openness, agreeableness, and conscientiousness (Jang, Livesley, & Vernon, 2006), as well as attitudes and belief

systems, such as political ideology (Alford, Funk, & Hibbing, 2005); religiosity (Waller, Kojetin, Bouchard, & Lykken et al., 1990); and attitudes toward some social issues and recreational activities (Olson, Vernon, Aitken, & Jang, 2001). Critics have asserted that similar environments, rather than genetics, account for these perceived likenesses between monozygotic twins (Beckwith & Morris, 2008).

18. Even in scholarly parlance, analysts frequently use intuitions, predispositions, and sentiments interchangeably, so the distinctions in the literature—which does, after all, cover works written over several centuries—are not clear-cut. All phenomena can be thought of as content-laden domains that are evolutionarily programmed and get triggered by external stimuli. But the content is there waiting to be activated by the external environment.

References

Alford, J., Funk, C., & Hibbing, J. (2005). Are political orientations genetically transmitted? *American Political Science Review, 99*, 153–167.

Anscombe, G. E. M. (1958). Modern moral philosophy. *Philosophy, 33*, 1–19.

Arnold, B. (1989). *The pursuit of virtue: The union of moral psychology and ethics*. New York: Peter Lang.

Aronson, E. (1997). The theory of cognitive dissonance: The evolution and vicissitudes of an idea. In C. McGarty & A. Halsam (Eds.), *The message of social psychology: Perspective on mind in society* (pp. 20–35). Cambridge: Blackwell.

Batson, C. (1991). *The altruism question: Toward a social-psychological answer*. Hillsdale, NJ: Lawrence Erlbaum.

Batson, C. (1992). Experimental tests for the existence of altruism. In D. Hull, M. Forbes, & K. Okruklik (Eds.), *Proceedings of the 1992 Biennial Meeting of the Philosophy of Science Association* (Vol. 2, pp. 69–78). East Lansing, MI: Philosophy of Science Association.

Beckwith, J., & Morris, C. (2008). Twin studies of political behavior: Untenable assumptions? *Perspectives on Politics, 6*(4), 785–791.

Bentham, J. (2002). *Works*. Charlottesville, VA: InteLex Corp.

Crisp, R., & Slots, M. (Eds.). (1997). *Virtue ethics*. Oxford: Oxford University Press.

Coles, R. (1967). *Children of crisis: A study of courage and fear*. Boston, MA: Little-Brown.

Damasio, A. (1994). *Descartes' error: Emotion, reason, and the human brain*. New York: Putnam.

De Waal, F. (1996). *Good natured: The origins of right and wrong in humans and other animals*. Cambridge, MA: Harvard University Press.

Dennis, J., Easton, D., & Easton, S. (1969). *Children in the political system: Origins of political legitimacy*. New York: McGraw-Hill.

Dworkin, R. (1977a). Is law a system of rules? In R. Dworkin (Ed.), *The philosophy of law* (pp. 38–65). Oxford: Oxford University Press.

Dworkin, R. (1977b). *The philosophy of law*. Oxford: Oxford University Press.

Dworkin, R. (1996).Taking Rights Seriously. In J. Arthur (Ed.), *Morality and moral controversies* (4th ed., pp. 120–126). Upper Saddle River, NJ: Prentice-Hall.

Festinger, L. (1957). *A theory of cognitive dissonance*. Stanford, CA: Stanford University Press.

Geertz, C. (1973). *The interpretation of cultures: Selected essays*. New York: Basic Books.

Gilligan, C. (1993). *In a different voice: Psychological theory and women's development*. Cambridge, MA: Harvard University Press.

Goodall, J. (1986). *The chimpanzees of Gombe: Patterns of behavior*. Cambridge, MA: Harvard University Press.

Greene, J. D., & Haidt, J. (2002). How (and where) does moral judgment work? *Trends in Cognitive Sciences, 6*, 517–523.

Greenstein, F. (1965). *Children and politics*. New Haven, CT: Yale University Press.

Haidt, J. (2001). The emotional dog and its rational tail: A social intuitionist approach to moral judgment. *Psychological Review, 108*, 814–834.

Haidt, J., & Bjorklund, F. (2008). Social intuitionists reason, as a normal part of conversation. In W. Sinnott-Armstrong (Ed.), *Moral psychology* (Vol. 2, pp. 181–217). Cambridge, MA: MIT Press.

Haidt, J., & Joseph, C. (2007). The moral mind: How five sets of innate moral intuitions guide the development of many culture-specific virtues, and perhaps even modules. In P. Carruthers, S. Laurence, & S. Stich (Eds.), *The innate mind* (Vol. 3, pp. 367–391). Oxford: Oxford University Press.

Hart, H. L. A. (1977). Positivism and the separation of law and morals. In R. Dworkin (Ed.), *The philosophy of law* (pp. 17–37). Oxford: Oxford University Press.

Hauser, M. (2006). *Moral minds: How nature designed our universal sense of right and wrong*. New York: Ecco.

Hume, D. (1998). *An enquiry concerning the principles of morals*. Oxford: Oxford University Press. (Original work published 1751)

Hursthouse, R. (1999). *On virtue ethics*. Oxford: Oxford University Press.

Hutcheson, F. (1968). *A system of moral philosophy*. New York: A.M. Kelley. (Original work published 1755)

Jang, K., Livesley, W., & Vernon, P. (2006). Heritability of the big five personality dimensions and their facets: A twin study. *Journal of Personality, 64*(3), 577–592.

Kagan, J. (1978). *The growth of the child: Reflections on human development*. New York: Norton.

Kagan, J. (1998). *Three seductive ideas*. Cambridge, MA: Harvard University Press.

Kahneman, D., Knetsch, J., & Thaler, R. (1991). The endowment effect, loss aversion, and status quo bias. *Journal of Economic Perspectives, 5*(1), 193–220.

Kahneman, D., Slovic, P., & Tversky, A. (1982). *Judgment under uncertainty: Heuristics and biases*. Cambridge: Cambridge University Press.

Kant, I. (1991). *The metaphysics of morals*. Cambridge: Cambridge University Press. (Original work published 1797)

Kohlberg, L. (1981). *The philosophy of moral development*. San Francisco, CA: Harper and Row.

Lakoff, G. (1996). *Moral politics*. Chicago: University of Chicago Press.

Lakoff, G., & Johnson, M. (1980). *Metaphors we live by*. Chicago: University of Chicago Press.

Lakoff, G., & Johnson, M. (1999). *Philosophy in the flesh: The embodied mind and its challenges to Western thought*. New York: Basic Books.

Latané, B., & Darley, J. M. (1970). *The unresponsive bystander: Why doesn't he help?* New York: Appleton-Century-Crofts.

Lyons, D. (1965). *Forms and limits of utilitarianism*. Oxford: Oxford University Press.

Marcus, G. (2004). *The birth of the mind: How a tiny number of genes creates the complexities of human thought*. New York: Basic Books.

Margolis, H. (1984). *Selfishness, altruism and rationality*. Chicago: University of Chicago Press.

Mikhail, J. (2007). Universal moral grammar: Theory, evidence and the future. *Trends in Cognitive Sciences, 11*(4), 143–152.

Monroe, K. (1996). *The heart of altruism: Perceptions of a common humanity*. Princeton, NJ: Princeton University Press.

Monroe, K. (2004). *The hand of compassion: Portraits of moral choice during the Holocaust*. Princeton, NJ: Princeton University Press.

Monroe, K. (Ed.). (2011). *Science, ethics, and politics: Conversations and investigations.* Boulder, CO: Paradigm.

Monroe, K. (2012). *Ethics in a time of terror and genocide: Identity and moral choice.* Princeton, NJ: Princeton University Press.

Monroe, K., Chiu, W., Martin, A., & Portman, B. (2009). What is political psychology? *Perspectives on Politics, 7*(4), 859–882.

Montesquieu C. S. (1748). *The Spirit of the Laws.* Geneva: Crowder, Wark, and Payne. (Original work published 1748)

Nisbett, R., & Ross, L. (1980). *Human inference: Strategies and shortcomings of social judgment.* Englewood Cliffs, NJ: Prentice-Hall.

Rose, C. (Host). (2010, September 30). The deciding brain [Brain Series, Episode 11]. In C. Rose (Executive producer), *Charlie Rose,* PBS, New York.

Rosen, F. (2003). *Classical utilitarianism from Hume to Mill.* New York: Taylor & Francis.

Sapolsky, R. (2002). *A primate's memoir.* New York: Touchstone Books.

Scherer, K. R. (2003). Introduction: Cognitive components of emotion. In R. J. Davidson, K. R. Scherer, & H. Goldsmith (Eds.), *Handbook of the Affective Sciences* (pp. 563–571). Oxford: Oxford University Press.

Schnall, S., Haidt, J., Clore, G., & Jordan, A. (2008). Disgust as embodied moral judgment. *Personality and Social Psychology Bulletin, 34,* 1096–1109.

Smart, J., & Williams, B. (1973). *Utilitarianism: For and against.* Cambridge: Cambridge University Press.

Smith, A. (1976). *The theory of moral sentiments.* Indianapolis, IN: Liberty Classics. (Original work published 1759)

Statman, D. (Ed.). (1997). *Virtue ethics.* Oxford: Oxford University Press.

Sun, R. (2009). Motivational representations within a computational cognitive architecture. *Cognitive Computation, 1,* 91–103.

Sun, R., & Wilson, N. (2011). Motivational processes within the perception-action cycle. In V. Cutsuridis, A. Hussain, & J. G. Taylor (Eds.), *Perception-action cycle: Models, architectures and hardware.* Springer, Berlin.

Sunstein, C. (2005). Moral heuristics. *Behavioral and Brain Sciences, 28*(4), 531–542.

Taylor, S. (2002). *The tending instinct: How nurturing is essential to who we are and how we live.* New York: Holt.

Taylor, T. (1992). *The anatomy of the Nuremberg trials: A personal memoir.* New York: Knopf.

Toulmin, S. (1958). *The uses of argument.* New York: Cambridge University Press.

Vaillant, G. (2002). *Aging well: Surprising guideposts to a happier life from the landmark Harvard study of adult development.* Boston: Little, Brown.

Waller, N. G., Kojetin, B. A., Bouchard, T. J., Jr., Lykken, D. T., & Tellegen, A. (1990). Genetic and environmental influences on religious interests, attitudes, and values: A study of twins reared apart and together. *Psychological Science, I,* 1-5.

Williams, B. (1981). *Moral luck.* Cambridge: Cambridge University Press.

Wilson, J. (1993). *The moral sense.* New York: The Free Press.

Zajonc, R. (1980). Feeling and thinking: Preferences need no inferences. *American Psychologist, 35*(2), 151–175.

IV RELIGION

8 Psychological Origins and Cultural Evolution of Religion

Scott Atran

He that is not with me is against me; and he that gathereth not with me scattereth abroad.
—Jesus of Nazareth, Matthew 12:30

And the Lord said unto the servant, Go out into the highways and hedges and compel them to come in.
—Luke 14:23

[A]n advancement in the standard of morality and an increase in the number of well-endowed men will certainly give an immense advantage. ... [A] tribe including many members who, from possessing in a high degree the spirit of patriotism, fidelity, obedience, courage, and sympathy, were always ready to give aid to each other and to sacrifice themselves for the common good, would be victorious over most other tribes.
—Charles Darwin, *The Descent of Man*

8.1 Introduction

Ever since Edward Gibbon's *Decline and Fall of the Roman Empire (1776–1789)*, written in Britain as the American republic was born, scientists and secularly minded scholars have been predicting the ultimate demise of religion. But in many places around the globe, religious fervor is increasing. At the beginning of the twenty-first century, new religious movements continued to arise at a furious pace—perhaps at the rate of two or three per day (Lester, 2002). There are now more than two billion self-proclaimed Christians (about one-third of humanity), a quarter of whom are Pentecostals or charismatics (people who stay in mainstream Protestant and Catholic churches but have adopted Pentecostal practices like healings, speaking in tongues, casting out demons, and laying hands upon the sick).

The Winner's Chapel, a Pentecostal church that celebrates newfound market wealth and success, is less than twenty years old but already has tens of thousands of members in thirty-two African countries. During the same period, the Falun Gong, a Buddhist offshoot that the Chinese government estimated at 70 million adherents at the turn of the century (Kahn, 1999), continues to grow (Moore, 2009), and Islamic revivalist movements have spread across the Muslim world.

The United States—the world's most economically powerful and scientifically advanced society—is also one of the world's most professedly religious societies. Evangelical Christians and fundamentalists include about 25 percent of Americans, and together with charismatics constitute about 40 percent of the American population. According to a June 2008 Pew poll, more than 90 percent believe in God or a universal spirit, including one in five of those who call themselves atheists (Pew, 2008). More than half of Americans surveyed pray at least once a day, but about three-quarters of religious believers in America think that there is more than one way to interpret religious teachings and that other religions can also lead to eternal salvation (Salmon, 2008). Even in France, the most secular of societies, two-thirds of the population believe in "a spirit, god or life force" (European Commission, 2005).

An underlying reason for religion's endurance is that science treats humans and intentions only as incidental elements in the universe, whereas for religion they are central. Science is not particularly well suited to deal with people's existential anxieties—death, deception, sudden catastrophe, loneliness, or longing for love or justice. Religion thrives because it addresses people's yearnings and society's moral needs.

Although science may never replace religion, science can help us understand how religions are structured in individual minds (brains) and across societies (cultures) and also, in a strictly material sense, why religious belief endures. Religion does not appear to be either a naturally selected adaptation of our species nor innate in us. All of its cognitive and social components are universally found in various mundane thoughts and activities. Nevertheless, since the Upper Paleolithic, several naturally selected elements of human cognition have tended to converge in the normal course of social interaction to produce a near-universal family of phenomena that most people recognize as religion. Roughly, religion exploits ordinary cognitive processes to passionately display costly commitment to supernatural agents that are invoked to deal with otherwise unsolvable existential problems, such as death and deception.

8.2 Evolutionary Enigmas

The origin of large-scale cooperative societies is an evolutionary problem because people frequently cooperate with genetic strangers of unknown reputation whom they will never meet again, and whose cheating, spying, lying, and defection cannot be monitored and controlled (Fehr & Fischbacher, 2003). Kinship cannot drive cooperation among genetic strangers, although "imagined kinship"—a cultural manipulation of kin psychology and terminology ("brotherhood," "motherland," etc.)—can be a potent psychological mobilizer of groups formed through other means (Johnson, 1987). Difficulty in maintaining trust based on accurate information about reputation is progressively less reliable the larger the group (reliability of information and mutual trust for an n-person group tends to decline as a function of $1/n$; Nowak & Sigmund, 1998). Moreover, "Watch my back and I'll watch yours," and other sensible reciprocity strategies in politics and economics, fail to account for cooperation with people you only meet once, such as aiding a stranger who asks for directions or a dollar for food, taking the hand of a lost child or a disabled person to cross the street, leaving a tip at a roadside restaurant you'll never visit again, or risking life and limb to save someone you don't know (Henrich, McElreath, Barr, Ensimger et al., 2006).

Studies show that people are usually quite suspicious of strangers who belong to an entirely different social milieu or cultural group and don't even bother to find out about their reputation or the history of their past transactions. The default assumption is usually zero-sum: anything I do for the other person is likely to be bad for me (Thompson & Hastie, 1990). But other studies show that invoking God or other supernatural concepts leads to reduced cheating and greater generosity between anonymous strangers (Bargh & Chartland, 1999). The key to the difference in behavior seems to be belief that there's a God watching over to make sure everyone stays honest (Norenzayan & Shariff, 2008).

Thus, as societies grow, it can be harder to enforce moral and altruistic norms, and to punish free-riders on the public good. This in turn can make such societies less cohesive and less able to compete with other expanding societies. Moral deities define the sacred boundaries of societies and the taboos you can't transgress. If you really believe in these moral gods, then the problem of punishment becomes easier, as you punish yourself for misbehavior. Consider the power of taking an oath before God among the Pashtun, traditionally dominant tribes of Afghanistan and of today's

Taliban, as a way of ensuring that a potential liar or thief would think twice before acting for fear that a supernatural entity would always know what went on and exact retribution, even if no other person on earth could possibly spot the transgression:

[B]ecause the role of honor is so serious ... particularly because oaths required of men who are accused of dishonorable acts (such as theft) come from potentially dubious sources, there are a host of supernatural consequences that will rain down upon the perjurer should a man lie. These include becoming poor, having your opinions disregarded, having your body become pale and ugly, seeing your land and livestock lose their productivity, as well as endangering the life of your children (Barfield, 2003, p. 12).

But the natural origin of gods and religious traditions is itself an evolutionary puzzle, because all religions require costly commitment to beliefs that violate basic tenets of rational inference and empirical knowledge necessary for navigating the world; for example, the belief in sentient but bodiless beings or beings with bodies that defy gravity and can pass through solid walls (Atran & Norenzayan, 2004). In all religions, there are bodiless but sentient souls and spirits that act intentionally, though not in ways empirically verifiable or logically understandable. For the Christian philosopher Sören Kierkegaard (1843/1954), true faith can only be motivated by "a gigantic passion" to commit to the "absurd." Abraham's willingness to sacrifice something more important to him than his own life—that of his only and beloved son—is exemplary: "For love of God is ... incommensurable with the whole of reality ... there could be no question of human calculation" (p. 25). Hundreds of millions of people across the planet celebrate Abraham's actions as noble and heroic, rather than murderous, evil, or insane.

Imagine creatures who consistently believed that the dead live on and the weak are advantaged over the strong, or that you can arbitrarily suspend the known physical and biological laws of the universe with a prayer. If people literally applied such prescriptions to the factual navigation of everyday life, they likely would be either dead or in the hereafter in short order—too short for most individuals to reproduce and the species to survive. The trick is in knowing how and when to suspend factual belief without countermanding the facts and compromising survival. But why take the risk of neglecting the facts at all, even in exceptional circumstances?

Religious practice is costly in terms of cognitive effort (maintaining both rational and nonrational networks of beliefs) as well as material sacrifice (ranging from human sacrifice to prayer time) and emotional expenditure (inciting fears and hopes). A review of anthropological literature on religious offerings concludes: "Sacrifice is giving something up at a cost. ... 'Afford it or not,' the attitude seems to be" (Firth, 1963). Indeed, what could be the calculated gain from:

- years of toil to build gigantic structures that house only dead bones (Egyptian, Mesoamerican, and Cambodian pyramids).
- giving up one's sheep (Hebrews) or camels (Bedouin) or cows (Nuer of Sudan) or chickens (Highland Maya) or pigs (Melanesian tribes, Ancient Greeks), or buffaloes (South Indian tribes).
- dispatching wives when their husbands die (Hindus, Inca, Solomon Islanders).
- slaying one's own healthy and desired offspring (the firstborn of Phoenicia and Carthage, Inca and Postclassical Maya boys and girls, children of South India's tribal Lambadi).
- chopping off a finger for dead warriors or relatives (Dani of New Guinea, Crow and other American Plains Indians).
- burning your house and all other possessions for a family member drowned, crushed by a tree, or killed by a tiger (Nāga tribes of Assam).
- knocking out one's own teeth (Australian aboriginals).
- making elaborate but evanescent sand designs (Navajo, tribes of Central Australia).
- being willing to sacrifice one's time, and sometimes even one's life, to keep Fridays (Muslims) or Saturdays (Jews) or Sundays (Christians) holy.

As billionaire Bill Gates pithily put it, "Just in terms of allocation of time resources, religion is not very efficient. There's a lot more I could be doing on a Sunday morning" (Keillor, 1999, citing Gates).

Evolution can't account for religion simply as a biological adaptation, naturally selected for some ancestral task that is "hard-wired" into us. Try to come up with an adaptive logic that generates a unitary explanation for all of the strange thoughts and practices above, or for just stopping whatever you're doing to murmur often incomprehensible words while gesticulating several times a day. And individual devotion to religion is quite variable, much wider than for other behaviors that likely did evolve as adaptations: like the ability to walk, to see, to think about other people thinking, or to imagine possible and counterfactual worlds that do not (yet) exist. There's no gene for the complex set of beliefs and behaviors that make up religion, any more than there's a gene for science; nor is there likely any genetic complex with lawlike or systematic qualities that is responsible for most religious belief or behavior (Atran, 2002).

The author's research suggests that the two evolutionary enigmas, large-scale cooperation and costly religion, have resolved one another through a process of cultural coevolution (Atran & Henrich, 2010). This coevolution centers on the concept of costly communal commitment to absurd, counterintuitive ideas that have no consistent logical or empirical connection to everyday reality.

8.3   Born to Believe: The Storytelling Animal

Humans are cause-seeking, purpose-forming, storytelling animals. As the eighteenth-century Scottish philosopher David Hume (1758/1955) famously noted, we find patterns in nature and look for the "hidden springs and principles" that bring those patterns to life and let us navigate them. We can't help doing this; Nature made us so.

Our mental facility for confabulation is strikingly evident in experiments with split-brain patients, where the left hemisphere (which controls language and the right hand) is physically cut off from the right hemisphere (which controls the left hand) (Gazzaniga, Ivry, & Mangun, 1998). In this classic study, images were selectively presented to each hemisphere: the left was shown a chicken claw, the right a snow scene. Patients were presented with an array of objects and asked to choose an object associated with the image they were shown. A representative answer was that of a patient who chose a snow shovel with his left hand and a chicken with the right. Asked why he chose these items, he (that is, the left-hemisphere story spinner) said: "Oh, that's simple. The chicken claw goes with the chicken and you need the shovel to clean out the chicken shed." Michael Gazzaniga, who carried out the experiment, observes, "The left brain weaves its story in order to convince itself and you that it is in full control" (Gazzaniga, Ivry & Mangun, 1998).

For the most part, our penchant for storytelling is a beneficial thing. Because the world we live in does have lots of recurring and sequential patterns, like sunsets and seasons and birth and death, knowing these regularities helps us to better survive and reproduce. Hunters and gatherers were able to systematically track game and store the resources of good times in anticipation of bad times. Agriculturalists were capable of subsisting on the land they settled because they could understand and manage the seasonal flooding of a river, the water needs and life cycles of plants, and the motivations of enemies to plunder the crops. We moderns can enjoy machines that work for us because we can represent the patterns we find in nature, manipulate their causes in our imagination, and create new applications that help us function in the world.

But we humans also expect to find patterns and underlying principles where there are none. Flip a coin that comes up ten heads in a row and most people believe there's a better than even chance that the next flip will be tails. "There's a sucker born every minute," quipped P. T. Barnum, who might have been talking about people in a casino who rush into the seat of any player who has had a long string of bad luck at the slot machine. Even apparent lack of regularity tends to be overgeneralized into a causal pattern: "Lightning never strikes twice in the same place," goes the age-old untruth.

This built-in "flaw" of our causal understanding is rarely catastrophic, and is usually far outweighed by the benefits. That's pretty standard for most of our evolutionary endowments. For example, we humans are also prone to bad backs and to choking when we eat and speak. Thank goodness, though, for upright posture, which frees our hands and widens the horizon of sight. Good, too, that we have speech, which vastly expands our field of communication and the sharing of thoughts. The creations of evolution resemble more the works of an amateur tinkerer than a trained engineer. Perfect may be better than good in an ideal world. But in the real world, good only has to beat out less good by a little to get to the top of the heap.

And so we tell ourselves stories. We look for and impose purpose on the events of our life by weaving them into a meaningful narrative with the hand of intentional design. Our constantly cause-seeking brain makes it hard to believe that "stuff just happens," especially if the stuff happens to us. This view is backed up by a recent experiment in which people were asked what patterns they could see in arrangements of dots or stock market information (Whitson & Galinsky, 2008). Before asking, the experimenters made half their participants feel a lack of control, either by giving them feedback unrelated to their performance or by having them recall experiences where they had lost control of a situation. The results were remarkable. When people felt a lack of control, they would fall back on preternatural and supernatural explanations. That would also suggest why religions enjoy a revival during hard times (Norris & Inglehart, 2004).

The greater the impact of events on our lives, the greater the drive to impose meaning on those events. Terrible, senseless accidents can never just be senseless accidents, our mind's voice tells us. When psychologist Jesse Bering and his students carried out interviews with atheists, it became clear that they often tacitly attribute purpose to significant or traumatic moments in their lives, as if some agency were intervening to make them happen (Bering, 2009). It seems that's just the way our brains are wired: atheists can muzzle some if its expression, but even they can't seem to completely stop it in themselves.

A parent may spend years trying to find some significance in his child's untimely death. An earthquake victim is likely to want a story to explain why she survived and those around her didn't. And there are those ready to help find a story: After a massive earthquake in Pakistan's Azad Kashmir, the author witnessed jihadi groups riding Pakistani military vehicles and blasting over loudspeakers what their banners read: "This happened because you have turned from God's path. We can help you find the true path." In early 2010, the Rev. Pat Robertson similarly declared that the massive Haitian earthquake happened because Haitians had long ago sworn "a pact with the devil" (Smith, 2010).

8.4 Supernatural Agents

Religions invariably center on supernatural agent concepts, such as gods, angels, ancestor spirits, demons, and jinns. Granted, nondeistic theologies, such as Buddhism and Taoism, doctrinally eschew personifying the supernatural or animating nature with supernatural causes (Pyysiäinen, 2003). Nevertheless, people who espouse these faiths routinely entertain belief in an array of minor buddhas that behave "counterintuitively" in ways that are inscrutable to factual or logical reasoning. Buddhist Tibetan monks and Japanese Samurai warriors ritually ward off malevolent deities by invoking benevolent spirits, and conceive altered states of nature as awe-inspiring.

Mundane concepts of *agent* (intentional, goal-directed actor) are central players in what psychologists refer to as folk psychology, specifically the *theory-of-mind module*, or ToM. ToM is a species-specific cognitive system devoted to "mind reading"—that is, making inferences about the knowledge, beliefs, desires, and intentions of other people's minds. Recent brain-imaging (fMRI) studies show that people's statements about God's level of involvement in social events, as well as the deity's purported emotional states, reliably engage ToM-related prefrontal and posterior regions of the brain that appeared most recently in human evolution (Kapoggianis, Darbey, Sua, & Zambonia et al., 2009).

One plausible hypothesis is that notions of agent evolved in humans to respond automatically under conditions of uncertainty to potential threats (and opportunities) by intelligent predators (and protectors; Guthrie, 1993).

From this evolutionary vantage, the proper evolutionary domain of *agent* encompasses animate objects, but its actual domain inadvertently extends to moving dots on computer screens, voices in the wind, faces in clouds, eyes in the shadows, and virtually any complex design or uncertain circumstance of unknown origin (Sperber, 1996). For example, in Miami many people claimed to have spotted the Holy Virgin in windows, curtains, and television afterimages as long as there was hope of keeping young Elián González from returning to godless Cuba. On the day of the World Trade Center attacks, newspapers showed photos of smoke billowing from one of the towers that seemed "to bring into focus the face of the Evil One, complete with beard and horns and malignant expression, symbolizing to many the hideous nature of the deed that wreaked horror and terror upon an unsuspecting city" ("Bedeviling," 2001).

A number of studies reveal that children and adults spontaneously interpret the contingent movements of dots and geometrical forms on a screen as interacting agents, whether as individuals or groups, with distinct goals and internal goal-directed motivations (Bloom & Veres, 1999). In the 1940s, Heider and Simmel made a silent cartoon animation in which two triangles and a circle move against and around each other and a diagram of a house. People who watched almost always made up a social plot in which the big triangle was seen as an aggressor (Heider & Simmel, 1944). Young children spontaneously over-attribute agency to all sorts of entities (clocks, clouds), and may thus be predisposed to construct agent-based representations of many phenomena (Keleman, 2004).

Why? Because, from an evolutionary perspective, it's better to be safe than sorry regarding the presence of agents under conditions of uncertainty. Such reliably developing programs provide efficient reaction to a wide—but not unlimited—range of stimuli that would have been statistically associated with the presence of dangerous agents in ancestral environments. Mistakes, or "false positives," would usually carry little cost, whereas a true response could provide the margin of survival. This was true at least until these supernatural agents were selected by cultural evolution to begin demanding costly actions, under threat of divine punishment, or offers of sublime rewards, or until they evoked hostility in the followers of another god.

This cognitive proclivity would favor emergence of malevolent deities in all cultures, just as a countervailing Darwinian propensity to attach to protective caregivers would favor conjuring up benevolent deities. Thus, for the Carajá Indians of Central Brazil, intimidating or unfamiliar regions of the local ecology are religiously avoided: "The earth and underworld are inhabited by supernaturals. ... There are two kinds. Many are amiable and beautiful beings who have friendly relations with humans ... others are ugly and dangerous monsters who cannot be placated. Their woods are avoided and nobody fishes in their pools" (Lipkind, 1940). Similar descriptions of supernaturals appear in ethnographic reports throughout the Americas, Africa, Eurasia, and Oceania.

Our brains may be trip-wired to spot lurkers (and to seek protectors) where conditions of uncertainty prevail—when startled, at night, in unfamiliar places, during sudden catastrophe, in the face of solitude, illness, prospects of death, etc. Given the constant menace of enemies within and without, concealment, deception, and the ability to generate and recognize false beliefs in others would favor survival. In potentially dangerous or uncertain circumstances, it would be best to anticipate and fear the worst of all likely possibilities: the unseen presence of a deviously intelligent agent that might just want your head as a trophy. Unfortunately, a worst-case analysis also fosters unnecessary wars.

Humans habitually "trick and tweak" their own innate releasing programs, as when people become sexually aroused by makeup (which artificially highlights sexually appealing attributes), fabricated perfumes, or undulating lines drawn on paper or dots arranged on a computer screen—that is, pornographic pictures (Hamilton & Orians, 1965). Horror movies, for example, play off hair-trigger agency detection to catch our attention and build suspense. Much of human culture—for better or worse—can be arguably attributed to focused stimulation and manipulation of our species' innate proclivities (Sperber, 1996). Such manipulations can serve cultural ends far removed from the ancestral adaptive tasks that originally gave rise to them, although manipulations for religion often centrally involve the collective engagement of existential needs (wanting security) and anxieties (fearing death).

8.5 The Appeal of the Absurd: Counterintuitive Beliefs

How do our minds make an *agent* concept into a god? And why do people work so hard against their preference for logical explanations to maintain two views of the world: the real and the unreal, the intuitive and the counterintuitive?

Whatever the specifics, certain beliefs can be found in all religions that fit most comfortably with our mental architecture. Thus, anthropologists and psychologists have shown that people attend to, and remember, things that are unfamiliar and strange, but not so strange as to be impossible to assimilate. Ideas about God or other supernatural agents tend to fit these criteria. They are "minimally counterintuitive," as anthropologist Pascal Boyer (1994) calls them: weird enough to get your attention and lodge in memory but not so weird as to be rejected altogether, like a burning bush that speaks, a frog that transforms into a prince, or women who turn into blocks of salt.

Together with psychologists Ara Norenzayan and colleagues, the author studied the idea of minimally counterintuitive agents (Norenzayan, Atran, Faukner, & Schaller, 2006). College students as well as Maya Indians were presented with lists of fantastical creatures and asked to choose the ones that seemed most "religious." The convincingly religious agents, the students said, were not the most outlandish—not the turtle that chatters and climbs or the squealing, flowering marble—but those that were just outlandish enough: giggling seaweed, a sobbing oak, a talking horse. Giggling seaweed meets the requirement of being minimally counterintuitive. So does a God who has a human personality except that he knows everything, or a Spirit with a mind but no body.

What goes for single or simple religious beliefs and utterances also goes for more complex collections of religious beliefs (Barrett & Nyhof, 2001). The Bible, for example, is a succession of mundane events—walking, eating, sleeping, dreaming, copulating, dying, marrying, fighting, suffering storms and drought—interspersed with just a few counterintuitive occurrences, such as miracles and appearances of supernatural agents. The same is true of the Koran, t e Hindu Veda, the Maya Popul Vuh, or any other cultural corpus of religious beliefs and narratives.

Religious beliefs are counterintuitive because they purposely violate what studies in cognitive anthropology and developmental psychology indicate are universal expectations about the world's everyday structure, including basic categories of *intuitive ontology*—the ordinary ontology of the everyday world that is built into the language learner's semantic system—such as *person, animal, plant,* and *substance* (Atran, 1989). Studies reveal that children across cultures do not violate such categorical constraints in learning the meaning of words. But in many religions, though never in reality, animals can conceive of the distant past and far future, plants can walk or talk, and mountains and lakes can have feelings, wishes, beliefs, and desires.

All the world's cultures have religious myths that are attention-arresting because they are counterintuitive, in the technical sense of violating intuitive ontology. Still, people in all cultures also recognize these beliefs to be counterintuitive, whether or not they are religious believers. In our society, Catholics and non-Catholics alike are undoubtedly aware of the difference between Christ's body and ordinary wafers, or between Christ's blood and ordinary wine. Catholics are no more crazed cannibals for their religious beliefs than are Muslims sick with sex when they invoke the pretty virgins floating in paradise.

Reasoning and inference in the communication of many religious beliefs is cognitively designed never to come to closure, but to remain open-textured. To claim that one knows what Judaism or Christianity is truly about because one has read the Bible, or what Islam is about because one has read the Koran, is to believe there is an essence to religion and religious beliefs. But psychological science (and the history of exegesis) demonstrates that this claim is false.

Polls suggest that 30 to 40 percent of all religious Christians say they believe that the Bible is the literal word of God (Newport, 2007), and about 50 percent of religious Muslims say they believe this of the Koran (Pew Research Center, 2007). But deeper study shows that evangelicals and other self-styled "fundamentalists" who espouse belief in a stable text as literal words of God do not ascribe fixed meanings to them. Indeed, it is literally impossible for normal human minds to do so. As for "one infinite, omnipotent, and eternal God," observed Thomas Hobbes (1651/1901), even the religiously enlightened "choose rather to confess He is incomprehensible and above their understanding than to define His nature by 'spirit incorporeal,' and then confess their definition to be unintelligible."

Instead of fixing meaning, studies show that people use the words to evoke many different ideas to give sense and significance to various everyday contexts (Barrett & Keil, 1996; Malley, 2004). Consider a telling example of how religious beliefs are actually processed by human minds. Many argue that the Ten Commandments mean exactly the same thing today as they did when Moses received them on Mount Sinai (Schlesinger, 1999). That's hardly likely, given changes in social conditions and expectations over the last two and half millennia. Thus, failure to heed the commandments to honor the Sabbath or forswear blasphemy merited capital punishment in ancient Israel, but no one in our time and in our society is condemned to death for frolicking on the Sabbath or cursing their Maker.

One study by students in the author's class on evolutionary psychology compared how people interpret the Ten Commandments. The study found that only autistics—who take social cues, including what is said, strictly at face value—produced consistently recognizable paraphrases: for example,

"Honor the Sabbath" might be rendered as "Honor Sunday" (or Saturday). In contrast, university students and members of Jewish and Christian Bible study groups produced highly variable interpretations that third parties from those groups could not consistently identify with the original commandments: "Honor the Sabbath" might be interpreted as "Don't work so hard" or "Take time for your family" (Atran & Norenzayan, 2004). Despite people's own expectations of consensus, interpretations of the commandments showed wide ranges of variation, with little evidence of consensus. (Unlike religious utterances, control phrases such as "two plus two equals four" or "the grass is green" do pass intact from mind to mind).

Rather, religion is psychologically "catchy"—cognitively contagious— because its miraculous and supernatural elements grab attention, stick in memory, readily survive transmission from mind to mind, and so often win out in the cultural competition for ideas that the collectivity can use. Like other human productions that are easy to think and good to use, religious beliefs spontaneously reoccur across cultures in highly similar forms despite the fact that these forms are not evolved by natural selection or innate in our minds. Religion is no more an evolutionary adaptation, as such, than are other near-universals like calendars, maps, or boats.

8.6 The Tragedy of Cognition

Core religious beliefs minimally violate ordinary notions about how the world is, with all of its inescapable problems, to produce surprising but easy to remember supernatural worlds that treat existential problems, like fear of death and worry about deception and defection: for example, a world with beings that resemble us emotionally, intellectually, and physically except they can move through solid objects and are immortal, like angels, ancestral spirits, and souls.

It's not enough for an agent to be minimally counterintuitive for it to earn a spot in people's belief systems. Mickey Mouse and the talking teapot are minimally counterintuitive, but people don't truly believe in them. An emotional component is needed for belief to take hold. If your emotions are involved, then that's the time when you're most likely to believe whatever the religion tells you should be believed. Religions stir up emotions through their rituals—swaying, singing, bowing in unison during group prayer and other ceremonial rituals, sometimes working people up to a state of physical arousal that can border on frenzy. And religions gain strength during the natural heightening of emotions that occurs in times of personal crisis, when the faithful often turn to shamans or priests. The most intense emotional crisis, for which religion can offer powerfully comforting answers, is when people face mortality.

"I simply can't build up my hopes on a foundation consisting of confusion, misery, and death," wrote Anne Frank, a Jewish teenager doomed to die in a Nazi concentration camp:

I hear the ever approaching thunder, which will destroy us too; I can feel the sufferings of millions; and yet, if I look up into the heavens, I think that it will all come right, that this cruelty will end, and that peace and tranquility will return again. ... I want to be useful or give pleasure to people around me who yet don't really know me. I want to go on living after my death! (Frank, 1944/1993)

There's no rational or empirically evident way to escape our eventual death. The tragedy of human cognition—of our ability to imagine the future—is that there is always before us the looming reality of our own demise, whereas the prospect of death for other creatures only arises in actual circumstances, in the here and now. Evolution has endowed all its sentient creatures with the mental and physical means to try to do everything in their power to avoid death, yet practically all humans soon come to understand through everyday reason and evidence that they can do nothing about it in the long run. Cross-cultural experiments and surveys indicate that people more readily accept the truth of narratives containing counterintuitive elements, including miracles, when they are reminded of death through images or descriptions (Norenzayan & Hansen, 2006), or when facing danger or insecurity, as with pleas of hope for God's intervention during wartime (Argyle & Beit-Hallahmi, 2000).

Fear of death, then, is an undercurrent of belief. The spirits of dead ancestors, ghosts, immortal deities, heaven and hell, the everlasting soul: the notion of spiritual existence after death is at the heart of almost every religion. Believing in God and the afterlife is a way to make sense of the brevity of our time on earth, to give meaning to this short existence.

8.7 Costly Commitment

We are a cultural species, evolved to have faith in culture. Unlike other animals, humans rely heavily on acquiring behavior, beliefs, motivations, and strategies from others in their group. These psychological processes have been shaped by natural selection to focus our attention on those domains and those individuals most likely to possess fitness-enhancing information (Henrich & Gil-White, 2001).

Like our own distant ancestors, contemporary foragers routinely process plant foods to remove toxins, often with little or no conscious knowledge of what happens if you don't process the food (Beck, 1992). Such foods often contain low doses of toxins that cause little harm for months or even years, and don't badly damage the food's flavor. But these toxins can accumulate and eventually cause severe health problems and death. A naive learner who favors her own experience of eating the foods without performing the arduous and time-consuming processing will do less work in the short run, but possibly die earlier in the long run. *Placing faith in traditional practices, without understanding why, can be adaptive.* Similarly, manufacturing complex technologies or medicines often involves a sequence of important steps, most of which cannot be skipped without producing something shoddy. We also have faith that any electronics we buy won't blow up in our faces, and that the buildings we go in won't collapse.

But here's the rub: because of language, it's simple for people to lie. You just have to say one thing and do another. Politicians do it often enough; so do shysters and frauds like Bernie Madoff. With evolution of language, faith in culturally transmitted information became vulnerable to exploitation, particularly by successful and prestigious individuals who could now transmit practices or beliefs they themselves may not hold. Language makes deception easy and cheap. Before language, learners observed and inferred people's underlying beliefs or desires by their behavior. Those wishing to deceive would have to actually perform an action to transmit it.

Suckers, though, really aren't born every minute—at least when it comes to most things that are important. It usually takes a lot of work, like cozying up to a victim, or special circumstances, like a distraction, to lure people into a con. That's because humans are universally endowed with a couple of important ways of deciding whether or not to trust what others say and do.

The first way of figuring out who to trust is by rational reasoning: considering if a person's expressed beliefs, and the actions they imply, are logically in line with their other beliefs and empirically consistent with prior actions (Byrne & Whiten, 1998). If the political candidate who expresses belief in the sanctity of the family also expresses belief in free love or turns out to be an adulterer, then you might conclude that the candidate's belief in family sanctity is suspect. (There are various psychological biases and heuristics that impinge on academic notions of pure rationality, but these are orthogonal to the basic distinction between rational and religious beliefs; Tversky & Kahneman, 1974).

The second way to spot deception is by looking for costly commitment: that is, seeing if the person expressing a belief is willing to commit himself to act on those beliefs. Studies show that young children usually refuse to taste a new food offered by a stranger unless the stranger eats it first (Harper & Sanders, 1975). If the food were really bad or dangerous, then it would have been too costly to the stranger to eat it. Developmental studies of the cultural transmission of altruistic giving show that neither preaching nor exhortation to charitable giving are effective without opportunities to observe costly giving by the models (Henrich & Henrich, 2003). Similarly, studies of beliefs about the existence of entities like intangible germs and angels show that children only subscribe to those agents whom adults seem to endorse through their daily actions, and seem rather skeptical of unseen and supernatural agents that are unendorsed in our culture, like demons and mermaids (Harris, 2006). Interviews with a racially diverse sample of parents from highly religious Christian, Jewish, Mormon, and Muslim families reveal that parents see religion holding their families together on a virtuous life course primarily because of costly investments in "practicing (and parenting) what you preach" (Marks, 2004).

A person who is being courted usually knows that a suitor is really in love if the suitor courts with extravagant disregard for expense and time, and struggles against all comers and through thick and thin. That kind of "irrational" and costly display is hard to fake (Frank, 1988). It makes little sense unless the suitor is sincerely in love beyond reason, at least during courtship, because it's unreasonable for anyone to believe that his current honey is the most attractive and understanding person in the world and that no other will ever come along to compare.

Preposterous beliefs fail the test of rational reasoning, which puts them at a disadvantage relative to mundane beliefs in terms of the learner's own commitment to figuring out the social relevance and implications of a belief. But religious beliefs overcome this disadvantage through ritual acts of costly commitment. These extravagant acts are designed to convince learners to pay attention to what socially successful actors say and do, to try to understand the underlying meanings and motivations for that success, and if possible to emulate it.

Mundane beliefs can be undermined by reasoned argument and empirical evidence. Not so religious beliefs, which are rationally inscrutable and immune to falsification: you can't possibly disprove that God is all-seeing or show why he's not more likely to see just 537 things at once. Experiments suggest that once people sincerely commit to religious belief, attempts to undermine those beliefs through reason and evidence can stimulate believers to actually strengthen their beliefs (Festinger, Riecken, & Schachter, 1956). For the believer, a failed prophecy may just mean there is more learning to do and commitment to make.

Preposterous beliefs involve more costly displays of commitment, and require more costly commitments from believers, than do mundane beliefs. Devotion to supernatural agents tends to spread in a population to the extent they elicit costly commitments, usually in the form of ritual ceremonies, offerings, and sacrifices (Tremlin & Lawson, 2006). When participants in costly rituals demonstrate commitment to supernatural beliefs, then observers who witness these commitments are more inclined to trust and follow the participants and so enhance group solidarity and survival. Consider, again, taking an oath before God among the Pashtun in terms of the ritual that underscores its seriousness, which involves time and labor and intense scrutiny by others:

Because swearing a false oath within village or farmstead is believed to bring illness or reduce the productivity of the land, the ritual must be held away from human habitation. Because the surface of the land is presumed to be polluted by contact with thieves, adulterers, and criminals, a half-meter deep hole is dug and a copy of the Quran laid in it. After undergoing ritual ablutions, the accused places his right arm on the Quran and swears by Allah that what he is saying is true and fair. Since at least some of the evil consequences that can befall a false swearer apply equally to a false accuser, demanding an oath unfairly also has serious consequences and may impugn the honor of the accuser (Barfield, 2003).

An examination of eighty-three utopian communes of the nineteenth century indicates that religious groups with more costly rituals were more likely to survive over time than religious groups with fewer rituals. Members and leaders often explicitly acknowledged that costly demands increased their religious engagement (Sosis & Bressler, 2003). Among Israeli kibbutzim (socialist agricultural settlements operating on the principle of labor according to ability with benefits according to need), groups with more religious rituals showed higher levels of cooperation than other religious and secular groups with fewer rituals (Sosis & Ruffle, 2003). Religious kibbutzim also economically outperform secular kibbutzim (Fishman & Goldschmidt, 1990).

Religious beliefs, then, are more likely to spread in a population and promote cooperation through mutual commitment than secular beliefs alone. Religious trust generally carries over to other beliefs and actions that the participants may affiliate with their ritualized religious beliefs, including cooperative works, charity, economic exchange, and warfare. As cooperating groups increase in size and expand, they come into conflict, competing with other groups for territory and other resources. The growing scope and cost of religious beliefs is both cause and consequence of this increasing group cooperation, expansion, and competition.

8.8   Religious Rituals and the Rise of Civilizations

Collective commitment to the absurd is the greatest demonstration of group love that humans have devised. Unlike other creatures, humans define the groups to which they belong in abstract terms. Often they kill and die not in order to preserve their own lives or those of the people they love, but for the sake of an idea—the conception they have formed of themselves. Call it love of group or God, it matters little in the end (Atran, 2011; see Durkheim, 1912/1995). This is "the privilege of absurdity; to which no living creature is subject, but man only'" of which Hobbes wrote in *Leviathan* (1651/1901; see Gray, 2010).

It is in religious rituals that supernatural agents, through their surrogates and instruments, most strongly insinuate themselves in people's affections. The ceremonies repetitively occur to make highly improbable, and therefore socially unmistakable, displays of mutual commitment. Within the congregation's coordinated bodily rhythms (chanting, swaying) and submission displays (bowing, prostrating) individuals commune with, and signal giving over part of their being to, the intensely felt existential yearnings of others. This demonstration, in turn, conveys the intention or promise of self-sacrifice by and toward others (charity, care, defense, support) without any specific person or situation necessarily in mind. Profession of religious belief and adherence to its costly rituals is a convincing statement of open-ended social commitment.

Archaeological research that focuses on the coevolution of ritual and society reveals that religious rituals became much more formal, elaborate, and costly as societies developed from foraging bands to chiefdoms to states (Marcus & Flannery, 2004). In ancient Mexico, for example, the nomadic egalitarian patterns of hunter-gatherer bands (before 4000 B.P., or before the present) selected for informal, unscheduled rituals from which no one was excluded. Much the same goes for contemporary hunter-gatherers, such as the Kung of Africa's Kalahari Desert, whose primary religious rituals, such as trance dancing, include everyone in the camp and are organized in an ad hoc manner depending on the contingencies of rainfall, hunting prospects, illnesses, and so forth (Lee, 1979).

With the establishment of permanent villages and multivillage chiefdoms (4000–3000 B.P.), main rituals were scheduled by solar and astral events and managed by social achievers ("big men" and chiefs). This also appears to be the case for predynastic Egypt (6000–5000 B.P.) and China (4500–3500 B.P.), as well as for nineteenth-century chiefdoms of Native North Americans. After the state formed in Mexico (2500 B.P.), most important rituals were performed by a class of trained full-time priests, materially subsidized by society, using religious calendars and occupying temples built at enormous cost in labor and lives. This is also true for the state-level societies of ancient Mesopotamia (after 5500 B.P.) and India (after 4500 B.P.), which, like their Mesoamerican counterparts, also practiced costly and fearsome human sacrifice.

Consider also what is arguably the first comparative study of society over time. In the fourteenth century, the great historian Ibn Khaldūn examined different waves of Islamic invasion of the North African Maghreb and the ensuing fate of their dynasties. He found that "dynasties of wide power and large royal authority have their origin in religion based on ... truthful propaganda [that is, with demonstrated commitment]. ... Superiority results from group feeling ... individual desires come together in agreement, and hearts become united. ... Mutual cooperation and support flourish. As a result, the extent of the state widens, and the dynasty grows" (Khaldūn, 1377/1989). Contemporary studies indicate that Islam's spread into Kenya and other parts of sub-Saharan Africa is associated with Muslim rituals (fasting, abstention from alcohol, from adultery, and from eating pork, etc.), drawing people into tighter networks of trust that facilitate trade and economic success (Ensminger, 1997). Similar considerations apply to the current growth of the Protestant evangelical movement in Africa, Asia, and Latin America (Freston, 2001).

Religious beliefs and obligations mitigate self-interest and reinforce trust in cultural norms, like the Ten Commandments or the "golden rule," by conferring on them supernatural authorship, or *sacredness*, and by enforcing them through supernatural punishment, or *divine retribution*. Sacred beliefs and values are invariably associated with taboos: nonnegotiable prohibitions on beliefs and behaviors that transgress the sacred. Punishment for transgressing taboos provides concrete markers and proof of the meaning and importance of what is sacred for society. Together, sacred values and taboos bound moral behavior at the most basic level of conduct in society (sex, diet, dress, greeting) and at the most general level (warfare, rule, work, trade). Along with religious rituals and insignia, these bounds strongly identify one cultural group as different from another, reinforcing interactions with one's "own kind" while distancing the group from others.

The ancient Hebrew Kingdom of Judah, to take an example, used circumcision, dietary laws, prohibition against Sabbath work, and other ritual rules and commandments to mark off their belief in the ineffable one true God with no name. This enabled the alliance of Hebrew hill tribes to set themselves apart from coastal peoples (Philistines, Canaanites) and to pull themselves together to withstand conquest and fragmentation by stronger invaders (Egyptians, Babylonians; Sweeney, 2001). Violating the Sabbath and idolatry were considered the gravest of norm violations and punishable by death. These were the most arbitrary markers of collective identity relative to the concrete needs of social life shared with other groups (in contrast to taboos against stealing, adultery, murder, and the like). Willful disregard of them was considered a strong signal of personal sin and rebellion. If left unchecked, rebellion could spread to the whole body politic and spell chaos and ruin, especially in a competitive situation of constant warfare against other groups: "For rebellion is as the sin of witchcraft, and stubbornness is as iniquity and idolatry" (Samuel 15:23).

For the Hebrews, as for the Catholic Church and its Holy Inquisition in its fight for survival against the Protestant Reformation, intolerance for religious deceivers "who fake miraculous signs" or who secretly entice defection "to serve other gods" was extreme: "the dreamers of dreams shall be put to death," and it is incumbent on those who find them out to assist in their execution, whether "your brother … your son or daughter … the wife … your friend" (Deuteronomy 13:6, 13:15). This religious prescription expresses the primacy of large-scale corporate society over kin and tribal loyalties and is a hallmark of state-level and trans-state societies. Recall that *Islam* means "submission" of family and tribe to the larger community under God.

Thus, in the course of human history, moral religions requiring costly commitments made large-scale cooperation possible between genetic strangers, people who weren't kin and kith, especially cooperation to compete in war. A quantitative cross-cultural analysis of 186 societies found that the larger the group, the more likely it culturally sanctioned deities directly concerned with human morality (Roes & Raymond, 2003). Another survey of 60 societies reveals that males in warring societies endure the costliest religious rites (genital mutilation, scarification, etc.), which "ritually signal commitment and promote solidarity among males who must organize for warfare" (Sosis, Kress, & Boster, 2007).

Recent surveys and experiments by colleagues Jeremy Ginges, Ian Hansen, and Ara Norenzayan with Palestinians and Israeli settlers in the West Bank and Gaza show that frequency of involvement in religious rituals predicts support for martyrdom missions (for Palestinians, suicide attacks; for settlers, support for the 1993 Hebron mosque massacre by a religious immigrant). This relation is independent of mere expressions of religious devotion (amount of prayer). Similar findings were obtained for representative samples of Indian Hindus, Russian Orthodox, Mexican Catholics, British Protestants, and Indonesian Muslims. Greater ritual participation predicts both declared willingness to die for one's god or gods and belief that other religions are responsible for problems in the world (Ginges, Hansen, & Norenzayan, 2009).

To the extent that oppression of religious minorities and leaders results in broadcasting their costly commitments, then such oppression can help rather than hinder the growth of their message and following. Historical studies suggest that early Christianity spread to become the majority religion in the Roman Empire through displays of costly commitment, such as martyrdom and charity: for example, risking death by caring for non-Christians infected with plague (Stark, 1997). In the case of the civil rights movement, the mediatization of police brutality ultimately worked to the good, by sensitizing America's political establishment and mainstream population to the oppressive plight of minorities.

So publicizing martyrdom actions is a good idea if you want the attraction of martyrdom to increase, but a bad idea if you want it to end.

8.9 Cultural Group Selection

Our species' heavy reliance on social learning spontaneously gives rise to norms and informal institutions (stable equilibria), which vary in their group-level competitive properties. Ecological and social pressures, especially with the spread of agriculture, favor norms and institutions that strengthen and extend the social spheres of cooperation and trust while sustaining internal harmony. Deep commitments to certain kinds of religious beliefs and practices can cement both adherence to prosocial norms and a willingness to sanction norm violators, thereby increasing group solidarity and competitiveness with other groups. Religious beliefs and practices, like group-beneficial norms, can spread by competition among social groups in several ways, including warfare, economic production, and demographic expansion. Such cultural representations can also spread through more benign interactions, as when members of one group preferentially acquire behaviors, beliefs, and values from more successful groups. These processes of cultural group selection have both theoretical (Boyd & Richerson, 2002) and empirical grounding (Henrich, 2004).

To illustrate cultural group selection both via the imitation of more prestigious groups and via direct economic competition, consider the well-documented case of three adjoining populations: the Itzà Maya of Guatemala's Petén lowlands, Spanish-speaking Ladino immigrants from diverse regions, and Q'eqchi Maya who arrived in clusters of families and neighbors from the highlands (Atran, Medin, Ross, & Lynch et al., 2002). Among the Itzà, one important predictor of sustainability is their consensus on supernatural (as opposed to human) forest preferences. This cultural consensus about which species are most valuable and worthy of protection accords well with the anthropogenic character of the forest in the Classic era of Maya civilization. The researchers' hypothesis is that spirit preferences represent a summary of experience accumulated over generations. Itzà believe spirits to be "guardians" of the forest. Spirits help people who do not harm the survival prospects of certain species (as spirits see those prospects). Hurting the forest can result in accidents, illness, and worse, such as punishment. This research team has witnessed Itzà, bitten by deadly pit vipers, refuse to be taken for anticoagulant treatment until they venture into the forest to ask spirits for guidance or forgiveness. It matters little if supernatural threats are real or not: if people believe in them, threats of punishment become real deterrents (Durkheim, 1912/1995).

Evidence indicates that much of this knowledge is being transmitted to Ladinos. Experimental elicitations show that Itzà knowledge predicts relative success in short and long-term agro-forestry. By attending to Itzà models of behavioral success in agro-forestry, and to Itzà stories that embed that behavior in context, prestigious Ladinos have managed to acquire a subset of Itzà knowledge of the ecological relationships between humans, animals, and plants. Social network analysis suggests how this knowledge and practice has spread through the Ladino community. It seems the initial lack of any communal Ladino religion or corporate structures, combined with the uncertainty created by immigration into a novel environment, made Ladinos open to learning from Itzà (Atran & Medin, 2008).

In contrast to Ladinos, migrant Q'eqchi, who have strong and highly ritualized religious institutions, pay little heed to Itzà. The Q'eqchi retain allegiances only to the spirits of their native highlands and have no knowledge of Itzà beliefs. Q'eqchi send delegations back to the highlands to consult deities there when they have agricultural troubles in the lowlands. Q'eqchi mental models of the forest are correspondingly poor, as are their associated agro-forestry practices, which are commercially oriented and unsustainable.

These divergent beliefs mean the Q'eqchi are now spreading more rapidly than the other two groups. In fact, Q'eqchi practices are well adapted to present "open-commons" conditions in Guatemala, which encourage massive immigration from the overcrowded highlands into the ecologically fragile lowlands. There is little incentive to avoid destructive practices: if one part of the forest is destroyed, Q'eqchi simply migrate. In this context, Itzà practices are currently maladaptive. By making costly commitments to preserve the forest, Itzà make it easier for the highly ritualized, corporately disciplined Q'eqchi to exploit it. Thus, Itzà may be subsidizing their own cultural extinction in the competition among ethnic groups.

The cultural evolution of the interrelationship between religious beliefs and costly rituals and devotions is apparent in the study of eighty-three utopian communes in the nineteenth century noted above (Sosis & Bressler, 2003). Religious groups with more costly rituals were more likely to survive over time than religious groups with fewer costly rituals. Differential group survival yielded an increase in the mean number of costly rituals per group over time. The above theory and evidence suggests that such rituals and devotions likely generated greater commitment and solidarity within groups. Indeed, members and leaders in the study explicitly acknowledged that costly demands increased members' religious commitment.

The relation of rituals to prosocial behavior toward in-group members is demonstrated in a variety of ways. Among Israeli kibbutzim (cooperatives), individuals from religious kibbutzim cooperated more in behavioral experiments than those from non-religious ones, with increased cooperativeness of religious members attributed to greater ritual participation (Sosis & Ruffle, 2003). Religious kibbutzim also economically outperform secular ones (Fishman & Goldschmidt, 1990). Surveys of Palestinians and Israeli settlers in the West Bank and Gaza reveal that a person's frequency of attendance at religious services predicts support for martyrdom missions. This relation is independent of time spent in prayer. Similar findings emerge for representative samples of religious Indians, Russians, Mexicans, British and Indonesians: greater ritual attendance predicts both declared willingness to die for one's deities, and belief that other religions are responsible for problems in the world (Ginges et al., 2009). Finally, a study of 60 small-scale societies reveals that males from groups in the most competitive socioecologies (measured by those with frequent warfare) endure the costliest rites (genital mutilation, scarification, etc.), which "ritually signal commitment and promote solidarity among males who must organize for warfare" (Sosis et al., 2007).

Cultural group selection shapes religious beliefs and rites to manipulate our psychology and increase solidarity and commitment. Such patterns, observed across history and in the anthropological record, re-emerge in today's terrorist groups (Atran, 2010). Even avowedly secular national and transnational movements retain many agentive (anthropomorphic) and transcendental (sacred) aspects of traditional religions: nations ritually mourn, rejoice, and demand sacrifice, and the "naturalness" of causes that defy prior human history (universal justice, equality, liberty) is anything but empirically or logically self-evident. Since this chapter argues that sociopolitical complexity coevolved with both commitment-inducing rituals and beliefs in high moralizing gods, these efforts also dovetail with recent work indicating that cultural selection, driven by differences in sociopolitical complexity, is crucial to understanding the global distribution and diversity of languages (Currie & Mace, 2009).

8.10 Conclusion

In sum, religion, as an interwoven complex of rituals, beliefs, and norms, plausibly arises from a combination of (1) the mnemonic power of counterintuitive representations, (2) our evolved willingness to put faith in culturally acquired beliefs rooted in the commitment-inducing power of devotions and rituals, and (3) the selective effect on particular cultural complexes created by competition among societies and institutions (Atran & Henrich, 2010). None of these evolved for religion, per se. The mnemonic power of minimally counterintuitive representations appears to be a byproduct of our evolved expectations about how the world works and our fitness-enhancing requirement to pay attention to anomalies. The faith we sometimes place in culture over our own experience and intuitions is a cognitive adaptation, resulting from our long dependence on vast bodies of complex cultural knowledge. Reliance on costly displays evolved to provide partial immunity against manipulation. The power of rhythm and synchrony in ritual to build solidarity likely arises from our imitative and ToM abilities. Cultural evolution, driven by competition among groups, exploits each of these cognitive processes to fashion sets of counterintuitive beliefs, rituals, and norms that spread by inter-group transmission, conquest, or reproductive differentials. As a result, for large-scale societies, these complexes tend to include potent supernatural agents that monitor and incentivize actions, expanding the sphere of cooperation, galvanizing solidarity in response to external threats, deepening faith, and sustaining internal harmony (Atran & Henrich, 2010).

Significant advances in the study of religious cognition, the transmission of culture, and the evolution of cooperation are all relatively recent. Bringing these new insights, in combination with older ideas, to bear on phenomena as complex as moralizing religions and large-scale societies will be an ongoing challenge. The argument and evidence presented here provide a plausible scenario showing how synthetic progress is possible. More rigorous study is needed regarding the evolved psychology and cultural processes associated with the role of counterintuitive agents and costly rituals in scaling up the scope of trust and exchange; of sacred values and taboos in sustaining large-scale cooperation against external threats; and of maintaining social and political causes that defy self-interest (Atran, Axelrod, & Davis, 2007; Ginges, Atran, Sachdeva, & Medin, 2011). Empirical research that combines in-depth ethnography with both cognitive and behavior experiments among diverse societies, including those lacking a world religion, is crucial to understanding how religion influences our cognition, decision making, and judgment. The formal modeling of cultural evolutionary processes should be combined with historical and archeological efforts to apply these emerging insights to broad patterns of history. These joint efforts should further illuminate the origins and development of religions, and the cooperation and conflicts they engender. There may be no more urgent study needed in the world today.

References

Argyle, M., & Beit-Hallahmi, B. (2000). *The social psychology of religion.* New York: Routledge.

Atran, S. (1989). Basic conceptual domains. *Mind & Language, 4,* 7–16.

Atran, S. (2002). *In gods we trust: The evolutionary landscape of religion.* Oxford: Oxford University Press.

Atran, S. (2010). *Talking to the enemy: Violent extremism, sacred values, and what it means to be human.* London: Penguin.

Atran, S. (2011, March 29). Why war is never really rational. *Huffington Post.* Retrieved from http://www.huffingtonpost.com/scott-atran/libya-war-obama_b_842298.html.

Atran, S., Axelrod, R., & Davis, R. (2007). Sacred barriers to conflict resolution. *Science, 317,* 1039–1040.

Atran, S., & Henrich, J. (2010). The evolution of religion: How cognitive byproducts, adaptive learning heuristics, ritual displays, and group competition generate deep commitments to prosocial religions. *Biological Theory, 5,* 18–30.

Atran, S., & Medin, D. (2008). *The native mind and the cultural construction of nature.* Cambridge, MA: MIT Press.

Atran, S., Medin, D., Ross, N., Lynch, E., Vapnarsky, V., Ucan Ek', E., et al. (2002). Folkecology, cultural epidemiology, and the spirit of the commons: A garden experiment in the Maya lowlands, 1991–2001. *Current Anthropology, 43,* 421–450.

Atran, S., & Norenzayan, A. (2004). Religion's evolutionary landscape: Counterintuition, commitment, compassion, communion. *Behavioral and Brain Sciences, 27,* 713–770.

Barfield, T. (2003, June 26). Afghan customary law and its relationship to formal judicial institutions. Washington, DC: United States Institute for Peace. Retrieved from http://www.usip.org/files/file/barfield2.pdf.

Bargh, J., & Chartland, T. (1999). The unbearable automaticity of being. *American Psychologist, 54,* 462–479.

Barrett, J., & Keil, F. (1996). Conceptualizing a nonnatural entity: Anthropomorphism in god concepts. *Cognitive Psychology, 31,* 219–247.

Barrett, J., & Nyhof, M. (2001). Spreading nonnatural concepts. *Journal of Cognition and Culture, 1,* 69–100.

Beck, W. (1992). Aboriginal preparations of Cycad seeds in Australia. *Economic Botany, 46,* 133–147.

Bedeviling: Did Satan rear his ugly face? (2001, September 14). *Philadelphia Daily News.* Retrieved from http://phillydailynews.newspaperdirect.com.

Bering, J. (2009, October 9). God's in Mississippi, where the getting' is good [blog post]. *Scientific American.* Retrieved from www.scientificamerican.com.

Bloom, P., & Veres, P. (1999). The perceived intentionality of groups. *Cognition, 71,* B1–B9.

Boyd, R., & Richerson, P. (2002). Group beneficial norms can spread rapidly in a structured population. *Journal of Theoretical Biology, 215,* 287–296.

Boyer, P. (1994). *The naturalness of religious ideas: A cognitive theory of religion.* Berkeley, CA: University of California Press.

Byrne, R., & Whiten, A. (1998). *Machiavellian intelligence: Social expertise and the evolution of intellect in monkeys, apes, and humans.* Oxford: Oxford University Press.

Currie, T. E., & Mace, R. (2009). Political complexity predicts the spread of ethnolinguistic groups. *Proceedings of the National Academy of Sciences of the United States of America, 106*(18), 7339–7344.

Darwin, C. (1871). *The descent of man, and selection in relation to sex.* London: John Murray.

Durkheim, E. (1995). *Elementary forms of religious life*. New York: Free Press. (Original work published 1912)

Ensminger, J. (1997). Transaction costs and Islam: Explaining conversion in Africa. *Journal of Institutional and Theoretical Economics, 153*, 4–28.

European Commission. (2005, June 1). Social values, science and technology. *Special Eurobarometer* (225)63.1. Retrieved from http://ec.europa.eu/public_opinion/archives/ebs/ebs_225_report_en.pdf.

Fehr, E., & Fischbacher, U. (2003). The nature of human altruism. *Nature, 425*, 785–791.

Festinger, L., Riecken, H., & Schachter, S. (1956). *When prophecy fails*. Minneapolis: University of Minnesota Press.

Firth, R. (1963). Offering and sacrifice. *Journal of the Royal Anthropological Institute, 93*, 12–24.

Fishman, A., & Goldschmidt, Y. (1990). The orthodox Kibbutzim and economic success. *Journal for the Scientific Study of Religion, 29*, 505–511.

Frank, A. (1993). *Anne Frank: The diary of a young girl*. New York: Bantam. (Originally published in 1944)

Frank, R. (1988). *Passions within reason*. New York: Norton.

Freston, P. (2001). *Evangelicals and politics in Asia, Africa, and Latin America*. Cambridge: Cambridge University Press.

Gazzaniga, M., Ivry, R., & Mangun, G. (1998). *Cognitive neuroscience: The biology of the mind*. New York: Norton.

Gibbon, E. (1993). *The decline and fall of the Roman Empire (1776–1789)* London: International Book Co.

Ginges, J., Atran, S., Sachdeva, S., & Medin, D. (2011). Psychology out of the laboratory: The challenge of violent extremism. *American Psychologist, 66*(6), 507–519.

Ginges, J., Hansen, I., & Norenzayan, A. (2009). Religious and popular support for suicide attacks. *Psychological Science, 20*, 224–230.

Gray, J. (2010, December). The privilege of absurdity [Review of S. Atran, *Talking to the enemy*]. *Literary Review*. Retrieved from http://www.literaryreview.co.uk/gray_12_10.html.

Guthrie, S. (1993). *Faces in the clouds: A new theory of religion*. Oxford: Oxford University Press.

Hamilton, W., & Orians, G. (1965). Evolution of brood parasitism in Altricial birds. *Condor, 67*, 361–382.

Harper, L., & Sanders, K. (1975). The effect of adults' eating on young children's acceptance of unfamiliar foods. *Journal of Experimental Child Psychology, 20,* 206–214.

Harris, P. (2006). Germs and angels: The role of testimony in young children's ontology. *Developmental Science, 9,* 76–96.

Heider, F., & Simmel, M. (1944). An experimental study of apparent behavior. *American Journal of Psychology, 57,* 243–249.

Henrich, J. (2004). Cultural group selection, coevolutionary processes and large-scale cooperation. *Journal of Economic Behavior & Organization, 53,* 3–35.

Henrich, J., & Gil-White, F. (2001). The evolution of prestige: Freely conferred deference as a mechanism for enhancing the benefits of cultural transmission. *Evolution and Human Behavior, 22,* 165–196.

Henrich, N., & Henrich, J. (2003). *Why humans cooperate: A cultural and evolutionary explanation.* Oxford: Oxford University Press.

Henrich, J., McElreath, R., Barr, A., Ensimger, J., Barrett, C., Bolyanatz, A., et al. (2006). Costly punishment across human societies. *Science, 312,* 1767–1770.

Hobbes, T. (1901). *Leviathan.* New York: E. P. Dutton. (Original work published 1651)

Hume, D. (1955). *An inquiry concerning human understanding.* New York: Bobb-Merrill. (Original work published 1758)

Ibn Khaldūn, A. (1989). *The Muqaddimah: An introduction to history.* Princeton, NJ: Princeton University Press. (Original work published 1377)

Johnson, G. (1987). In the name of the Fatherland: An analysis of kin term usage in patriotic speech and literature. *International Political Science Review, 8,* 165–174.

Kapoggianis, D., Barbey, A. K., Sua, M., Zambonia, A., Kruegera, F., & Grafman, J. (2009). Cognitive and neural foundations of religious belief. *Proceedings of the National Academy of Sciences of the United States of America, 106,* 4876–4881.

Kahn, J. (1999, April 27) Notoriety now for movement's leader. *New York Times.* Retrieved from http://www.nytimes.com/1999/04/27/world/notoriety-now-for-movement-s-leader.htm.

Keillor, G. (1999, June 14). Faith at the speed of light. *Time.* Retrieved from http://www.time.com/time/magazine/article/0,9171,991211,00.html.

Keleman, D. (2004). Are children "intuitive theists"? Reasoning about purpose and design in nature. *Psychological Science, 15,* 295–301.

Kierkegaard, S. (1954). *Fear and trembling and the sickness unto death.* Garden City, NY: Doubleday. (Original work published 1843)

Lee, R. (1979). *The! Kung San: Men, women and work in a foraging society*. Cambridge: Cambridge University Press.

Lester, T. (2002, February). Supernatural selection. *Atlantic Monthly*. Retrieved from www.theatlantic.com.

Lipkind, W. (1940). Carajá cosmography. *Journal of American Folklore, 53*, 248–251.

Malley, B. (2004). *How the Bible works: An anthropological study of evangelical biblicism*. Landham, MD: AltaMira Press.

Marcus, J., & Flannery, K. (2004). The coevolution of ritual and society. *Proceedings of the National Academy of Sciences of the United States of America, 101*, 18257–18261.

Marks, L. (2004). Sacred practices in highly religious families: Christian, Jewish, Mormon, and Muslim perspectives. *Family Process, 43*, 217–231.

Moore, M. (2009, April 24) Falun Gong "growing" in China despite 10-year ban. *Telegraph*. Retrieved from http://www.telegraph.co.uk/news/worldnews/asia/china/5213629/Falun-Gong-growing-in-China-despite-10-year-ban.html.

Newport, F. (2007). *One-third of Americans believe the Bible is literally true: High inverse correlation between education and belief in a literal Bible*. Princeton, NJ: Gallup News Service. Retrieved from http://www.gallup.com/poll/27682/OneThird-Americans -Believe-Bible-Literally-true.apx.

Norenzayan, A., Atran, S., Faukner, J., & Schaller, M. (2006). Memory and mystery: The cultural selection of minimally counterintuitive narratives. *Cognitive Science, 30*, 531–553.

Norenzayan, A., & Hansen, I. (2006). Belief in supernatural agents in the face of death. *Personality and Social Psychology Bulletin, 32*, 174–187.

Norenzayan, A., & Shariff, A. (2008). The origin and evolution of religious prosociality. *Science, 322*, 58–62.

Norris, P., & Inglehart, R. (2004). *Sacred and secular: Religion and politics worldwide*. Cambridge: Cambridge University Press.

Nowak, M., & Sigmund, K. (1998). Evolution of indirect reciprocity by image scoring. *Nature, 393*, 573–577.

Pew Forum on Religious & Public Life. (2008, June). U.S. Religious Landscape Survey. Retrieved from http://religions.pewforum.org/reports#.

Pew Research Center. (2007, May 22). *Muslim Americans: Middle class and mostly mainstream*. Retrieved from http://pewresearch.org/assets/pdf/muslim-americans.pdf.

Pyysiäinen, I. (2003). Buddhism, religion, and the concept of "God." *Numen, 50,* 147–171.

Roes, F., & Raymond, M. (2003). Belief in moralizing gods. *Evolution and Human Behavior, 24,* 126–135.

Salmon, J. (2008, June 24). Most Americans believe in higher power, poll finds. *The Washington Post.* Retrieved from http://www.washingtonpost.com.

Schlesinger, L. (1999). *The Ten Commandments: The significance of God's laws in everyday life.* New York: HarperCollins.

Smith, B. (2010, January 13). Robertson: Haiti "cursed" since Satanic pact. *Politico.* Retrieved from http://www.politico.com

Sosis, R., & Bressler, E. (2003). Cooperation and commune longevity: A test of the costly signaling theory. *Cross-Cultural Research, 37,* 211–239.

Sosis, R., Kress, H., & Boster, J. (2007). Scars for war: Evaluating signaling explanations for cross-cultural variance in ritual costs. *Evolution and Human Behavior, 28,* 234–247.

Sosis, R., & Ruffle, B. (2003). Religious ritual and cooperation: Testing for a relationship on Israeli religious and secular Kibbutzim. *Current Anthropology, 44,* 713–722.

Sperber, D. (1996). *Explaining culture: A naturalistic approach.* Oxford: Wiley-Blackwell.

Stark, R. (1997). *The rise of Christianity.* New York: Harper Collins.

Sweeney, M. (2001). *King Josiah of Judah.* Oxford: Oxford University Press.

Thompson, L., & Hastie, R. (1990). Social perception in negotiation. *Organizational Behavior and Human Decision Processes, 47,* 98–123.

Tremlin, T., & Lawson, E. T. (2006). *Minds and gods: The cognitive foundations of religion.* Oxford: Oxford University Press.

Tversky, A., & Kahneman, D. (1974). Judgment under uncertainty: Heuristics and biases. *Science, 185,* 1124–1131.

Whitson, J., & Galinsky, A. (2008). Lacking control increases illusory pattern perception. *Science, 322,* 115–117.

9 Religion: From Mind to Society and Back

Ilkka Pyysiäinen

9.1 Introduction

Exploring the cognitive basis of the social sciences and trying to ground the social in the cognitive requires taking an explicit stance on reduction(ism) as discussed in philosophy of science. In social science and the humanities, the question of reductionism has been especially salient in the study of religion. This chapter begins with a philosophical analysis of reduction; after that, two relatively new research programs in the study of religious thought and behavior are discussed: the standard model of the cognitive science of religion and approaches based on gene-culture coevolutionary theories. Finally, the question of reductionism is addressed and the possibility of combining multilevel explanations of religious phenomena is evaluated.

9.2 Reductionism and Its Discontents

In the study of religion, *reduction* has usually meant elimination, in the sense that religion is an entity-like whole and sociological and psychological explanations are regarded as eliminating it by the mere act of explaining it. In other words, an explanation is thought to be a substitute for the *explanandum,* or what was to be explained (see Edis, 2008, p. 112). Religion is "explained away" because the proper explanandum is lost when all questions about religion are reduced to sociological or psychological questions (see Idinopulos & Yonan, 1994; Pyysiäinen, 2004b, pp. 1–27). These arguments involve some deep philosophical problems, as will be discussed in what follows.

Philosophers of science distinguish between theory reduction and elimination. In the classical account of theory reduction proposed by Ernst Nagel (1961), reducing an upper-level theory to a lower-level theory means that all its laws can then be deduced from the lower-level theory, of which it is a special case (see McCauley, 2007). Nothing is thereby eliminated. Here, as in the so-called New Wave models of scientific reduction (e.g., Bickle, 1998), the distinction between theory succession over time within some science and the relations of theories from different sciences at some particular point in time has been obscured. Successional and cross-scientific relations are not distinguished from each other; all illustrations of theory elimination are the result of theory succession within a particular science, rather than of cross-disciplinary theory reduction (McCauley, 2009).

There is nothing wrong in principle with the strategy of searching for lower-level mechanisms to account for upper-level patterns. The crucial issue is how we distinguish between different levels and how we understand the relations between lower- and upper-level theories and their application in different sciences (see Craver, 2007; Craver & Bechtel, 2007; Bechtel, 2009; McCauley, 2009; Hedström & Ylikoski, 2010). An important question is whether it is possible to distinguish between different ontological levels in a general manner, irrespective of the context of particular studies and experiments (Bechtel, 2009; McCauley, 2009).

In a classic paper, Oppenheim and Putnam (1958) distinguished between the levels of subatomic particles, atoms, molecules, cells, organisms, and society. The list is not exhaustive and involves many problems, however. It ignores such levels as neural networks and families, for example, while families are not part of galaxies, although they are below galaxies in the taxonomy of levels (Craver, 2007). McCauley thus argues that different levels might be better differentiated on the basis of their relative complexity, rather than the size of their respective objects. Higher levels are more organized and involve more complex causal mechanisms. The altitude of the level of analysis is directly proportional to the complexity of the systems it deals with. The altitude of the level is also inversely proportional to the size of the domain of the events in question: psychology, for example, is a higher level than biology, and it deals with only some of the phenomena within the realm of biology. Finally, the age of the relevant phenomena decreases bottom-up: the lower the level, the longer its phenomena have been around in the world (McCauley, 1986, 2007).

A more detailed description is provided by Craver (2007; see also Wimsatt, 1994), whose analysis focuses on the levels of causal-mechanical explanation. He first distinguishes between levels of science (e.g., products, units) and of nature (e.g., causation, size, composition). Levels of composition include the levels of mechanisms, among other things. Levels of mechanisms are context sensitive; there are no universally distinct levels. In causal-mechanical explanation, mechanisms that produce or support the phenomena to be explained are specified (McCauley, 1986; Wimsatt, 1994; Craver, 2007; Bechtel, 2008).

When we, for example, want to explain a particular type of religious experience, levels of sociocultural processes, emotions, and cognitive mechanisms as well as neural processes and their molecular basis are involved. At each level, the explanandum is different. There is, for instance, no direct molecular-level explanation of sociocultural phenomena (see Pyysiäinen, in press). It thus seems that a full explanation of a particular type of religious experience would have to specify mechanisms that extend from the molecular level all the way up to the sociocultural level. This, of course, is mostly a theoretical ideal; in practice, only a multidisciplinary research group could even try to achieve something like this.

This also raises the question of the nature of relationships between different levels. Craver and Bechtel (2007), for example, argue that there is no inter-level causation. Causation takes place within levels; between levels, there are constitutional relationships. What is often regarded as top-down *causation* does not involve top-down *causes*; there are only mechanistically mediated effects, which are hybrids of constitutive and intra-level causal relations (Craver & Bechtel, 2007, pp. 561–62; Bechtel, 2009). Molecular-level processes, for instance, do not cause cultural-level phenomena, the constitution of which yet also includes molecular-level processes (see Sun, Coward, & Zenzen, 2005).

From the point of view of the humanities and social sciences, it is often argued that lower-level explanations are uninformative and only provide unnecessary details (see Garfinkel, 1981). If we want to explain why person X performed ritual Y, lower-level explanations are of no help because they account for different kinds of explananda. It is one thing to explain the molecular basis of phenomena such as memory, for example, and quite another to explain intentional behavior (Kamppinen, 2009; cf. Eronen, 2009).This does not imply that lower-level explanations are somehow mistaken and "reductionist," in the negative sense of the term. Lower-level mechanisms are part of the constitution of religious experiences, although there is no direct causal relationship between them and sociocultural phenomena.

Mechanistic explanation refers to a strategy of specifying the mechanism responsible for producing or sustaining a phenomenon. Such explanation consists of describing the parts, operations, and organization of a mechanism and showing how the mechanism realizes the phenomenon to be explained (Bechtel, 2008, p. 49; Craver, 2007, p. 5). Mechanistic explanations do not imply any "covering laws"; it is enough that an explanatory generalization is stable, in the sense that the specified relation between the cause and the effect holds under a range of conditions, which are generally not universal (Craver, 2007, p. 99). One way to describe a mechanistic explanation is to say that a certain mechanism produces a probability distribution over possible outcomes and shows that the explanandum is an instance of one of those possible outcomes (Railton, 1978, as applied in Craver, 2007, p. 40). A typical explanandum is the activity or behavior of the mechanism; the *explanans* is a description of the internal organization or structure of the system in terms of lower-level entities and activities.

According to Craver, causality means causal relevance in the sense that any given X is causally relevant with regard to Y if an "ideal intervention I on X is such a change in the value of X that it also changes Y *only via* the change in X (Craver, 2007, pp. 95–96; see also Woodward, 2003). The ideal intervention need not be actually made by humans; it suffices that the intervention is possible in principle. It thus is possible to say what would have happened if the cause of the event had been manipulated by an ideal intervention. The variables in question must have measureable values but it is enough that the intervention or manipulation is conceptually possible (Woodward, 2003, pp. 94, 114, 127–33).

Apart from philosophy, the idea of mechanistic explanation has recently also received attention in the social sciences; it has been discussed mostly by social scientists themselves, whereas biological and neuroscientific explanations have been studied by philosophers of science (Hedström & Ylikoski, 2010). Rather than providing explicit methodological principles for empirical research, mechanistic accounts of explanation and causality offer a way of exploring and conceptualizing what is involved in a scientific explanation of religion. Such analyses could help empirical scientists avoid fruitless debates and dead ends, like the one on reductionism, in the study of religion.

9.3 Two Traditions in the Study of Religion

9.3.1 The Standard Model of the Cognitive Science of Religion
During the past two decades, two new research programs on religious phenomena have emerged: the "standard model" of the cognitive science of religion (Boyer, 2005b), and approaches based on the so-called gene-culture coevolutionary theories of cultural transmission, which draw from

methodology in population genetics (Boyd & Richerson, 1985; Henrich & Boyd, 2002; Richerson & Boyd, 2005; Bell, Richerson, & McElreath, 2009; Henrich & Henrich, 2007; Henrich, 2009; Henrich, Ensminger, McElreath, Barr et al., 2010a). Within the standard model, religion is explained by cognitive mechanisms, whereas coevolutionary theories largely focus on modeling the effects of various biases in social behavior. In the following, we refer to these latter theories as *coevolutionary theories*, although this may not be entirely accurate.

Differences notwithstanding, there is also some overlap between these two approaches; their relative potential has recently been briefly evaluated using modeling (Claidière & Sperber, 2007; see also Boyer, 1994, pp. 263–96). A closely related line of research concerns the evolution of religion as a by-product of or adaptation for cooperation (see Boyer & Bergstrom, 2008; Sosis, 2009; Pyysiäinen & Hauser, 2010). Representatives of different new (and old) disciplines now seek interdisciplinary cooperation in explaining religious phenomena, as testified by, for example, the launching of the new journal *Religion, Brain, and Behavior* in 2010 (see Stausberg, 2009).

The cognitive and evolutionary approaches have meant a new turn in the study of religion, traditionally governed by humanistic attempts at interpreting the meaning of religious "symbols" either philosophically or anthropologically (cf. Sperber, 1975). Explanatory strategies have been accused of reductionism on the grounds that they neglect the meaning religious beliefs and practices have for people (e.g., Eliade, 1969; Geertz, 1973; Clarke & Byrne, 1993; Stark & Finke, 2001; cf. Holland & Quinn, 1987). As Boyer (1994, p. 295) puts it, the lack of humanistic significance may be the price we have to pay for causal relevance.

The cognitive science of religion emerged in the 1990s as a study of the mental representation of ritual structures (Lawson & McCauley, 1990) and of the role of implicit cognition in cultural transmission (Boyer, 1994; see also Barrett, 2004; Tremlin, 2006; Pyysiäinen, 2008, 2009). Boyer critically examines the idea of "exhaustive cultural transmission," in which people supposedly have religious and other cultural ideas because these are learned from culture. Boyer shows that not all ideas are based on such cultural transmission; instead, much of our mental contents are based on intuitive inferences and assumptions that are not learned or even conscious, yet nevertheless guide behavior. Thus, many culturally and cross-culturally recurrent ideas are based on the intuitive ways the human mind works, not on explicit transmission (Boyer, 1994, pp. 87, 263–266). Communication processes are not semiotic but inferential. In communication, we do not produce coded versions of mental representations that could then simply be decoded by the receiver, a point made *in extenso* by Sperber and Wilson (1986). Rather, we produce material tokens (utterances, texts, etc.) that then modify the audience's mental representations in roughly predictable ways (Boyer, 1994, p. 284).

A central role is here played by the idea of intuitive ontology, the human tendency to assign entities into categories which then allow for implicit inferences concerning the properties of the entities in question. If, for example, an entity can cry, it obviously belongs to the category of persons and thus has the standard properties of persons. Such assumptions as, for example, "entities in the category of persons can think" need not be culturally transmitted; they are implicitly assumed (Boyer, 1994, pp. 91–124). Consequently, models of cultural transmission that neglect the role played by implicit cognition and the inferential nature of communication are incomplete (Boyer, 1994, pp. 270–78). The observation that human intuitions canalize the spread of cultural ideas is the basis for the "epidemiology of beliefs" (or of representations) that Boyer's work, which is the study of the differential spread of ideas in populations, represents (Sperber, 1985, 1996, 2006; Sperber & Hirschfeld, 2004).

With regard to religion, the most important cognitive mechanisms are those dedicated to the representation of agency (see Pyysiäinen, 2009). Agents are organisms whose behavior can be predicted by attributing to them conscious beliefs and desires (Dennett, 1993, pp. 15–17). Agency consists of the components of animacy (liveliness, self-propelledness), and of mentality (beliefs and desires) (Pyysiäinen, 2009, pp. 13–14). Animacy refers to liveliness expressed as self-propelledness. *Liveliness* is a compositional concept and covers such things as, for example, reproduction, certain stability in the face of small changes, metabolism, interaction with the environment, and the fact that components of living systems depend on one another in preserving identity. Self-propelled movement seems to be a crucial factor in animacy. Only moving entities can have a mind and consciousness, which thus may have their root in motor action (see Gallese & Metzinger, 2003; Pyysiäinen, 2010c).

Mentality, for its part, has probably evolved from action control that has been detached from actual bodily movements. When any given sensory input can be responded to in differing ways, the organism in question has internal states, in the sense of possible motor actions from which the organism has to choose one to execute. Originally merely reactive structures thus can lead to the emergence of adaptive controllers that can handle tasks with a large number of degrees of freedom. Such a system is able to plan ahead, relying on its internal world model and by temporarily disconnecting the motor output (Cruse, 2003; see also Hurley, 2008). This has also led from implicit responding to explicit knowledge (Sun, 2002; Hélie & Sun, 2010).

Mentality refers to intentional states (beliefs and desires). Humans have the ability to attribute intentional states to others, which is referred to as a "theory of mind," "mentalizing," or "mindreading" (Nichols & Stich, 2003; Saxe, Carey, & Kanwisher, 2004). *Mindreading* means attributing intentional states to others; these cannot be perceived, but only *inferred* from various kinds of cues. Humans not only attribute intentions to others but can also form recursive chains of intentional attitudes, so that one attitude is embedded in another. Dennett (1993) refers to this as "orders of intentionality." I may believe that John *wants* Mary to *understand* that Bill *believes* Linda *loves* Bill, for example (Pyysiäinen, 2009, p. 19).

There is now evidence that such brain areas as the superior temporal sulcus and the temporoparietal juncture are most responsive when subjects view real people (and not, for example, animated cartoon characters). It has been shown that an animated figure that closely resembles but is not quite human is experienced as disturbing and even felt to be physically revolting (Hari & Kujala, 2009, p. 460). There seem to be neural circuits that respond specifically to patterns of intentional *biological* motion (Mar, Kelley, Heatherton, & Macrae, 2007).

However, it may also be that the temporoparietal junction is sensitive not only to biological motion, but also to any stimuli which otherwise signal intentional activity (Gallagher, Happé, Brunswick, Fletcher et al., 2000). In a recent experiment, for example, animal and non-living stimuli were processed by subjects in different parts of the brain in both blind and sighted persons; the categorization between living things and non-living kinds thus is part of the hard-wired organization of brain and independent of visual experience of apparently intentional motion (Mahon, Anzellotti, Schwarzback, & Zampini, 2009).

As to the supernaturalness of agents, this is based on violations of intuitive expectations related to intuitive ontological categories. (For more on this issue, also see chapter 8 by Scott Atran in this volume.) Boyer's theory of counterintuitiveness of representations explains the nature and construction of supernatural representations in terms of the human cognitive architecture. It also explains the differential spread of these representations by the constraints provided by attention and the human memory systems: representations that only involve one violation of intuitive expectations are at once salient and yet relatively easy to process in mind (Boyer, 2001, pp. 85–87; Barrett, 2000, pp. 21–30). Therefore, they are likely to be remembered and thus also to be passed on to others (see Barrett, 2004; but cf. Shtulman, 2008).

Human agent detection is "hypersensitive" in the sense that even minimal cues can trigger the postulation of agency (Guthrie, 1993; Barrett, 2000, pp. 31–32; see also Saler & Ziegler, 2006). Agent-causality is our preferred mode of explanation because it is better to mistake a non-agent for an agent than the other way around (Guthrie, 1993; Barrett, 2000). This might be because, for our ancestors, the most important threat has been other human agents (Alexander, 1979, pp. 222–24). We are therefore inclined to see agents even where there are none. The cognitive mechanism of agency detection produces a lot of false positives, which is not counterproductive insofar as we are able to discard the false positives in the face of further evidence. This is the "better safe than sorry" principle (Guthrie, 1993): it is better to be prepared to deal with (a possibly hostile) agent than mistake an agent as a non-agent. This fact about human cognition helps explain why beliefs about supernatural agents are felt to be intuitively compelling and thus are easy to spread.

9.3.2 Coevolutionary Theories

Coevolutionary theorists' focus on modeling behavioral traits in populations in no way denies the importance of cognitive factors; they are merely regarded as insufficient as such (e.g., Gervais & Henrich, 2010). Cultural learning, for example, takes place when individuals imitate others, which leads to a cumulative "ratchet effect": once a new cultural trait emerges, it can no longer be lost if the population is large enough to make the trait persist. In small, isolated populations, cultural achievements can be lost if the necessary experts die without being replaced by new successors (Tomasello, 1999; Henrich, 2004).

Coevolutionary theories emphasize five factors that shape the selection of ideas for cultural transmission (Boyd & Richerson, 1985; Richerson & Boyd, 2005):

- random variation caused by accidental errors in transmission
- loss of variants in small samples, comparable to genetic drift in biology
- guided variation, or adjusting phenotypes to environment through learning and rational calculation
- biased transmission, in the sense that cultural transmission favors certain variants. This includes:
  - *direct bias:* intrinsic value of a variant
  - *frequency-dependent bias:* people often imitate the majority

- *indirect bias:* choosing a variant because it co-occurs together with a favorable variant (e.g., choosing prestigious individuals as role models)
- *natural selection:* some variants have an effect on the inclusive fitness of people.

Henrich and Henrich (2007, pp. 64–68) distinguish between prestige bias, conformist bias, and punishing behaviors as evolved behavioral strategies of trying to cope with differing kinds of situations. *Prestige bias* means that people are selective regarding the sources from whom they adopt ideas and beliefs. We use various kinds of cues related to skill, success, and prestige to figure out which individuals are most likely to have useful ideas, beliefs, values, preferences, or strategies to be gleaned through observation. Social norms are preferentially learned from individuals who share the same ethnic markers, such as dialect or dress, for instance. Prestigious individuals are often trusted and imitated even when their domain of prestige is unrelated to the opinion domain in question (Henrich & Henrich, 2007, pp. 11–12, 20).

Conformist bias means that people often copy the behaviors, beliefs, and strategies of the majority. When, for example, the accuracy of information acquired through individual learning decreases, reliance on conformist transmission increases. *Informational conformism* means that people actually change their beliefs and opinions, while *normative conformism* means that people only alter their superficial behavior but not their true beliefs and opinions. Conformist cultural transmission can maintain behaviors only when they are neutral, not too costly, or if the costs are ambiguous (Henrich & Henrich, 2007, pp. 22–30, 66). Conformism works because it makes intra-group cooperation possible, and groups that are capable of cooperation tend to outperform in selection groups that are not (Richerson & Boyd, 2005; Bell et al., 2009; Henrich et al., 2010a). Prestige bias can either support conformism or compete with it, depending on the status of the prestigious persons imitated. Both biases, however, lead members of a social group to adopt similar mental representations.

Punishing refers to any behavior by which people signal disapproval of others' behavior or thinking and force them to pay a cost for their norm violation. People shun norm violation because they do not want be punished. Thus, they may behave in accordance with social norms even if they do not truly believe in the rightness of these norms. Punishing is a cheap strategy because it is possible to punish violators of a norm even without adhering to the norm itself. It works especially when it is not necessary to punish violators too often (Henrich & Henrich, 2007, pp. 64–67).

When this does not work, another option is to punish those who refuse to punish violators. As norm violation is rare, and failing to punish violators rarer still, it is necessary to punish those who refuse to punish only when the first two conditions are met. Punishing those who refuse to punish is cheap in the sense that it is not needed too often. This strategy can get people "locked in" on almost any social norm. The norm itself can be beneficial, neutral, or even outright harmful, and yet become stabilized (Henrich & Henrich, 2007, pp. 66–68). One example is the notorious practice of giving daughters clitoridectomies. Families that refuse to do this get a bad reputation in the social group and thus run the risk of not receiving aid from others when in need (Henrich & Henrich, 2007, p. 71). Conformist bias, prestige bias, and punishment thus help explain why explicit commitment to theological doctrines can survive, although the doctrines are not intuitively compelling or cognitively optimal (Henrich & Henrich, 2007; see also Pyysiäinen, 2010a; cf. Whitehouse, 2004).

Henrich and Boyd (2002) argue that the crucial issue in cultural transmission is selection: some ideas and practices survive because of the aforementioned biases in cultural transmission. Although Sperber's "attractors" are also important, the final outcomes are determined by selection alone. By *attractors*, Sperber means representations that have a great *relative probability* of surviving in cultural transmission because these representations are experienced as intuitively natural (Sperber, 1996). Henrich and Boyd (2002) accept that social learning requires certain "innate" expectations about objects in the environment and the nature of relationships among them; they, however, argue that this does not render population dynamic models of cultural transmission unnecessary. Selection actually tends to trump attraction.

Claidière and Sperber (2007) agree that distribution-based transmission biases and content-based attraction both play a role in explaining cultural stability and evolution. But while Henrich and Boyd (2002) understand attraction deterministically as the variable size of the departure from the model always in one and the same direction, Sperber sees attraction as the *relative probability* of departing from the model in one direction rather than another. Only when selection is quite strong and probabilistic attraction quite weak can attraction be ignored. Generally, when there is both attraction and selection, both contribute to the evolution of the population. Although imitation may well play a major role in the propagation of culture, for example, it cannot be considered faithful enough to explain its stability (Claidière & Sperber, 2010). Recently, Lindeyer and Reader (2010) have provided empirical evidence for this among animals on the basis of experimental studies with zebra fish. The claim of people imitating prestigious individuals leaves open the question of *why* some individuals are regarded as prestigious.

Henrich et al. (2010a), for their part, argue that the emergence of sedentary culture and growing populations brought along new kinds of culturally transmitted norms and values; these, then, made fairness with regard to strangers not only possible but also important. Extending the demand for cooperation to non-kin was not simply an erroneous extension of an intuitive kin-selection to non-kin but, rather, was an idea forced upon people by the fact that they had to learn to live in large groups of potential cooperators (see also Norenzayan & Shariff, 2008).

People living in small communities lack the kind of market integration and "doctrinal" religion (Whitehouse, 2004) that characterize modern Western culture; consequently, they lack the generalized ideas of fairness that are typical of large-scale communities (see also Pyysiäinen, 2010b). Thus, the question is whether sociocultural change can also change human intuitions about fairness by introducing culturally transmitted norms and values, or whether social change only brings along new kinds of incentives for using intuitive ideas about fairness and cooperation (Baumard, 2010).

9.3.3 The Evolution of Religion

In the standard model of the cognitive science of religion, beliefs about supernatural agents are evolutionary by-products of human cognition (Boyer, 2001, 2005b; Kirkpatrick, 2006; McKay & Dennett, 2009; Pyysiäinen & Hauser, 2010; for a review, see Boyer & Bergstrom, 2008). Supernatural beliefs persist because we intuitively connect complex order with an intelligent agent as the source of that order, and because certain kinds of event structures spontaneously trigger intuitions about intentional agency even when no agents are in fact perceived (Guthrie, 1993; Pyysiäinen, 2009; Rosset, 2008; Kelemen & Rosset, 2009; Waytz, Gray, Epley, & Wegner, 2010; see also Saler and Ziegler, 2006). There is no specifically religious cognition, let alone a module. Ordinary cognitive mechanisms that mediate religious representations also make possible many other representations, such as in imagination, fiction, lying, and so forth.

Representations of supernatural agents rely on the same mechanisms for agent representation as any agent representation; all that is needed are the ability to reason about agents that are not immediately present, and metarepresentation of beliefs about beliefs (see Pyysiäinen, 2009). According to the standard model, religion is too heterogeneous a category to have emerged as an adaptation (Boyer, 1994; Kirkpatrick, 2006). Trying to explain "religion" is like trying to explain all white objects (Boyer, 1994, p. 32): the explanandum is simply too multifaceted to be captured by any single "magic bullet" explanation (Boyer, 2001). The concept of religion is a scholarly construct, not a natural kind.

There is also no evidence of any such genetic variation among early humans that would have allowed natural selection to favor religion. All this is not, however, to deny that religious beliefs and practices may have later been coopted for adaptive uses; once in place, religious beliefs and practices may have enhanced intra-group cooperation (see Norenzayan & Shariff, 2008; Pyysiäinen & Hauser, 2010).

The evolutionary adaptationist research program is less easily defined. Some of its representatives claim that religion is an adaptation for intra-group cooperation, since either religious practices or beliefs contribute to group cohesion. Such practices and beliefs effectively block potential free-riders from cheating and taking the benefits of a cooperating group without themselves cooperating (e.g., Bering & Johnson, 2005; Johnson & Bering, 2006; Johnson & Krüger, 2004). Bering (2006) argues that a cognitive system dedicated to forming illusory representations of psychological immortality and symbolic meaning evolved as a response to the unique selective pressures of the human social environment. Thus, afterlife beliefs are not direct products of natural selection, but have been selected for as an intuitive pattern of reasoning that does not depend on the presence of explicit religious concepts. The general idea of an afterlife is present in human cognitive structures from the start (Bering, 2002). In this view, religion is not a set of ideas that survive in cultural transmission because they parasitize other evolved cognitive structures. Rather, "a representational bias for envisioning personal immortality" has "impacted the net genetic fitness of individual humans in ancestral environments" (Bering, 2006).

The line of demarcation between adaptationism and the by-product view is somewhat muddled, however. Sosis, for example, argues that religion is an adaptive system similar to the respiratory, circulatory, or immune systems (2009, pp. 320–21). It consists of a recurrent set of core elements that form an appropriate unit of evolutionary analysis. The focus of analysis is in the coalescence of elements that form a religious system. Adaptations are "traits that exist because of a process of phenotypic modification by natural selection for a particular gene-propagating effect" (p. 321). Whether the religious system is an adaptation cannot be settled by studying its parts in isolation: the proper unit of adaptationist analysis is the entire system, not its constituent parts. Yet the actual constitution of the "religious system" has not been described in detail; in addition, the question of why the constituent parts recurrently coalesce waits for an explanation. It may be rather difficult to delineate any general or universal religious system where all components of religion are always copresent.

Sosis first points out that demonstrating that a trait is adaptive does not mean that it is an adaptation (2009, pp. 321–23, 328). Yet he claims that it has not been shown that religious beliefs and practices are inevitable spandrels or by-products of cognitive mechanisms. Even if hypersensitive agent detection were a by-product, this would not prove the religious system to be a by-product. The question of the religious system, "a seamless whole" with "functional goals" (pp. 322–328), as an adaptation cannot be settled by studying its parts in isolation (cf. Shariff, 2008, p. 119). Yet Sosis argues that in the most likely scenario cognitive, emotional, and behavioral elements were exapted for use in the religious system (2009, pp. 323–24). The coopting of preexistent structures for novel solutions is a hallmark of evolutionary adaptations. The religious system is an adaptation if the exapted mechanisms have been modified by the new socioecological niche created by religion.

Studying the actual evolutionary history of religion has turned out to be difficult. The by-product view and context-sensitive, multilevel mechanistic explanation both by and large avoid the conceptual and empirical problems of the adaptationist view. Therefore, the adaptationists have recently come closer to the by-product view (e.g., Wilson, 2008; see Wilson, 2002) in saying that what is called religion is a set of cognitions and behaviors made possible by non-religious cognitive and emotional mechanisms, which is precisely the by-product thesis. The adaptationist account only adds to this the claim that, once in place, religious cognition and behavior have become the targets of selection. Whether the relevant process is one of natural selection or gene-culture coevolution, or both, is still under debate.

One option is that we are dealing with a Baldwin effect: learned behaviors (i.e., religion) have affected the direction and rate of evolutionary change by natural selection (see Depew, 2003; Pyysiäinen, 2006). Learning and behavioral flexibility can amplify and bias natural selection because they enable individuals to modify the context of natural selection that affects their offspring, who thus face new kinds of selection pressures (Deacon, 1998, p. 322; see also Dennett, 1991, pp. 184–87). Thus, as cultural change is more rapid than biological evolution, religion may not have shaped the human brain; instead, religion may have evolved guided by the fact that forms of religion that are easier to adopt have survived better than others. Religion has not become genetically assimilated but, rather, coevolved with the human brain (see Pyysiäinen, 2006, pp. 213–17; Godfrey-Smith, Dennett, & Deacon, 2003, p. 112).

9.4 On Reducing Religion

The issue under debate is whether what is known as religion can be reduced to certain "innate" cognitive biases (Boyer, 1994), or whether macro-level cultural factors also have explanatory power, or at least call for an interpretive approach (Geertz, 1973; Malley, 1997). In particular, can certain doctrinal forms of religion change individuals' intuitive ideas about norms and values, or does change occur only in the kinds of social contexts available for reflective reasoning (see Pyysiäinen, 2004a)? The question is difficult because "religion" is a heterogeneous category: it refers to a variety of beliefs, narratives, doctrines, experiences, institutions, and social hierarchies that anthropologists have been documenting and describing for decades (Morris, 1987; for a more theoretical argument, see Saler, 2000; Day, 2005). Can all these be reduced to certain cognitive mechanisms, and in what sense (ontological, epistemological, and so forth)?

It seems that we also need the higher-level story in order to be able to discuss religions as a set of sociocultural phenomena, because this level involves questions that cannot be directly answered by referring to the lower levels where the explanandum is different (see, e.g., Kuorikoski & Ylikoski, 2010). Boyer's "three page short history of all religions ever" (2001, pp. 326–28) does not satisfy historians of religion, although it might satisfy a cognitive scientist. It is not clear how differing forms of religiosity really could be explained with reference to cognitive mechanisms only because, for example, different people have varying credal attitudes toward one and the same god concepts (Gervais & Henrich, 2010). Even if the human cognitive architecture canalizes the cultural spread of religious ideas, it may still be that differing sociocultural contexts also have an effect on the cultural selection of beliefs and practices (see Griffiths & Machery, 2008).

In the standard model from cognitive science, religion is defined as a set of behaviors and thoughts that have arisen as by-products of certain cognitive functions. Such cognitive mechanisms as the theory of mind canalize the spread of cultural traditions, making certain representations more likely than others to be passed on to others both horizontally and vertically (within and between generations; Atran, 2002).

In other words, the stability and recurrence we observe in religious traditions are the result of the architecture of the human mind. Religious social institutions are the outcome of the way the human mind works. The philosophical question of the ontological reduction of culture to the mind has not received much attention from the cognitive scientists of religion; instead, they provide methodological arguments for the proper way of studying religion (see Visala, 2009). Boyer (1993, pp. 6, 9–13), for example, argues that the assumption of culture as a specific ontological level does not produce explanatory schemes that could not be produced simply by studying people's actual ideas and actions. What is in this way explained are trends in populations, not the motives or intentions of individual persons (Boyer, 2001). Explanations do not target the "proximate" mechanisms underlying individual behavior but, instead, the "ultimate" evolutionary causes of the spread of religious ideas (Ariew, 2003; see Pyysiäinen, in press).

Coevolutionary theorists, for their part, start from observed behaviors and try to account for their persistence using mathematical modeling of cultural selection (e.g., Henrich & Henrich, 2007, Henrich et al., 2010a). Henrich and Henrich notwithstanding, there is little in the way of research regarding empirical applications of this theoretical perspective to religion. The same holds true for the standard model of the cognitive science of religion, which has by and large focused on explaining religion in the abstract, not on explaining the actual evolution of religious thinking and behavior. Thus, we face three problems: first, the actual evolutionary history of religion has so far not been described; second, the role of sociocultural factors in shaping human religiosity has not been carefully analyzed by the cognitive scientists of religion; third, the neural implementation of religious cognition is thus far poorly understood (see Schjoedt, 2009; Persinger, 2009).

However, the way social cognition is neurally implemented is an area of vigorous research in socio-cognitive neuroscience (for a review, see Hari & Kujala, 2009). This new research paves the way for multilevel explanation that takes into account neural networks, psychology, and social reality. The nature of the relationship between neural, psychological, and sociocultural explanations is a difficult question, though (see Kuorikoski & Ylikoski, 2010). The standard model of the cognitive science of religion seems to accord with New Wave reductionism: that is, the traditional cultural and hermeneutical explanations or interpretations of religious thought and behavior can be eliminated, once we have the proper cognitive scientific explanations of religious thought, behavior, and experience (see Boyer, 2001, pp. 326–28).

Many social scientists object to such methodological individualism, insisting that social action, structures, and institutions are real and can be used as explanatory factors (see Kincaid, 1996; Schmaus, 2010; Hedström & Ylikoski, 2010). They are not reducible to what an individual alone thinks and does. If groups are reduced to cognizing individuals, the question arises why the level of individuals or of cognition should be real, whereas groups are not; the levels of intentional behavior and of cognitive process are reducible in principle to neural events, after all. So, if one insists on the reduction of groups and social institutions to cognizing individuals, why not go all the way down to neurochemistry, or even its microphysical basis (see Huebner, 2008)?

The answer is that different-level explananda require different-level explanations.

Durkheim, for instance, distinguished between psychological explanations of belief in the efficacy of rituals, and sociological explanations of the macro-level effects of rituals. In each case, both the explanandum and the explanans are different. Although society exists only in and through individuals, "social facts" form a level of analysis that is not reducible to mental mechanisms (Durkheim, 1912/1965, p. 389). Social facts are distinct from their individual manifestations and should only be explained by other social facts, not by individual psychology (Durkheim, 1895/1966). In his later works on suicide and on religion, Durkheim did use psychological concepts and explanations; he was not as much an anti-psychologist as he is often considered to have been, although he was anti-reductionist (Schmaus, 2010).

Very much the same holds true of coevolutionary theories. Not only do they refer to a different explanans as compared to the standard model of the cognitive science of religion, but the explanandum is different, because mere cognitive explanations are deemed insufficient (psychological vs. social). The coevolutionary theories are supposed to capture phenomena that the cognitive explanations ignore. The persistence of god beliefs, for example, cannot be explained merely by the cognitive form of these representations, because people in one religious group usually do not believe in the gods of other traditions, even though the representations of alien gods are structurally identical with the god(s) one believes in (Gervais & Henrich, 2010).

The coevolutionary theories explain these kinds of phenomena by referring to social processes such as imitation, conformism, and punishing. Coevolutionary theories are not anti-psychological, however; they only claim that cognitive explanations alone are not enough. There are phenomena they cannot capture, while coevolutionary theories can. Representatives of the standard cognitive model, for their part, argue that coevolutionary theories leave unexplained why, for example, some individuals are regarded as worthy of imitation. Cognitive explanations thus might help make coevolutionary theories more robust by providing an account of the underlying cognitive processes. Thus, these two research programs are complementary rather than mutually exclusive.

However, for those who endorse a hermeneutical understanding of religion and argue against reductionism, both the cognitive science of religion and coevolutionary theories are reductionist: they do not capture the specifically "religious" nature of religion (e.g., Eliade, 1969; Geertz, 1973; cf. Martin, 2000; Pyysiäinen, 2004b, pp. 1–27). Theologians and some scholars of religion think that in both the Durkheimian tradition and the cognitive science of religion, religion is reduced to something non-religious; thus its essential nature is lost (see, e.g., Pihlström, 2002, 2005). But, from the point of view of multilevel mechanistic explanation, there is no one fundamental level at which all questions should be settled. Things can be viewed at differing mechanistic levels; at each level the explanandum is different. This calls into question the claim that religion is a *sui generis* category and can only be studied as such. It is well-nigh impossible to specify what it actually takes to view religion only "from inside" (see Pyysiäinen, 2004b, 67–80).

9.5 General Discussion and Conclusion

Cognitive and coevolutionary explanations of religious phenomena have had to face two objections. First, many humanists and social scientists argue that cultural and social phenomena are not reducible to psychology, and that psychology thus is irrelevant for social science. Second, some scholars of religion claim that social scientific explanations of religion are reductionist and that the specifically religious nature of beliefs and practices is lost when they are studied merely as social phenomena (Eliade, 1969; Stark & Finke, 2001 see also Pihlström, 2002, 2005).

Such anti-reductionist arguments concerning religion and the study of religion are based on an implicit or explicit presupposition that the proper level of analysis is simply given, whether it is the level of religion as *sui generis*, the social or cultural systems, and so forth. In the background is the (implicit) belief that reality shows itself to us as if sliced up into neat categories from the start. Once such a "myth of the given" (Sellars, 1980) is accepted in the study of religion, approaches at differing levels are deemed either irrelevant or as actually distorting a proper understanding of religion. But how can we objectively decide what counts as the proper level?

This chapter has tried to show how a multilevel causal-mechanical view of explanation can help solve the problem. As "religion" is a heterogeneous and vague category without any essence, not a natural kind, it is questionable that any one approach could show us religion from the inside, while all others only view it from outside. The question also emerges how we can generalize from sets of experiments or models to "religion" at large (see Saler, 2000; Henrich et al., 2010b). Things have gotten all the more complicated with the emergence of cognitive scientific and biological perspectives in the study of religion, which have inspired historians of religion and scholars engaged in cultural studies to emphasize the "cultural construction" of both science and the concept of religion (e.g., Rydving, 2008; cf. Sørensen, 2008; Sutcliffe, 2008).

In emphasizing that phenomena can be explained at differing levels, multilevel mechanistic explanations can help avoid misleading questions about reductionism. Religious phenomena can be explained at differing mechanistic levels, and at each level, the explanandum is different. There are no "magic bullet" explanations at once capturing everything that is worth explaining in religion. "Religion" is a heuristic term that directs our attention to certain kinds of recurrent phenomena that can be explained at differing levels of mechanisms.

The practical question then emerges how explanations at differing levels of mechanisms are to be coordinated. It is not enough that representatives of different disciplines politely listen to each other in seminars; what is needed is rather a forgetting of the traditional disciplinary boundaries (see Boyer, 2005a). The standard model of the cognitive science of religion and coevolutionary theories are not two alternatives between which we should make a choice; instead, they are complementary in the sense that they operate at differing levels of mechanisms. Whether attraction or selection is a stronger force is an empirical question that has not been conclusively settled so far (*pace* Claidière & Sperber, 2007, 2010). Instead of principled debates about which approach is to be favored, we rather need exploration on their mutual relationship and a research program that takes into account both cognitive processes and biases in the cultural transmission of beliefs and practices.

Acknowledgment

I want to thank Petri Ylikoski for helpful comments.

References

Alexander, R. D. (1979). *Darwinism and human affairs*. Seattle: University of Washington Press.

Ariew, A. (2003). Ernst Mayr's "ultimate/proximate" distinction reconsidered and reconstructed. *Biology and Philosophy, 18,* 553–565.

Atran, S. (2002). *In gods we trust: The evolutionary landscape of religion*. Oxford: Oxford University Press.

Barrett, J. L. (2000). Exploring the natural foundations of religion. *Trends in Cognitive Sciences, 4,* 29–34.

Barrett, J. L. (2004). *Why would anyone believe in God?* Walnut Creek, CA: AltaMira Press.

Baumard, N. (2010, 24 April). Are variations in economic games really caused by culture? [Blog post]. *International Cognition and Culture*. Retrieved from http://www .cognitionandculture.net

Bechtel, W. (2008). *Mental mechanisms: Philosophical perspectives on cognitive neuroscience*. New York: Routledge.

Bechtel, W. (2009). Looking down, around, and up: Explanation in psychology. *Philosophical Psychology, 22*(5), 543–564.

Bell, A. V., Richerson, P. J., & McElreath, R. (2009). Culture rather than genes provides greater scope for the evolution of large-scale human prosociality. *Proceedings of the National Academy of Sciences of the United States of America, 106*(42), 17671–17674.

Bering, J. M. (2002). Intuitive conceptions of dead agents' minds: The natural foundations of afterlife beliefs as phenomenological boundary. *Journal of Cognition and Culture, 2,* 263–308.

Bering, J. M. (2006). The folk psychology of souls. *Behavioral and Brain Sciences, 29,* 453–462.

Bering, J. M., & Johnson, D. D. P. (2005). "Oh Lord ... you perceive my thoughts from afar": Recursiveness and the evolution of supernatural agency. *Journal of Cognition and Culture, 5,* 118–142.

Bickle, J. (1998). *Psychoneural reduction: The new wave*. Cambridge, MA: MIT Press.

Boyd, R., & Richerson, P. J. (1985). *Culture and the evolutionary process*. Chicago: The University of Chicago Press.

Boyer, P. (1993). Cognitive aspects of religious symbolism. In P. Boyer (Ed.), *Cognitive aspects of religious symbolism* (pp. 4–47). Cambridge: Cambridge University Press.

Boyer, P. (1994). *The naturalness of religious ideas: A cognitive theory of religion.* Berkeley, CA: University of California Press.

Boyer, P. (2001). *Religion explained: The evolutionary origins of religious thought.* New York: Basic Books.

Boyer, P. (2005a). Ten problems for integrated behavioural science: How to make the social sciences relevant. Global Fellows Seminars, University of California, Los Angeles. Retrieved from http://escholarship.org/uc/item/46h429tx

Boyer, P. (2005b). A reductionistic model of distinct modes of religious transmission. In H. Whitehouse & R. N. McCauley (Eds.), *Mind and religion: Psychological and cognitive foundations of religiosity* (pp. 3–29). Walnut Creek, CA: AltaMira Press.

Boyer, P., & Bergstrom, B. (2008). Evolutionary perspectives on religion. *Annual Review of Anthropology, 37,* 111–130.

Claidière, N., & Sperber, D. (2007). The role of attraction in cultural evolution (Reply to J. Henrich & R. Boyd, "On modeling cognition and culture"). *Journal of Cognition and Culture, 7*(1–2), 89–111.

Claidière, N., & Sperber, D. (2010). Imitation explains the propagation, not the stability of animal culture. *Proceedings of the Royal Society B: Biological Sciences, 277*(1681), 651–659.

Clarke, P. B., & Byrne, P. (1993). *Religion defined and explained.* New York: St. Martin's Press.

Craver, C. (2007). *Explaining the brain: Mechanisms and the mosaic unity of neuroscience.* New York: Oxford University Press.

Craver, C., & Bechtel, W. (2007). Top-down causation without top-down causes. *Biology and Philosophy, 22,* 547–563.

Cruse, H. (2003). The evolution of cognition—a hypothesis. *Cognitive Science, 27,* 135–155.

Day, M. (2005). The undiscovered and undiscoverable essence: Species and religion after Darwin. *Journal of Religion, 85*(1), 58–82.

Deacon, T. W. (1997). *The symbolic species: The co-evolution of language and the brain.* New York: Norton.

Dennett, D. C. (1993). *The intentional stance.* Cambridge, MA: MIT Press.

Dennett, D. C. (1991). *Consciousness explained.* Harmondsworth, UK: Penguin.

Depew, D. J. (2003). Baldwin and his many effects. In B. H. Weber & D. J. Depew (Eds.), *Evolution and learning: The Baldwin effect reconsidered* (pp. 3–30). Cambridge, MA: MIT Press.

Durkheim, É. (1965). *The elementary forms of the religious life* (J. W. Swain, Trans.). New York: Free Press. (Original work published 1912)

Durkheim, É. (1966). *The rules of sociological method* (Ed. G. E. G. Catlin, Trans. S. A. Solovay & J. H. Mueller). New York: Free Press, Collier Macmillan. (Original work published 1895)

Edis, T. (2008). *Science and nonbelief.* Amherst, NY: Prometheus Books.

Eliade, M. (1969). *The quest: History and meaning in religion.* Chicago: University of Chicago Press.

Eronen, M. I. (2009). Reductionist challenges to explanatory pluralism: Comment on McCauley. *Philosophical Psychology, 22*(5), 637–646.

Gallagher, H. L., Happé, F., Brunswick, N., Fletcher, P. C., Frith, U., & Frith, C. D. (2000). Reading the mind in cartoons and stories: An fMRI study of "theory of mind" in verbal and nonverbal tasks. *Neuropsychologia, 38*, 11–21.

Gallese, V., & Metzinger, T. (2003). Motor ontology: The representational reality of goals, actions and selves. *Philosophical Psychology, 16*(3), 365–388.

Garfinkel, A. (1981). *Forms of explanation.* New Haven, CT: Yale University Press.

Geertz, C. (1973). *The interpretation of cultures.* New York: Basic Books.

Gervais, W. M., & Henrich, J. (2010). The Zeus problem: Why representational content biases cannot explain faith in gods. *Journal of Cognition and Culture, 10*, 383–389).

Godfrey-Smith, P., Dennett, D., & Deacon, T. W. (2003). Postscript on the Baldwin effect and niche construction. In B. H. Weber & D. J. Depew (Eds.), *Evolution and learning: The Baldwin effect reconsidered* (pp. 107–112). Cambridge, MA: MIT Press.

Griffiths, P. E., & Machery, E. (2008). Innateness, canalization, and "biologicizing the mind." *Philosophical Psychology, 21*(3), 397–399.

Guthrie, S. (1993). *Faces in the clouds.* Oxford: Oxford University Press.

Hari, R., & Kujala, M. (2009). Brain basis of human social interaction: From concepts to brain imaging. *Physiological Reviews, 89*, 453–479.

Hedström, P., & Ylikoski, P. (2010). Causal mechanisms in the social sciences. *Annual Review of Sociology, 36*, 49–67.

Hélie, S., & Sun, R. (2010). Incubation, insight, and creative problem solving: A unified theory and a connectionist model. *Psychological Review, 117*(3), 994–1024.

Henrich, J. (2004). Demography and cultural evolution: How adaptive cultural processes can produce maladaptive losses—The Tasmanian case. *American Antiquity, 69*(2), 197–214.

Henrich, J. (2009). The evolution of costly displays, cooperation, and religion: Credibility enhancing displays and their implications for cultural evolution. *Evolution and Human Behavior, 30,* 244–260.

Henrich, J., & Boyd, R. (2002). On modeling cognition and culture: Why cultural evolution does not require replication of representations. *Journal of Cognition and Culture, 2*(2), 87–112.

Henrich, N., & Henrich, J. (2007). *Why humans cooperate: A cultural and evolutionary explanation.* Oxford: Oxford University Press.

Henrich, J., Ensminger, J., McElreath, R., Barr, A., Barrett, C., Bolyanatz, A., et al. (2010a). Markets, religion, community size, and the evolution of fairness and punishment. *Science, 327,* 1480–1484.

Henrich, J., Heine, S. J., & Norenzayan, A. (2010b). The weirdest people in the world? *Behavioral and Brain Sciences, 33*(2–3), 61–83.

Holland, D., & Quinn, N. (Eds.). (1987). *Cultural models in language and thought.* Cambridge: Cambridge University Press.

Huebner, B. (2008). Do you see what we see? An investigation of an argument against collective representation. *Philosophical Psychology, 21*(1), 91–112.

Hurley, S. (2008). The shared circuits model (SCM): How control, mirroring, and simulation can enable imitation, deliberation, and mindreading. *Behavioral and Brain Sciences, 31*(1), 1–58.

Idinopulos, T. A., & Yonan, E. A. (Eds.). (1994). *Religion and reductionism.* Leiden, The Netherlands: Brill.

Johnson, D., & Bering, J. (2006). Hand of God, mind of man: Punishment and cognition in the evolution of cooperation. *Evolutionary Psychology, 4,* 219–233.

Johnson, D., & Krüger, O. (2004). The good of wrath: Supernatural punishment and the evolution of cooperation. *Political Theology, 5,* 159–176.

Kamppinen, M. (2009). Rationality, religion and intentional systems theory: From objective ethnography to the critical study of religious beliefs. *Method and Theory in the Study of Religion, 21*(3), 274–284.

Kelemen, D., & Rosset, E. (2009). The human function compunction: Teleological explanation in adults. *Cognition, 111,* 138–143.

Kincaid, H. (1996). *Philosophical foundations of the social sciences: Analyzing controversies in social research.* Cambridge, MA: Cambridge University Press.

Kirkpatrick, L. A. (2006). Religion is not an adaptation. In P. McNamara, *Where God and science meet: How brain and evolutionary studies alter our understanding of religion. Vol. 1: Evolution, genes, and the religious brain* (pp. 159–180). Westport, CT: Praeger.

Kuorikoski, J., & Ylikoski, P. (2010). Explanatory relevance across disciplinary boundaries: The case of neuroeconomics. *Journal of Economic Methodology*, *17*(2), 219–228.

Lawson, E. T., & McCauley, R. N. (1990). *Rethinking religion: Connecting cognition and culture*. Cambridge: Cambridge University Press.

Lindeyer, C. M., & Reader, S. M. (2010). Social learning of escape routes in zebrafish and the stability of behavioural traditions. *Animal Behaviour*, *79*, 827–834.

McCauley, R. N. (1986). Intertheoretic relations and the future of psychology. *Philosophy of Science*, *53*, 179–199.

McCauley, R. N. (2007). Reduction: Models of cross-scientific relations and their implications for the psychology-neuroscience interface. In P. Thagard, *Philosophy of psychology and cognitive science* (pp. 105–158). Amsterdam, The Netherlands: Elsevier.

McCauley, R. N. (2009). Time is of the essence: Explanatory pluralism and accommodating theories about long-term processes. *Philosophical Psychology*, *22*(5), 611–635.

McKay, R. T., & Dennett, D. C. (2009). The evolution of misbelief. *Behavioral and Brain Sciences*, *32*(6), 493–561.

Mahon, B.Z., S. Anzellotti, J. Schwarzback, & M. Zampini. (2009). Category-specific organization in the human brain does not require visual experience. *Nature, 63* (August 20), 397–405.

Malley, B. (1997). Causal holism in the evolution of religious ideas: A reply to Pascal Boyer. *Method and Theory in the Study of Religion, 9*, 389–399.

Mar, R. A., Kelley, W. M., Heatherton, T. F., & Macrae, C. N. (2007). Detecting agency from the biological motion of veridical *vs* animated agents. *Social Cognitive and Affective Neuroscience, 2*, 199–205.

Martin, M. (2000). *Verstehen: The uses of understanding in social science*. New Brunswick, NJ: Transaction Publishers.

Morris, B. (1987). *Anthropological studies of religion: An introductory text*. Cambridge: Cambridge University Press.

Nagel, E. (1961). *The structure of science: Problems in the logic of scientific explanation*. London: Routledge & Kegan Paul.

Nichols, S., & Stich, S. P. (2003). *Mindreading: An integrated account of pretence, self-awareness, and understanding other minds*. Oxford: Oxford University Press.

Norenzayan, A., & A.F. Shariff. (2008). The origin and evolution of religious prosociality. *Science, 322* (3 October), 58–62.

Oppenheim, P., & Putnam, H. (1958). Unity of science as a working hypothesis. In H. Feigl, M. Scriven, & G. Maxwell (Eds.), *Minnesota studies in the philosophy of science, Vol. 2* (pp. 3–36). Minneapolis: University of Minnesota Press.

Persinger, M. (2009). Are our brains structured to avoid refutations of belief in God? An experimental study. *Religion, 39,* 34–42.

Pihlström, S. (2002). Pragmatic and transcendental arguments for theism. *International Journal for Philosophy of Religion, 51,* 195–210.

Pihlström, S. (2005). A pragmatic critique of three kinds of religious naturalism. *Method and Theory in the Study of Religion, 17*(3), 177–218.

Pyysiäinen, I. (2004a). Intuitive and explicit in religious thought. *Journal of Cognition and Culture, 4*(1), 123–150.

Pyysiäinen, I. (2004b). *Magic, miracles, and religion: A scientist's perspective.* Walnut Creek, CA: AltaMira Press.

Pyysiäinen, I. (2006). Amazing grace: Religion and the evolution of the human mind. In P. McNamara, *Where God and science meet: How brain and evolutionary studies alter our understanding of religion. Vol. 1: Evolution, genes, and the religious brain* (pp. 209–225). Westport, CT: Praeger.

Pyysiäinen, I. (2008). Introduction: Religion, cognition, and culture. *Religion, 38*(2), 101–108.

Pyysiäinen, I. (2009). *Supernatural agents: Why we believe in souls, gods, and buddhas.* Oxford: Oxford University Press.

Pyysiäinen, I. (2010a). How religion resists the challenge of science. *Skeptical Inquirer, 34*(3), 39–41.

Pyysiäinen, I. (Ed.). (2010b). *Religion, economy, and cooperation.* Berlin: Mouton de Gruyter.

Pyysiäinen, I. (2010c). Believing and doing: Ritual action enhances religious belief. In A. W. Geertz & J. S. Jensen (Eds.), *Religious narrative, cognition and culture: Image and word in the mind of narrative.* London: Equinox.

Pyysiäinen, I. (in press). Mechanical explanation of ritualized behavior. In A. Geertz & J. Sørensen (Eds.), *Religious ritual, cognition and culture.* London: Equinox.

Pyysiäinen, I., & Hauser, M. (2010). The origins of religion: Evolved adaptation or by-product? *Trends in Cognitive Sciences, 14*(3), 104–109.

Railton, P. (1978). A deductive-nomological model of probabilistic explanation. *Philosophy of Science, 45,* 206–226.

Richerson, P., & Boyd, R. (2005). *Not by genes alone: How culture transformed human evolution.* Chicago: University of Chicago Press.

Rosset, E. (2008). It's no accident: Our bias for intentional explanations. *Cognition*, *108*, 771–780.

Rydving, H. (2008). A Western folk category in mind? *Temenos: Nordic Journal of Comparative Religion*, *44*(1), 73–99.

Saler, B. [1993] (2000). *Conceptualizing religion: Immanent anthropologists, transcendent natives, and unbound categories*. New York: Berghahn Books.

Saler, B., & Ziegler, C. A. (2006). Atheism and the apotheosis of agency. *Temenos: Nordic Journal of Comparative Religion*, *42*(2), 7–41.

Saxe, R., Carey, S., & Kanwisher, N. (2004). Understanding other minds: Linking developmental psychological and functional neuroimaging. *Annual Review of Psychology*, *55*, 87–124.

Schjoedt, U. (2009). The religious brain: A general introduction to the experimental neuroscience of religion. *Method and Theory in the Study of Religion*, *21*, 310–339.

Schmaus, W. (2010). Durkheim and psychology. In I. Pyysiäinen (Ed.), *Religion, economy, and cooperation* (pp. 99–126). Berlin: Mouton de Gruyter.

Shariff, A. (2008). One species under God? Sorting through the pieces of religion and cooperation. In J. Bulbulia, R. Sosis, E. Harris, R. Genet, C. Genet, & K. Wyman (Eds.), *The evolution of religion: Studies, theories, & critiques* (pp. 119–125). Santa Margarita, CA: Collins Family Foundation.

Sellars, W. (1980). Behaviorism, language and meaning. *Pacific Philosophical Quarterly*, *61*, 3–25.

Shtulman, A. (2008). Variation in the anthropomorphization of supernatural beings and its implications for cognitive theories of religion. *Journal of Experimental Psychology*, *34*(5), 1123–1138.

Sørensen, J. (2008). Cognition and religious phenomena—A response to Håkan Rydving. *Temenos: Nordic Journal of Comparative Religion*, *44*(1), 111–120.

Sosis, R. (2009). The adaptationist–by-product debate on the evolution of religion: Five misunderstandings of the adaptationist program. *Journal of Cognition and Culture*, *9*(3–4), 315–332.

Sperber, D. (1975). *Rethinking symbolism*. Cambridge: Cambridge University Press.

Sperber, D. (1985). Anthropology and psychology: Towards an epidemiology of representations. *Man (N.S.)*, *20*, 73–89.

Sperber, D. (1996). *Explaining culture: A naturalistic approach*. Oxford: Blackwell.

Sperber, D. (2006). Conceptual tools for a naturalistic approach to cultural evolution. In S. C. Levinson & P. Jaisson (Eds.), *Evolution of culture: A Fyssen Foundation symposium* (pp. 147–165). Cambridge, MA: MIT Press.

Sperber, D., & Wilson, D. (1986). *Relevance: Communication and cognition*. Cambridge, MA: Harvard University Press.

Sperber, D., & Hirschfeld, L. A. (2004). The cognitive foundations of cultural stability and diversity. *Trends in Cognitive Sciences*, *8*(1), 40–46.

Stark, R., & Finke, R. (2001). *Acts of faith: Explaining the human side of religion*. Berkeley, CA: University of California Press.

Stausberg, M. (Ed.). (2009). *Contemporary theories of religion*. London: Routledge.

Sun, R. (2002). *Duality of mind: A bottom-up approach toward cognition*. Mahwah, NJ: Lawrence Erlbaum.

Sun, R., Coward, L. A., & Zenzen, M. J. (2005). On levels of cognitive modeling. *Philosophical Psychology*, *18*(5), 613–637.

Sutcliffe, S. J. (2008). "Two cultures" in the study of religion? A response to Håkan Rydving. *Temenos: Nordic Journal of Comparative Religion*, *44*(1), 101–110.

Tomasello, M. (1999). *The cultural origins of human cognition*. Cambridge, MA: Harvard University Press.

Tremlin, T. (2006). *Minds and gods: The cognitive foundations of religion*. Oxford: Oxford University Press.

Waytz, A., Gray, K., Epley, N., & Wegner, D. M. (2010). Causes and consequences of mind perception. *Trends in Cognitive Sciences*, *14*(8), 383–388.

Whitehouse, H. (2004). *Modes of religiosity: A cognitive theory of religious transmission*. Walnut Creek, CA: AltaMira Press.

Wilson, D. S. (2002). *Darwin's cathedral: Evolution, religion and the nature of society*. Chicago: University of Chicago Press.

Wilson, D. S. (2008). Evolution and religion: The transformation of the obvious. In J. Bulbulia, R. Sosis, E. Harris, R. Genet, C. Genet, & K. Wyman (Eds.), *The evolution of religion: Studies, theories, & critiques* (pp. 23–29). Santa Margarita, CA: Collins Family Foundation.

Wimsatt, W. (1994). The ontology of complex systems: Levels, perspectives, and causal thickets. *Canadian Journal of Philosophy*, *20*(Suppl.), 207–274.

Visala, A. (2009). *Religion explained? A philosophical appraisal of the cognitive science of religion*. Unpublished Ph.D. thesis. University of Helsinki, Helsinki, Finland.

Woodward, J. (2003). *Making things happen: A theory of causal explanation*. Oxford: Oxford University Press.

10 Ritual, Cognition, and Evolution

Harvey Whitehouse

10.1 Introduction

Social scientists have long recognized ritual to be a universal and ancient feature of human societies that influences the scale, structure, and cohesiveness of cultural groups as well as the various forms of competition (including violent conflict) that divide them (Ibn Khaldūn, 1958; Robertson Smith, 1889/2002; Frazer, 1922; Durkheim, 1912/2008; Weber, 1947). Recent convergences and developments in cognitive science and evolutionary theory point to new directions for interdisciplinary research on this topic. Such approaches focus attention on developmental and proximate causes (Tomasello, 1999; Boyer & Lienard, 2006); social consequences and functions (e.g., Atkinson & Whitehouse, 2010; Cohen, Montoya, & Insko, 2006); and processes of natural and cultural selection within the constraints of phylogeny or history (e.g., Wilson, 2002; Turchin, 2006).

Rituals are commonly credited with all kinds of functions: supernatural, symbolic, expressive, social, and so on. But how they serve these functions is opaque, inasmuch as the causal link between socially stipulated procedures and their putative end goals (if any) is opaque. *Teleological opacity* of this kind is one of the hallmark features of ritualized behavior. Social anthropologists have often observed that ritual participants are powerless to explain why they carry out their distinctive procedures and ceremonies, appealing only to tradition or to ancestors. But of considerable interest, too, is the fact that nobody has any difficulty understanding the anthropologist's question, when she asks what the rituals mean. People know that ritualized actions can be invested with functional and symbolic properties even though they may struggle on occasion to identify what those may be, often pointing the hapless researcher in the direction of somebody older or wiser. On other occasions, people have very strong intuitions about the meaning of a ritual, for instance the communicative function of a military salute, even though they are powerless to explain why that particular gesture rather than any other is the privileged method of express-ing respect in the military. What distinguishes rituals from other kinds of teleologically opaque behavior is that the relationship between actions and stated goals (if indeed they are stated at all) cannot *even in principle* be specified in physical-causal terms (Whitehouse, 2004, 2011; Legare & Whitehouse, under review; Sørensen, 2007). To seek out a practical ratio-nale is to misunderstand the very nature of ritualized behavior.

Much of the cultural knowledge we acquire in the course of socialization may be described as ritualized. Religious dogmas, embodied skills, social etiquette, clothing fashions, and even the rules of childhood games *may* be ascribed functions or purposes (whether in the process of teaching and learning, or in reflecting later on why our habits take the form that they do). Often these functions remain mysterious, however—behavioral norms copied simply without question. But whether or not we invest a ritual with a particular function or meaning, we do not consider it sensible to formu-late in physical-causal terms how that function is realized. This is the crux of the difference between rituals and purely practical actions.

Imagine the various ways in which swords can be used for practical ends. The long, sharp pointed blade is admirably designed for piercing and slashing, whether for the purposes of maiming and slaying foes on the battlefield or for a range of more prosaic tasks. These natural properties of sharp materials have been known to our ancestors stretching far back into the mists of prehistory. They are endlessly rediscovered by each new gen-eration of children as a consequence both of our teleological reasoning capacities and our rapacious exploration of object affordances in the envi-ronment from early infancy onwards, often under the guidance of more experienced individuals. But sharp objects can also be used in ways that it would be absurd to interpret in teleological terms. For instance, a sword may be used to confer a knighthood by tapping the flat edge of the blade on the candidate's shoulders. To make sense of this behavior, we must adopt a ritual stance rather than an instrumental one—we must abandon all hope of understanding what is happening in physical-causal terms and instead assume that whatever-it-is that requires us to observe this particular sequence of actions in this particular way derives from an altogether dif-ferent way of reasoning. Quite what that reasoning should be is seldom entirely clear: perhaps the actions of a ritual should be regarded as sym-bolic, perhaps as divinely sanctioned for reasons inscrutable to mere mortals, perhaps as the result of some unknowable mechanisms of super-natural causation, or perhaps for the sole reason that this is the proper or traditional way to behave. But whatever the consensus on such issues, nobody would think that the ritual is explainable in mechanistic terms. To interpret it that way would be to assert that this is not really a ritual at all.

This chapter is part of the treatment of religion in the current volume. Most *ritual* traditions, ancient and modern, postulate beliefs in one or more gods, and largely for this reason are commonly referred to as religions. Nevertheless, the "religion" label is slippery and is also used to refer to cultural traditions that entail beliefs in ancestors (the spirits of dead people), creator beings (not necessarily gods in the senses most commonly used), or various kinds of magic (whether or not requiring the intervention of supernatural agents). While there is nothing wrong with referring to ritual traditions that espouse such beliefs as religions, we cannot assume that the category "religion" has underlying coherence. In fact, it may comprise an arbitrary collection of ideas (e.g., about gods, ghosts, creation, magic, etc.) that have quite distinct and unrelated causes. Some ritual traditions (e.g., Maoism) eschew concepts commonly classed as religious. Still, secular rituals are a relatively recent cultural innovation, and even today remain the exception rather than the rule. It is possible that certain aspects of the psychology underlying ritualized behavior tend to prompt "religious" ideas, and may even provide some coherence to the category after all. This chapter considers that possibility in its concluding discussion.

10.2 Why Humans Have Rituals

Many species besides humans exhibit stereotypic behaviors that it is tempting to describe as rituals. Well-known avian examples include the exquisite choreography of mating swans and the ornate temple-like structures of the tropical bowerbird. Complex forms of courtship in birds have adaptive functions, acting for instance as hard-to-fake signals of fitness. Ritual-like behavior in animals may also contribute to the building of social networks or the inhibition of aggression within groups. Do human rituals serve similar functions?

From a gene's eye perspective, men and women should have very different mating strategies. Males should be opportunistic, females cautious and picky. It has been argued that much of the cultural achievements of men (in art, science, politics, and so on) serve the same biological function as the peacock's tail (Wight, 2007). They are hard-to-fake signals of an individual's genetic endowments. Does the same apply to human rituals? In many traditional religions, ritual is the exclusive province of men, with women commonly forming an audience. It is tempting to draw an analogy with the behavior of the bowerbird, designed to attract critical and discerning females. Avian displays have other functions than to attract mates, for instance to strengthen bonds between an established pair, thereby assisting cooperation in the rearing of offspring. Penguins and albatrosses (among many other species) pair bond through synchronous head-bobbing. Human rituals also commonly involve synchronous movement. Recent experiments suggest that this increases social attachment and cooperation (Wiltermuth & Heath, 2009), much as it appears to do in certain birds.

It is also possible that the human propensity for ritualized behavior emerged partly in response to more species-specific problems. With the move from jungle to savannah, our ancestors became increasingly omnivorous in their food-procurement strategies. Among the advantages of being a generalist would have been reduced vulnerability to food shortages resulting from climate change, disease, or competition from other species. Openness to trying out new potential foodstuffs, however, would have carried a greatly increased risk of imbibing toxins (Rozin, 1999). It has recently been argued that humans evolved a unique method of reducing such risks: the *hazard precaution system* (Boyer & Lienard, 2006). According to the theory, dubious objects and substances trigger a program of stereotyped actions involving cleaning and separating and a concern with symmetry, exactness, or boundary marking. This mechanism, it is suggested, evolved to protect us from contaminants by impelling us to take precautionary action when a risk is suspected. The neural systems responsible for producing hazard-precaution routines would seem to tragically malfunction in patients suffering from obsessive-compulsive disorder, but are quite useful when operating normally. According to Boyer and Lienard, a bizarre byproduct of the hazard-precaution system is that humans readily pick up behaviors, however random and unnecessary, which resemble the system's stereotyped outputs, primarily cultural rituals.

The byproduct theory of why humans have rituals has some appealing features. Indeed, it is quite possible that even the peacock's tail or the synchrony-cohesion arguments best explain cultural rituals as byproducts of mechanisms whose original adaptive functions have been lost or diminished. After all, performing rituals together may help us to attract a mate, but being an especially good performer in the church choir is only one of many potential clinchers in mate selection, and probably not a privileged one. We participate in rituals for many other reasons that have no adaptive benefits at the individual level, and may even carry significant costs. Likewise, the feel-good factor resulting from singing, dancing, and performing other synchronous actions in groups may promote bonhomie and help to motivate participation, but no more so than sharing a good joke or piece of juicy gossip. In sum, cognitive evolution can take us only part of the way to explaining the prevalence and diversity of rituals in our species.

To appreciate the prominence and heterogeneity of ritual in human societies, we must consider the adaptive benefits of rituals for social groups. The teleological opacity of ritual produces a potentially infinite universe of behavioral diversity. Human populations living side by side may have much in common, adopting the same basic techniques of production, using similar tools, exploiting similar natural resources and foodstuffs, living in similar kinds of houses, and so on. Indeed, at the level of practical affairs and day-to-day life, there may be little to tell them apart. People cannot distinguish themselves from their neighbors by continually inventing new ways of tackling the technical challenges of life. Useful inventions typically appear slowly, and their spread is difficult to control. But the arbitrariness of ritualized behavior makes it extraordinarily easy for a group to differentiate itself from others. For this potential to be realized, however, we need to understand another consequence of communal rituals: ingroup bonding.

Social scientists have long argued that rituals bind groups together. Recently, anthropologists and psychologists have assembled systematic evidence that ritual participation increases trust and cooperation among participants, by acting as a costly and therefore hard-to-fake signal of commitment to the group. An important part of the story, however, concerns the relationship between ritual meaning and social identity. Rituals can be invested with a great variety of potential meanings, emotions, moods, and associations (Geertz, 1973). The fact that the ritual actions are not transparently linked to any particular causal structure and function allows for many possible interpretations. Insofar as people reflect on exegetical matters (and, as noted above, this is not always the case) the resulting meanings may be quite idiosyncratic. But if interpreters do not know very much about what others are privately thinking, they can easily form the impression, however illusory, that what is personally meaningful and motivating about the ritual experience, for them, is shared by all other participants. This point too has long been recognized by social scientists, who argue that the common experience of publicly observable aspects of ritual (such as the actions and props) fosters the illusion of collective emotion and interpretation (Kertzer, 1988).

Rituals serve as admirable markers of group boundaries and also as mechanisms for increasing internal cohesion and shared identity within communities. But different forms of communal ritual can exploit these affordances in quite strikingly different ways and to varying degrees, as described in the following section.

10.3   Ritual, Group Formation, and Competition: The Imagistic Mode

Group formation is one of the most adaptive and yet also perhaps the most devastating of all human traits. Without groups, we could not wage wars, commit genocides, or colonize other people's lands. There is much to admire about human groupishness, insofar as it gives rise to acts of altruism, loyalty, camaraderie, heroism, and love. But these qualities typically extend only to the group (e.g., the family, tribe, or nation). Beyond the group, caution and suspicion reign, and when provoked by the members of rival coalitions we have a seemingly insatiable appetite for organized violence. Social cohesion whets the appetite for such conflict. As a simple rule of thumb, the more intensely we love our fellows, the more systematically and brutally we slay our foes.

Rituals play a crucial role in inter-group conflict and competition. This section begins with one of the most ancient ways in which rituals have been used to promote in-group cohesion and out-group hostility, namely through the performance of rare but highly traumatic rites that have enduring psychological effects on those who experience them. This syndrome has become known as the *imagistic mode of religiosity* (Whitehouse, 1995, 2000, 2004; Pachis & Martin, 2009). Imagistic rituals take many diverse forms, sometimes involving induction into the group (or *initiation*). They are found in all the world's most bellicose tribes and also in modern armies. There is evidence that they were performed at least as long ago as the Upper Paleolithic, and it is quite possible they date back much earlier still, helping to explain not only the success of human groups at war with each other but also why the spread of modern humans into new territories was so often accompanied by the extermination of rival species, from the large animals of Australasia and New Guinea to the Neanderthals of Europe.

It has long been appreciated that rare, traumatic rituals promote intense social cohesion, but efforts to tease apart the psychological mechanisms involved only really took off in the 1950s, much of the work inspired by Festinger's theory of *cognitive dissonance* (1957). Rituals *incur costs* (e.g., time, labor, and psychological endurance), often with the promise of only poorly defined or indeterminate rewards, and in some cases for no explicit purpose at all. In the case of initiations, the costs are typically extreme, for instance involving physical or psychological tortures, often of a terrifying kind. In a now classic application of Festinger's theory, Aronson and Mills (1959) demonstrated that the more severe the requirement for entry into an artificially created group, the greater would be the participants' liking for other group members. Their explanation for this was that our feelings toward the groups we join will never be wholly positive, and the experience of *disliking* aspects of the group will be *dissonant* with the experience of having paid a price to join; this dissonance could be resolved by downplaying the costs of entry, but the greater the severity of initiations into the group, the less sustainable that strategy will become. Under these circumstances, dissonance reduction will focus instead on generating more positive evaluations of the group.

More recent studies using psychological experiments, economic games, and cross-cultural surveys have shown repeatedly that within-group liking and out-group hostility are directly correlated (Cohen et al., 2006). As one game theorist neatly put it: "When Joshua killed twelve thousand heathen in a day and gave thanks to the Lord afterwards by carving the Ten Commandments in stone, including the phrase 'Thou shalt not kill,' he was not being hypocritical" (Ridley, 1996, p. 192).

For over two decades, our understanding of imagistic dynamics was based largely on a relatively small sample of detailed case studies (Whitehouse & Laidlaw, 2004; Whitehouse & Martin, 2004). But more systematic data on this topic are now being assembled. In a recent survey of 644 rituals selected from a sample of 74 cultures, Atkinson and Whitehouse found an inverse correlation between ritual frequency and levels of dysphoric arousal, with most rituals clustering around the two poles of the continuum (Atkinson & Whitehouse, 2010). In this survey, most low-frequency rituals involving intense dysphoric arousal were not used to mark entry into a group. They served a diversity of overt goals, such as communion with gods and spirits, honoring the dead, veneration of icons, the promotion of crop fertility, and so on. Moreover, extensive analysis of case studies has revealed considerable variability in the rationale assigned to them by participants (Whitehouse & Laidlaw, 2004; Whitehouse & Martin, 2004) and in some cases no rationale at all was provided (Barth, 1987; Humphrey & Laidlaw, 1994).

"Rites of terror" (Whitehouse, 1996) increase cohesion and tolerance *within groups,* but they also intensify feelings of hostility and intolerance *toward out-groups.* Cognitive dissonance does not appear to be the whole explanation. Two other factors also have crucial consequences for in-group cohesion and out-group hostility. One is *memory*: one-off traumatic experiences, especially ones that are surprising and consequential for participants, are remembered over longer time periods (and with greater vividness and accuracy) than less arousing events. Such memories have a canonical structure, sometimes referred to as "flashbulb memory" (Conway, 1995), specifying not only details of the event itself but what happened afterwards and who else was present. This last point is especially important in establishing the exclusivity of ritual communities: there is little scope for adding to or subtracting from ritual groups whose membership derives from uniquely encoded, one-off experiences.

The other factor is *interpretive creativity*. Since the procedures entailed in rituals are a matter of stipulation, and are not transparently related to overall goals (if indeed those goals are articulated at all), the meanings of the acts present something of a puzzle for participants. In the case of traumatic ritual experiences that are recalled for many months and years after the actual event, questions of symbolism and purpose are typically a major focus of attention. In a series of experiments using artificial rituals and varying levels of arousal, Richert, Whitehouse, and Stewart have shown that, after a time delay, the volume and specificity of spontaneous reflection on the meanings of rituals is substantially greater in high-arousal conditions than in controls (2005).

Similar effects have been found using field studies by systematically comparing the interpretive richness of people's accounts of rituals involving variable levels of arousal (Whitehouse, 1995; Xygalatas, 2007). Since rites of terror are typically also shrouded in secrecy and taboo, participants have little opportunity to compare the contents of their personal ruminations. As a result, they form the impression that their rich interpretations are shared by others undergoing the same experience, increasing the sense of camaraderie.

This heady cocktail of psychological mechanisms (cognitive dissonance, shared memory, and the illusion of common revelation) binds together small, exclusive communities of ritual participants. Groups formed in this way display high levels of trust, cooperation, and tolerance for fellow members. But there is also a darker side to this syndrome, which finds expression in out-group hostility. Comparative research, both ethnographic and historical, has revealed a strong correlation between rites of terror and chronic inter-group conflict and warfare (Cohen et al., 2006). Exactly why ritually induced cohesion produces intolerance toward out-groups requires further research.

Understanding small-group cohesion and out-group hostility within an evolutionary framework requires close attention not only to the proximate causes of the trait (in this case the psychological processes involved in rites of terror), but also the ultimate causes, namely *why* a given ritual syndrome spreads and persists over time. Cultural evolution is governed by many of the same fundamental principles as biological evolution, except that inheritance is by learning (rather than by genes); selection by consequences for cultural traits tends to be rapid; adaptive cultural mutations arise frequently (often as a result of deliberate innovation); and prior cultural forms are only loosely constraining (cultural revolutions do sometimes happen). Nevertheless, the study of how ritual variability affects the survival of cultural groups can be understood in the same basic terms that any evolutionary biologist would recognize. Specifically, we need to understand how changing features of a given group's ecology and resourcing needs might make the adoption of particular ritual forms *adaptive* (by contributing to group survival and reproduction over time), allowing also for the possibility of drift (random factors contributing to the ritual's persistence), and phylogeny (the constraints imposed by pre-existing ritual traditions).

Rites of terror (Whitehouse, 1996) and the intense cohesion these produce in small groups are typically an adaptation to conditions of tribal warfare (although also serving as a commitment mechanism for other dangerous pursuits, such as the hunting of large game). Activities involving high risk and temptation to defect (e.g., raiding, head-hunting, bride-capture, sectarian violence, and gangland disputes) would seem to be linked to the presence of low-frequency, dysphoric rituals (typically involving severe physical and psychological tortures). These patterns seem to have emerged and spread not only in simple societies (for instance, in sub-Saharan Africa, native America, Highland Philippines, Melanesia, Amazonia, etc.), but also play a prominent role in the formation of military cells in modern armies, terrorist organizations, and rebel groups (Whitehouse & McQuinn, in press). On this view, the imagistic mode is a kind of "gadget" for binding together military coalitions: in conditions of chronic warfare, groups lacking this gadget rapidly disappear and those possessing it systematically destroy, absorb, or become allied with their neighbors.

10.4 Ritual, Group Formation, and Competition: The Doctrinal Mode

Until just a few thousand years ago, group rituals were typically occasions for high excitement, but nowhere had people learned to regularize their rituals around daily or weekly cycles. High-frequency ritual (or *routinization*) is a hallmark of world religions and their offshoots, but is also characteristic of a great many regional religions and ideological move-ments. Routinized rituals play a major role in the formation of large-scale identities, enabling strangers to recognize each other as members of a common in-group, facilitating trust and cooperation on a scale that would otherwise be impossible. This syndrome has come to be known as the *doctrinal mode of religiosity* (Whitehouse, 1995, 2000, 2004). It heralds not only the first large-scale societies, but also the first complex political systems in which roles and offices are understood to be detachable from the persons who occupy them. To understand the proximate causes of these patterns, we need to return to the issue of how rituals are remembered.

When people participate in the same rituals on a daily or weekly basis, it is impossible for them to recall the details of every occasion. Instead they represent the rituals and their meanings as *types* of behavior—a Holy Com-munion or a call to prayer, for instance. Psychologists describe these rep-resentations as procedural scripts and semantic schemas. Scripts and schemas specify what typically happens in a given ritual and what is gener-ally thought to be its significance. In a group whose identity markers are composed mainly of scripts and schemas, what it means to be a member of the tradition is generalized beyond people of our acquaintance, applying to everyone who performs similar acts and holds similar beliefs. This route to the construction of communal identity, based on routinization, is a necessary condition for the emergence of *imagined communities* (Anderson, 1983)—large populations sharing a common tradition and capable of behaving as a coalition in interactions with non-members, despite the fact that no individual in the community could possibly know all the others, or even hope to meet all of them in the course of a lifetime.

Routinization has other important effects as well. For instance, it allows very complex networks of doctrines and narratives to be learned and stored in collective memory, making it relatively easy to spot unauthorized inno-vations. Moreover, routinization artificially suppresses reflection, in effect producing more slavish conformism to group norms. In one experiment, for instance, a group of thirty students performed an unfamiliar ritual twice a week for ten weeks and were then asked to post comments on the mean-ings of the ritual after each performance; reflexivity dramatically declined once the ritual had become a familiar routine (Whitehouse, 2004). Part of the reason seems to be that, having achieved procedural fluency, one no longer needs to reflect on *how* to perform the ritual, and this in turn makes one less likely to reflect on *why* one should perform it. Thus routinization would seem to aid the transmission of doctrinal orthodoxies, which are traditions of belief and practice that are relatively *immune to innovation* and in which unintended deviation from the norm is *readily detectable*.

Putting these things together, it would seem that routinized rituals provide a foundation for much larger-scale communities, capable of encompassing indefinitely many individuals singing from the same hymn sheet (both literally and metaphorically). Expanding the size of the in-group in this way has implications for the scale on which people can engage in cooperative behavior, extending both *trust* and *tolerance* even to strangers simply because they carry the insignia that display shared beliefs and practices. At the same time, however, the cohesion engendered through common membership in the tradition is less intensely felt than that accomplished in small groups undergoing rare and painful rituals together. In other words, as cohesion is expanded to encompass greater populations, it is also, in an important sense, spread more thinly. Some routinized traditions, however, manage to get the best of both worlds: a mainstream tradition, constructed around regular worship under the surveillance of an ecclesiastical hierarchy, may tolerate much more colorful local practices involving rare, dysphoric rituals (such as self-flagellation at Easter parades in the Philippines or walking on red hot coals among the Anastenaria of Northern Greece). While these localized practices undoubtedly produce highly solidary groups distinct from the mainstream tradition, the resulting cohesion can be projected onto the larger community, rejuvenating commitment to its unremitting regime of repetitive rituals (Whitehouse, 1995). Other patterns are also possible, however. One grand theorist of Muslim society, Ernest Gellner, showed that rural tribes bound together by high-arousal rituals formed the most formidable small military units in Islam, capable of periodically toppling urban elites, whose more routinized rituals and doctrinal beliefs failed to generate the kind of cohesion needed to mount an effective defense (Gellner, 1969). Other major patterns include periodic splintering and reformation (Pyysiäinen, 2004).

Although much work on these topics has been concerned with understanding the effects of psychological affordances, biases, and constraints, efforts are now being made to model the ultimate causes of patterns of religious group formation over time (Whitehouse, Bryson, Hochberg, & Kahn, in press). What factors favor the appearance and persistence of routinized rituals and the large-scale communities they engender? Some recent efforts to answer this question have focused on the first appearance of routinized collective rituals in human prehistory (Mithen, 2004; Johnson, 2004; Whitehouse & Hodder, 2010).

A watershed in the evolution of modes of religiosity seems to have occurred around 8,000 years ago at Çatalhöyük, in what is now Central Anatolia in Turkey. In the early layers of Çatalhöyük, the imagistic mode prevailed. There we find much evidence of low-frequency, high-arousal rituals, detectable from animal bones resulting from hunting and feasting activities, pictorial representations of major rituals, and human remains manipulated in elaborate mortuary practices. These practices would have produced highly cohesive groups necessary for coordinated hunting of large, dangerous animals. The boundedness of these groups may still be visible today in the massive trenches that appear to have divided communities in the earlier phases of settlement. But as hunting gradually gave way to farming, the need for such groups disappeared, and instead more day-to-day forms of cooperation across the settlement were required to sustain novel forms of specialized labor, reciprocity, pooling, and storage. Sustainable exploitation of the commons now required the dissolution of small-group boundaries and inter-group rivalry in favor of larger-scale forms of collective identity, trust, and cooperation extending to tens of thousands of individuals at the enlarged settlement.

This change in the scale of political association was facilitated by the appearance of the first-ever regular collective rituals, focused around daily production and consumption, and the spread of identity markers across the entire settlement, for instance in the form of stamp seals used for body decoration and more standardized pottery designs. The appearance and spread of routinized rituals seems to have been linked to the need for greater trust and cooperation when interacting with relative strangers. Consider the difficulties of persuading people you scarcely know that they should make long-term investments in your services based on a promise, or should pay taxes or tribute in return for protection or sustenance in times of need. In the absence of more detailed information about trustworthiness of prospective trading partners or remote governors to fulfill their part of any bargain, shared insignia proclaiming commitment to common beliefs and practices becomes a persuasive form of evidence. In such conditions, groups with routinized rituals capable of uniting large populations will tend to out-compete those who lack shared identity markers of this kind.

With the appearance of the first large-scale complex societies unified by routinized rituals, the dissociation between office and office-holder became more salient. In groups whose beliefs and practices are specified by generic scripts and schemas, we no longer represent our fellow members primarily as particular persons, but as incumbents of more generic qualities and roles (worshippers, imams, gurus, choir boys, etc.). The advent of routinized rituals heralds a fundamental shift from particularistic social relations to more universalistic conceptions of the social world, in which offices are understood as transcending the office holders, outliving them and regulating their behavior (Durkheim, 1912/2008). A natural corollary is that achieved status and power gives way to ascribed attributes that can be inherited. Principles of rank and royalty begin to override more personal qualities such as aggressivity or eloquence.

10.5 Modes of Ritual Domination and Cohesion

The shift from imagistic to doctrinal dynamics implies a change of strategy in the means of coercion. Anthropologists have distinguished two broadly contrasting modes of domination in political life (Sahlins, 1963; Bloch, 2008). First, there is the *transactional* strategy of persuasion and threat, the equivalent of "fighting and biting" among our non-human primate cousins. Individuals can garner resources through acquisitive strategies, for instance, based on the use of force (e.g., the exercise of superior individual physical strength and courage, or the command of a well-equipped militia); exceptional prosociality (e.g., acts of magnanimous generosity, or the espousal of doctrines seemingly contrary to the individual's self interest); the arts of diplomacy (e.g., protection of the group against perceived external threats using effective techniques of negotiation); or the manipulation of supernatural forces (e.g., holding sway with the gods or ancestors). All these methods of garnering power, status, and wealth are the outcomes of individual effort and are similar to (though arguably never identical to) the strategies of upward mobility found among other intelligent social animals, at least insofar as individual achievements cannot be passed on when the privileged individuals lose their grip or die. Thereupon, another achiever must succeed to the position of dominance.

Secondly, there is the *transcendental* route to leadership: the establishment of offices of an enduring kind, whose occupants are ascribed positions of superiority. Emperors, kings, chiefs, and popes (for instance) are, once installed, seen as inherently better than the rest of us by virtue of the offices they hold. Like any self-styled leader, such office-holders normally have reciprocal obligations to the subject population, but since their dominance is based on the rights of office rather than earned obligations, they can demand more than they replenish. Moreover, unlike the achiever, office-holders can pass on their dominions to successors (either by broadly democratic means or by principles of inheritance). Humans are unique among other animals in this respect. Inequalities in human societies are accomplished through a mixture of the transactional and the transcendental, but in widely varying degrees. For instance, New Guinea "big-men" are largely self-made and unable to pass on their networks to sons and nephews (Godelier, 1986; Whitehouse, 1991), whereas the inherited spiritual qualities (*mana*) of Polynesian chiefs enabled them to extract tribute by virtue of office and transmit their power, status, and wealth to succeeding generations, over time building ever larger empires (Feinberg & Watson-Gegeo, 1996). Yet even chiefly figures must be effective achievers if they are to consolidate or expand the jurisdiction of their authority. And some Melanesian big-men have succeeded in recruiting their sons to positions of influence. So we are dealing with differences of degree rather than kind in the emergence and reproduction of inequalities. But what drives the shift from achieved to ascribed forms of leadership is a stepping-up of the pace of ritual life and a reconceptualization of social relations in more abstract and formal terms.

The shift from transactional to transcendental modes of domination is accompanied by changes in social cohesion. Psychological studies suggest that when people think they share the same thoughts and emotions, they like each other more (Byrne, 1971). Arguably the most powerful glue that binds us to our fellows is the impression (sometimes only the illusion) of shared mental content, prompting not only greater liking for those who are like us, but greater confidence in their reliability. A possible evolutionary explanation for this finding is that shared mental content acts as a proxy for genetic relatedness; in other words, that we have an evolved propensity to treat people who share our memories, feelings, norms, values, etc. as *kin* (cf. Roscoe, 1993).

We identify shared mental content as a consequence of gathering information about each other via two broadly distinct channels: testimony and personal experience. Whereas our explicit beliefs about the qualities of people and places (including the dangers they pose) rely heavily on testimony, we tend to accord greater weight to experience (and the inferences derived from it). Actions, so the saying goes, speak louder than words. For instance, while we might base much of what we know about a partner on her self-presentation and the testimony of others, leading to a well-formed portrait at an early stage of the relationship, our confidence in the accuracy of our portrait only gathers momentum over many experiences of the partner's behavior over time (and our interpretations of that behavior, which presumably may be quite heavily colored by our more testimony-based beliefs about her).

Prior to the emergence of the doctrinal mode in human prehistory, group identity was forged entirely on the basis of directly shared experiences—including participation in rituals—that enriched our internally generated representations about co-participants and evinced high levels of confidence in their trustworthiness. Any set of memorable shared experiences could produce this effect, but the more representations a set of shared experiences can elicit over time (as a result of observation and private reflection rather than verbal testimony), the more confidence we have in the trustworthiness of co-participants and the more rigidly we adhere to the group's values and beliefs. The imagistic mode has long proven to be exceptionally effective at producing the illusion of shared mental content based on common experience. With the appearance of more routinized rituals, however, a new kind of group identity became possible based on semantic schemas and procedural scripts that could be generalized to any member of the in-group, even to complete strangers. Simply wearing a certain mode of dress or hairstyle now revealed a lot about a person's beliefs and practices. We could then make inferences on this basis about their trustworthiness, even people we had never met before. But just as this kind of cohesion could spread more widely, it was inevitably spread more thinly. Group identity cast in this generic mold, like testimony pertaining to people's personal character and history, may activate our evolved kin-detection heuristics, but only weakly, because it is not based on direct experience of the person in question.

10.6   General Discussion

Ritualized behavior is rooted in our evolved psychology, closely linked to our natural propensity to imitate trusted others. Understanding the nature, origins, and developmental pathways of this propensity is primarily a task for experimental psychology. Rituals also have some striking affordances for group-building, the variable exploitation of which is most productively understood within the framework of sociocultural evolution. Collective rituals come in two broad varieties: low-frequency/high-arousal (associated with the imagistic mode of group formation) and high-frequency/low-arousal (associated with the doctrinal mode). A large body of research has been conducted into the proximate causes of imagistic and doctrinal dynamics, including the accumulation of case studies, new field research, large-scale ethnographic surveys, and psychological experiments. Studies of ultimate causation have so far been devoted to the development of theoretical models, especially agent-based computational simulations, and to empirical studies focusing especially on the archeology of the Neolithic Middle East and the history of Europe and the Mediterranean.

More systematic research is needed into the role of ritual in the formation and regulation of human societies. There is much yet to discover about how people learn the rituals of their communities and how rituals promote social cohesion within the group and distrust of groups with different ritual traditions. Qualitative field research and controlled psychological experiments are needed in a wide range of societies to explore the effects of ritual participation on in-group cohesion and out-group hostility. Longitudinal databases would be particularly useful if we are to explore the evolution of ritual, resource extraction patterns, and group structure and scale over significant time periods. Further research is also needed to establish more clearly the relationship between ritual and religion, the topic with which this chapter concludes.

Many of the rituals recorded by anthropologists and historians are carried out with the ostensible purpose of maintaining relationships with gods, ancestors, and creator beings. Moreover, there appears to be a link between ritual action and magical thinking more generally. This raises the question whether rituals might serve in some way to bind together the disparate concepts that are commonly referred to as "religion." Even if the psychology required to build concepts of gods is quite different from that needed to conceive of an afterlife or a creator or a magical spell, nevertheless there might be some feature (or cluster of features) associated with ritualized action that tends to trigger all those features we commonly describe as religious, so lending the category some kind of coherence.

Perhaps rituals prompt us to think about magical causation and supernatural agency as a consequence of their causal opacity. The fact that we cannot specify any physical-causal link between ritual actions and hoped-for outcomes encourages us to postulate a supernatural link, whether quasi-mechanistic or agent-driven. There is some evidence that rituals proliferate around activities with uncertain outcomes. Athletes, for instance, are prone to performing rituals spontaneously when the risks of failure are most acute and the longing for success most intensely felt (Sørensen, 2007). Often these rituals take the form of appeals to supernatural agents, however vaguely specified. Ethnographers have long noted that rituals tend to accompany risky endeavors, but less so in pursuits entailing more predictable outcomes (Malinowski, 1935/2001, 1945/1992). But little is known about the psychological mechanisms linking teleological opacity with appeals to supernatural forces or agents.

Traditional rituals may also prompt us to think about dead ancestors and questions of origin and creation because of the prescriptive character of the behavior. Where causally opaque actions are culturally prescribed or normative, our ordinary intuitions about the intentionality of actors are disrupted. Clearly, the ritualized actions cannot be comprehended as the expression of intentional states internal to the actor, but derive at least partly from the intentional states of actors who came before and stretch back into the past (Whitehouse, 2004). Thus, rituals prime reflection on the minds of the dead—the group's ancestors —leading inexorably also to questions of origin and creation.

Ritual is popularly misconstrued as an exotic, even quirky topic—a facet of human nature that, along with beliefs in supernatural agents and magical spells, is little more than a curious fossil of pre-scientific culture, doomed to eventual extinction in the wake of rational discovery and invention. Nothing could be further than the truth. Humans are as ritualistic today as they have ever been. Even the most secular political systems ever devised, for instance under the sway of historical materialism and its vision of a communist utopia, were as devoted to ritual as any in human history. Each time a child is born, a new bearer of rituals from the past is created: another member of *Homo ritualis*.

Acknowledgment

This work was supported by an ESRC Large Grant (REF RES-060-25-0085) entitled "Ritual, Community, and Conflict."

References

Anderson, B. (1983). *Imagined communities: Reflections on the origin and spread of nationalism*. London: Verso.

Aronson, E., & Mills, J. (1959). The effect of severity of initiation on liking for a group. *Journal of Abnormal and Social Psychology, 59*, 177–181.

Atkinson, Q. D., & Whitehouse, H. 2010. The cultural morphospace of ritual form: Examining modes of religiosity cross-culturally. *Evolution and Human Behavior, 32*(1), 50–62.

Barth, F. (1987). *Cosmologies in the making: A generative approach to cultural variation in inner New Guinea*. Cambridge: Cambridge University Press.

Bloch, M. (2008). Why religion is nothing special but is central. *Philosophical Transactions of the Royal Society, 363*, 2055–2061.

Boyer, P., & Lienard, P. (2006). Why ritualized behavior? Precaution systems and action parsing in developmental, pathological and cultural rituals. *Behavioral and Brain Sciences*, *29*(6), 595–612.

Cohen, T. R., Montoya, R. M., & Insko, C. A. (2006). Group morality and intergroup relations: Cross-cultural and experimental evidence. *Personality and Social Psychology Bulletin*, *32*, 1559.

Conway, M. A. (1995). *Flashbulb memories*. Hillsdale, NJ: Lawrence Erlbaum.

Durkheim, E. (1912/2008). *The elementary forms of religious life*. New York: Free Press. (Original work published 1912)

Feinberg, R., & Watson-Gegeo, K. A. (Eds.). (1996). *Leadership and change in the Western Pacific*. London: Athlone.

Festinger, L. A. (1957). *A theory of cognitive dissonance*. Stanford, CA: Stanford University Press.

Frazer, J. (1922). *The golden bough*. London: Macmillan.

Geertz, C. (1973). *The interpretation of cultures*. New York: Basic Books.

Gellner, E. (1969). A pendulum swing theory of Islam. In R. Robertson (Ed.), *Sociology of religion: Selected readings*. Harmondsworth, UK: Penguin.

Godelier, M. (1986). *The making of great men*. Cambridge: Cambridge University Press.

Humphrey, C., & Laidlaw, J. (1994). *The archetypal actions of ritual: A theory of ritual illustrated by the Jain rite of worship*. Oxford: Clarendon Press.

Ibn Khaldūn, A. (1958) *The Muqaddimah: An introduction to history*. Princeton, NJ: Princeton University Press.

Johnson, K. (2004). Primary emergence of the doctrinal mode of religiosity in prehistoric southwestern Iran. In H. Whitehouse & L. H. Martin (Eds.), *Theorizing religions past: Historical and Archaeological Perspectives on Modes of Religiosity*. Walnut Creek, CA: AltaMira Press.

Kertzer, D. (1988). *Ritual, politics, and power*. New Haven, CT: Yale University Press.

Legare, C. H., & Whitehouse, H. (under review). The imitative foundations of cultural learning.

Malinowski, B. (1935/2001). *Coral gardens and their magic: A study of the methods of tilling the soil and of agricultural rites in the Trobriand Islands*. London: Allen and Unwin. (Original work published 1935)

Malinowski, B. (1945/1992). *Magic, science and religion and other essays*. Garden City, NY: Doubleday. (Original work published 1945)

Mithen, S. (2004). From Ohalo to Çatalhöyük: The development of religiosity during the early prehistory of western Asia, 20,000–7000 BC. In H. Whitehouse & L. H. Martin (Eds.), *Theorizing religions past: Historical and archaeological perspectives.* Walnut Creek, CA: AltaMira Press.

Pachis, P., & Martin, L. H. (Eds.). (2009). *Imagistic traditions in the Graeco-Roman world.* Thessaloniki, Greece: Vanias.

Pyysiäinen, I. (2004). Corrupt doctrine and doctrinal revival: On the nature and limits of the modes theory. In H. Whitehouse & L. H. Martin (Eds.), *Theorizing religions past: Historical and archaeological perspectives on modes of religiosity.* Walnut Creek, CA: AltaMira Press.

Richert, R. A., Whitehouse, H., & Stewart, E. E. A. (2005). Memory and analogical thinking in high-arousal rituals. In H. Whitehouse & R. N. McCauley (Eds.), *Mind and religion: Psychological and cognitive foundations of religiosity.* Walnut Creek, CA: AltaMira Press.

Ridley, M. (1996). *The origins of virtue.* London: Penguin Viking.

Robertson Smith, W. (1889/2002). *Religion of the Semites. Fundamental institutions. First series.* London: Adam & Charles Black. (Original work published 1889)

Roscoe, P. (1993). Amity and aggression: A symbolic theory of incest. *Man (N.S.)* 28:49–76.

Rozin, P. (1999). Food is fundamental, fun, frightening, and far-reaching. *Social Research, 66,* 9–30.

Sahlins, M. (1963). Poor man, rich man, big-man, chief: Political types in Polynesia and Melanesia. *Comparative Studies in Society and History, 5*(3), 285–303.

Sørensen , J. (2007). *A cognitive theory of magic.* Walnut Creek, CA: Altamira Press.

Tinbergen, N. (1963). On the aims and methods of ethology. *Zeitschrift für Tierpsychologie, 20,* 410–433.

Tomasello, M. (1999). *The cultural origins of human cognition.* Cambridge, MA: Harvard University Press.

Turchin, P. (2006). *War and peace and war: The rise and fall of empires.* New York: Penguin Group.

Weber, M. (1947). *The theory of social and economic organization.* Oxford: Oxford University Press.

Whitehouse, H. (1991). Leaders and logics, persons and polities. *History and Anthropology, 6*(1), 103–124.

Whitehouse, H. (1995). *Inside the cult: Religious innovation and transmission in Papua New Guinea.* Oxford: Oxford University Press.

Whitehouse, H. (1996). Rites of terror: Emotion, metaphor, and memory in Melanesian initiation cults. *Journal of the Royal Anthropological Institute, 4*, 703–715.

Whitehouse, H. (2000). *Arguments and icons: Divergent modes of religiosity.* Oxford: Oxford University Press.

Whitehouse, H. (2004). *Modes of religiosity: A cognitive theory of religious transmission.* Walnut Creek, CA: AltaMira Press.

Whitehouse, H. Bryson, J., Hochberg, M., & Kahn, K. (in press). The potential of modeling religion. *Method and theory in the study of religion.*

Whitehouse, H., & Hodder, I. (2010). Modes of religiosity at Çatalhöyük. In I. Hodder (Ed.), *Religion in the emergence of civilization: Çatalhöyük as a case study.* Cambridge: Cambridge University Press.

Whitehouse, H. (2011) The coexistence problem in psychology, anthropology, and evolutionary theory. *Human Development, 54*, 191–199.

Whitehouse, H., & Laidlaw, J. (2004). *Ritual and memory: Toward a comparative anthropology of religion.* Walnut Creek, CA: AltaMira Press.

Whitehouse, H., & Laidlaw, J. (2007). In H. Whitehouse & J. Laidlaw (Eds.), *Religion, anthropology and cognitive science.* Durham, NC: Carolina Academic Press.

Whitehouse, H., & Martin, L. (2004). *Theorizing religions past: Archaeology, history, and cognition.* Walnut Creek, CA: AltaMira Press.

Whitehouse, H., & McCauley, R. N. (2005a). *Mind and religion: Psychological and cognitive foundations of religiosity.* Walnut Creek, CA: AltaMira Press.

Whitehouse, H., & McCauley, R. N. (2005b). The psychological and cognitive foundations of religiosity. *Journal of Cognition and Culture, 5*, 1–2.

Whitehouse, H., & McQuinn, B. (in press). Ritual and violence: Divergent modes of religiosity and armed struggle. In M. Kitts, M. Juergensmeyer, & M. Jerryson (Eds.), *Oxford handbook of religion and violence.* Oxford: Oxford University Press.

Wight, R. (2007). *The peacock's tail and the reputation reflex: The neuroscience of art sponsorship.* The Arts & Business Lecture.

Wilson, D. S. (2002). *Darwin's cathedral: Evolution, religion, and the nature of society.* Chicago: University of Chicago Press.

Wiltermuth, S. S., & Heath, C. (2009). Synchrony and cooperation. *Psychological Science, 20*, 1–5.

Xygalatas, D. (2007). *Firewalking in northern Greece: A cognitive approach to high-arousal rituals* (Unpublished doctoral dissertation). Queen's University, Belfast, Northern Ireland.

V ECONOMICS

11 Cognitive Variables and Parameters in Economic Models

Don Ross

11.1 Introduction: Overview of Cognitive Variables in Economics

The current relationship between cognitive science and economics is complex, contested, and unsettled. An overview of it therefore cannot be a summary of points of consensus, but must indicate and explain points of disagreement. The task is complicated by the fact that the economics–cognitive science relationship inherits residues from a long history of interanimations between economics and psychology that have implicitly shaped the attitudes and methodological biases of practitioners. In consequence, particular economists' pronouncements of their views on the role of cognitive science are often less informative than what is revealed by their practices. Finally, at all times in the history of economics, an orthodoxy or mainstream methodology has coexisted with various dissenting heterodoxies. In this chapter, when it is said that economists of a given time and place thought or think such-and-such, this should be understood as referring to the orthodoxy of that time and place.

Let us begin with a broad sketch of the evolution of economists' attitudes toward psychology prior to the coming of the cognitive revolution in the 1960s. The earliest major thinkers on topics that preoccupy modern economists, in particular Smith and Hume, did not concern themselves with a boundary between economics and psychology because no traditions or institutions had yet stabilized such a frontier. Read from a time in which institutionalized disciplinary borders are being redrawn, Smith and Hume often seem more modern and sensible than their successors. However, as there is not space in this chapter to do any justice to the rich recent scholarship on Smith's thought, it will be set to one side, and the sketch will open with 1870, the beginning of the decade in which the early marginalists self-consciously separated economics from political philosophy.

The founders of modern economics, Walras, Jevons, and Menger, disagreed on various matters. However, all shared the view, doubted by some of their successors, that aggregate economic phenomena and regularities should be explained by reference to the actions and dispositions of individual people. That is, all were what we would now call *methodological individualists*.

Jevons (1871) exemplified what we might call the English philosophy, explicitly derived from Bentham's utilitarian psychology. According to this perspective, psychological mechanisms that are partly innate and partly adaptive endow people with natural, relative, subjective valuations of experiences. These valuations are then directly reflected in their choices among alternative consumption baskets. Two key general properties of the natural human psychology of valuation, according to Jevons, were that (1) wants are insatiable, so that the person who has fulfilled her basic needs will devote her energies to the pursuit of more complex consumption experiences; and (2) people attach decreasing marginal value to further increments of any given stream of consumption goods as their stock, their rate of flow of that stream, or both, increase. These principles lead directly to an emphasis on *opportunity cost*—analysis of the value of a choice in terms of the alternative possibilities it forecloses—that has characterized all subsequent mainstream economics.

Jevons was optimistic that empirical research in psychology would uncover the mechanisms that give rise to the principles of non-satiation and decreasing marginal utility, and also to the sources of variance in individual tastes. It is therefore appropriate to say that he anticipated eventual unification of economics with psychology; the need to drive economic analysis from the top-down logic of maximization of utility through minimization of opportunity cost was, on his view, a temporary artifact of ignorance of psychological mechanisms. One important caveat is necessary in this regard. Along with many of the first generation of modern economists, Jevons doubted that people's social and moral preferences are determined by mechanisms that empirical psychology would reveal. Such "higher wants" were taken by him to be functions of distinctive moral reasoning that people must learn from teaching and philosophical reflection. Thus Jevons and many of his immediate successors thought that economists should concentrate their attention on selfish consumption, not because they believed that people are narrowly selfish by nature, but because they thought that other-regarding behavior remained outside the scope of both economics and mechanistic psychology.

This historical English view of the relationship between economics and psychology contrasts with a rival Austrian one, articulated originally by Menger (1871/1981). Its mature exposition was given by von Mises (1949), and the boiled-down articulation and simplification through which most economists know it was provided by Robbins (1935). According to this view, the foundations of economics lie in broad psychological truisms that are discovered by introspection and logical reflection, and which are not susceptible to significant modification by discoveries of scientific psychology concerning their underlying basis. (Jevons's two principles above were held by Robbins to be among the truisms in question.) Robbins explicitly argued that economists in their professional capacity should ignore psychology. Though the psychology he had in mind, given the time at which he wrote, was behaviorist, his premises would apply equally to contemporary cognitive science. In addition, Robbins denied that economic analysis applies only to choices among "lower," material goods, arguing that opportunity cost influences the consumption of anything that is scarce, including sources of social, moral, and intellectual satisfaction.

The methodological views of the Austrians have enjoyed more success among economists than their first-order economic doctrines. The period from the end of the nineteenth century to the 1980s saw a steady retreat of the English view. In consequence, economists—with the notable exception of Herbert Simon—were not important contributors to, or an important audience for, the early stages of the cognitive revolution. However, the past four decades have witnessed an increasing resurgence of Jevons's position as so-called behavioral economists and their collaborators from psychology have empirically unearthed cognitive biases and framing effects that influence economic decisions (Kahneman, Slovic, & Tversky, 1982); posited computationally fast and frugal heuristics that allow people to exploit environmental regularities and ignore information that would otherwise overwhelm limited processing capacities (Gigerenzer, Todd, & ABC Research Group, 1999); and modeled endogenous, dynamic construction of preferences in response to experience (Lichtenstein & Slovic, 2006).

Behavioral economists regularly accuse their less psychologically informed colleagues of applying false models of economic agency that are derived from normative attributions of rationality rather than from empirically tested hypotheses about the actual cognitive resources with which evolution has equipped our species, and which learning and development can refine only to certain limits. (For a widely cited example of such criticism, see Camerer et al., 2005. Angner & Loewenstein, in press, is more comprehensive in coverage and somewhat more cautious in tone. Ariely, 2008, is one of many recent popular announcements of revolution against what is held to be orthodox dogma.) Ross (2008) sorts the discoveries that are held to challenge conventional economics into four sets: (1) findings that people don't reason about uncertainty in accordance with sound statistical and other inductive principles; (2) findings that people behave inconsistently from one choice problem to another as a result of various kinds of framing influences; (3) findings that people systematically reverse preferences over time because they discount the future hyperbolically instead of exponentially; and (4) findings that people don't act so as to optimize their personal expected utility, but are heavily influenced by their beliefs about the prospective utility of other people, and by relations between other peoples' utility and their own. Note that the findings in set (4) challenge mainstream economics only insofar as it is taken to be committed to the hypothesis that individual economic behavior is motivated by narrow selfishness; but this attribution applies only to the rhetoric of some well-known economists, rather than to any proper element of economic theory (Ross, 2005; Binmore, 2009). Findings in sets (1) through (3) challenge the idea that people's economic behavior conforms to normative ideals of rationality. It is natural to suppose that this might result from features of their brains or cognitive architectures that reflect adaptive sculpting by particular environments. This is why the results of behavioral economics are taken to imply the inclusion of cognitive processing variables in economic models.

The past decade has, in addition, seen the rise of the interdisciplinary field of neuroeconomics, which studies the mediation of choices and valuation of consumption alternatives by neural computations in specific brain areas (Glimcher, Camerer, Fehr, & Poldrack, 2009). Standard methodologies in neuroeconomics involve confronting monkeys and humans with reward choices while measuring their neural activity through, respectively, invasive single-cell recording and functional magnetic resonance imaging (fMRI). Neuroeconomists often interpret their work as realizing Jevons's anticipation of a time when psychologists would directly observe mechanisms responsible for varying subjective judgments of value.

Some aspects of Jevons's view have not enjoyed revival. No behavioral economists follow him in thinking that social motivations lie outside their domain. If anything, they tend to invert the Jevonsian picture: narrowly selfish conduct is typically held to be a learned, cognitive response to impersonal market institutions, while more primitive, automatic, and cognitively encapsulated neural-computational mechanisms incline people to sympathetic action for the benefit of perceived affiliates, and inherent anti-sympathy for and distrust of non-affiliates.

The side of opinion descended from the Austrian view has also evolved. In particular, the attempt to ground economic generalizations in introspectively accessible psychological truisms has been largely abandoned. Completing an effort begun by Pareto (1909/1971), Samuelson's revealed preference theory (RPT) derived the existence of sets of preferences mappable onto the real numbers by monotonic, complete, acyclical, and convex functions from observable schedules of aggregate demand (1938, 1947/1983). Samuelson would have preferred not to call these "utility" functions, but the lure of semantic continuity turned out to be decisive. Notwithstanding the semantic suggestion of "revealed" preference, his utility functions were intended as descriptions of actual and hypothetical behavior, not indicators of hidden inner states. It is common to attribute the motivation for this to the behaviorism and positivism that dominated the psychology and social science of the 1930s, 40s, and 50s, and certainly this influence played its part. However, the main impetus was the strongly felt doubt, clearly articulated by Robbins (1935) and also expressed by Pareto, Fisher and Hicks, that economic generalizations about aggregate responses to changes in supply, demand, or prices are hostage to contingencies about individuals that are discoverable by scientific psychology. Post-Samuelson economists began from observable aggregate demand. The central result of RPT was that, if aggregate demand has certain testable properties, then the existence of continuous preference fields is implied. The causal influences that stabilize such fields might or might not be properties of individual psychologies; the revealed preference theorist disavows professional interest in this question. Becker (1962) showed that the fundamental property of the standard model of the market—downward sloping demand for any good, given, constant real income—depends on no claim about the *computational* rationality of any agent; it depends only on the assumption that households with smaller budgets, and therefore smaller opportunity sets, consume less.

The dominance of RPT in economics was the primary reason for the initial disinterest shown by economists—outside of a small dissident minority championed by Simon (1969, 1972)—in the cognitive revolution. Two main developments that unfolded across economics during the 1970s undermined this cross-disciplinary isolationism.

The first of these developments was the extension of game theory (GT) to ever-widening subfields of microeconomics. Resistance to incorporating individual psychological variables in economic models was based to some extent on practical concerns about mathematical tractability: the constrained optimization methods used by economists to solve most problems did not readily allow for representation of idiosyncratic preference structures among different agents, of the sort that psychologically derived hypotheses about motivation tend to furnish. This was precisely the restriction lifted by GT. Furthermore, GT didn't merely make the introduction of psychological variables into economic models *possible*; in specific ways, it *invited* such introduction. Application of GT to economics required the enrichment of utility theory that von Neumann and Morgenstern (1944) and Savage (1954) provided in order to incorporate players' uncertainty about the valuations of and information available to other players. This enrichment was elucidated at every step by formalization of concepts drawn from everyday psychology. Thus the non-mathematical version of the vocabulary of GT came to be full of psychological notions, particularly beliefs, conjectures, and expectations. GT *can* be given a strictly behaviorist interpretation, according to which one uses it to guide inferences about players' stable behavioral orientations through observing which vectors of possible behavioral sequences in strategic interanimation are Nash equilibria. But the power of such inferences is often limited, because most games have multiple Nash equilibria. Efforts to derive stronger predictions led a majority of economic game theorists in the 1980s to follow the lead of Harsanyi (1967) in interpreting games as descriptions of players' *beliefs* instead of their *actions*. On this interpretation, a solution to a game is one in which all players' conjectures about one another's preferences and (conditional) expectations are mutually consistent. Such solutions are, in general, stronger than Nash equilibria, and hence more restrictive. The resulting "equilibrium refinement program" drew game theorists deep into evaluation of alternative, speculative computational models. This program has somewhat petered out over recent years after devolving into unproductive philosophical disputes over the meaning of rationality; however, along the way it familiarized many economists who had ignored Simon's work with issues surrounding computational tractability.

Second, the formal completion of general equilibrium theory by Arrow and Debreu (1954) required clearer formalization of the idea of the economic agent (Debreu, 1959). In particular, it was necessary to assume that the agents in general equilibrium models can rank all possible states of the world with respect to value, and that they never change their minds about these rankings. As argued by Ross (2005), nothing in this regimentation requires that agents be interpreted as perfectly coextensive with natural persons. However, a crucial intended point of general equilibrium theory since its inception had been to serve as a framework for thinking about the consequences of changes in exogenous variables, especially policy variables, for overall welfare. Improvement and decline in the feelings of natural persons about their well-being is what most people, as a matter of fact, mainly care about. Thus if the loci of preference fields in general equilibrium theory are not at least idealizations of natural persons, then it is not evident why efficiency, the touchstone of general equilibrium analysis, should be important enough to *warrant* touchstone status. Once one acknowledges pressure to include models of distinct human consumers in general equilibria, however, troubles arise. The so-called excess demand literature of the 1970s (Sonnenschein, 1972, 1973; Mantel, 1974, 1976; Debreu, 1974) showed that although all general equilibria are efficient, there is no unique one-to-one mapping between a given general equilibrium and a vector of individual demand functions. (Put more plainly, for a given set of demand functions, there is more than one vector of prices at which all demand is satisfied.) In tandem with the Lipsey-Lancaster (1956) theory of the second-best, the Sonnenschein-Mantel-Debreu theorem challenged the cogency of attempts by welfare economists to justify policy by reference to merely inferred (as opposed to separately and empirically observed) subjective preferences of consumers. Note that this problem arises whether one assumes an atomistic or an intersubjective theory of the basis of value. Nevertheless, the excess demand results shook the general postwar confidence that if one attended properly to the aggregate scale, then specific preferences of individuals could be safely ignored. Thus, we now find some leading economists (e.g., Bernheim & Rangel, 2008) devoting effort to incorporating psychological limitations and agent heterogeneity into normative models.

These developments have certainly not convinced all economists to join a "back to Jevons" movement. Gul and Pesendorfer (2008) recently published a manifesto against such incorporation. They deny as a matter of principle that anything can be learned about economics by studying neural information processing or cognitive structures. According to Gul and Pesendorfer, the task of positive economics is to predict choices as functions of shifts in incentives and opportunity sets. Such choices are abstract constructs rather than directly observable phenomena, as are all relationships into which they enter. Though Gul and Pesendorfer concede that the mechanics of choice presumably involve neural computations, they see this as no more relevant to the *economics* of choice than is the fact that every choice must occur at some specific set of geographical coordinates.

The intended value of such extreme abstraction lies in achieving generalizations of maximum scope. Economic models, on this conception, should be represented by reduced-form equations that leave psychological and neural processing variables inside black boxes.

To judge from responses to Gul and Pesendorfer's paper that have appeared to date, few economists endorse such strong disciplinary isolation. Harrison (2008), despite being sympathetic to the idea that economic models should include no more psychological variables than are necessary to best fit aggregate empirical choice data in large subject samples, argues that the idea that economists could black-box all processing variables in reduced-form equations is a theorist's illusion that could not be seriously entertained by any researcher with experience in modeling real experimental data from human subjects. However, the Gul and Pesendorfer critique is useful for establishing a general methodological standard by which to evaluate claims about the relevance of cognitive influences on economic relationships. The general goal of the economist is to propose and test models that feature patterns of choices as dependent variables, with incentives and budgets among the independent variables. Which additional independent variables, including variables for cognitive structures and processes, should be included in empirically adequate and predictively powerful generalizations is a matter to be determined by evidence, not a priori reflection. Most economists, including many who conduct experiments, probably agree with Gul and Pesendorfer that, in light of the value of parsimony in modeling, the burden of argument in any given case lies with the theorist who proposes additional variables or parameters.

This is not the view promoted by "radical" behavioral economists, who argue that the discipline stands in need of a "paradigm shift" following a long history of inadequate psychological realism. Popular accounts, and occasional scholarly publications (e.g., Zak, 2008), suggest that the need to condition economic models on cognitive or neural structures and processes follows directly from the fact that economic behavior must in every instance be produced by cognitive or neural computation. The invalidity of this general form of reasoning (which would license the conclusion that variables from fundamental physics must appear in every generalization of every science) should be obvious. Nevertheless, someone might reasonably wonder how there could be generalizations about people's responses to changes in prices or other incentives that are *not* based on generalizations about human cognitive architecture.

A partial answer to this question may lie in what Vernon Smith (2007) refers to as the *ecological* nature of economic rationality. Ecological rationality emphasizes the extent to which people, at least sometimes, approximate consistent, optimizing rationality in their choice behavior by means not of computational marvels they achieve with their raw cognitive or neural apparatus, but thanks to what Hutchins (1995) and Clark (1997) call "scaffolding." Social scaffolding consists of external structures in the environment that encode culturally accumulated information and constrain and channel behavior through processes that Grassé (1959) called *stigmergic*. Grassé's example was of termites coordinating their colony building actions by means of algorithms that take modifications to the environment made by other termites as inputs; the modifications themselves, in the contemporary terminology, are scaffolding. The social scaffolding elements that support stigmergy can become *cognitive* scaffolding when agents internalize them to cue thoughts through association. Sun (2006, p. 13) summarizes the idea as follows:

First of all, an agent has to deal with social and physical environments. Hence, its thinking is structured and constrained by its environments. ... Second, the structures and regularities of an environment may be internalized by agents, in the effort to exploit such structures and regularities to facilitate the attainment of needs. Third, an environment itself may be utilized as part of the thinking (cognition) of an agent ... and therefore it may be heavily reflected in the cognitive process of an agent.

Consider a simple example of social scaffolding first. Most people who regularly consume alcohol avoid becoming addicted to it. Some may do this by carefully choosing consumption schedules that allow them to maintain high equilibrium levels of the neurotransmitters serotonin and GABA, which inhibit control of consumption by midbrain dopamine systems that are vulnerable to obsessive recruitment of attention by short reward cycles (Ross, Sharp, Vuchinich, & Spurrett, 2008). This typically involves manufacture and maintenance of personal rules that, as Ainslie (2001) demonstrates, require complex cognitive self-manipulation, at least until they become habitual. Another person might avoid addiction without calling upon any effortful will power by becoming an airline pilot, thereby facing extreme sanctions for indulgence that are so strong as to prevent temptation from ever arising (Heyman, 2009). The first person thus relies upon inboard cognition to regulate consumption, while the second person is restricted by social scaffolding. To an economist persuaded by the Gul and Pesendorfer view, what is relevant about the two cases is captured by a single model of that portion of the population that does not drink more when the relative price of alcohol falls, and the cognitively interesting dynamics of the first case are left aside.

As an example of cognitive scaffolding, consider a person whose assets consistently increase, not because she knows how to gather information or think about financial markets, but because she has learned the simple rule "buy through whichever broker was used by the friend who suggests an asset purchase," and she has lots of friends who tell her about their investments. Her simple heuristic causes her to distribute her business across many brokers, and thus she hedges.

The phrase *scaffolding* has not yet entered the economics literature. However, economists often speak of institutions in a way that is general enough to incorporate the idea. An example that is prominent in the literature is Gode and Sunder's (1993) simulation of "zero intelligence" agents who participate in a double auction experiment subject to budget constraints and very simple rules. These rules are that sellers do not charge more than their marginal cost, and buyers do not make negative offers or offers above a fixed ceiling. Otherwise the agents bid randomly. The efficiency of these simulated markets matches that achieved by human subjects. This suggests that efficiency of outcomes in such markets may result from the ecological rationality of the institutional rules, rather than sophisticated "inboard" computations. Sunder (2003) discusses generalizations of this result, and further economic applications.

Widespread use of scaffolding to achieve economic rationality does not suggest that cognitive variables are irrelevant to economic modeling; quite the contrary. People making decisions by use of fast and frugal heuristics will tend to suffer catastrophic performance collapses if their environments change drastically, and this will only be successfully predicted by a model that captures the cognitive structures underlying their choices. What the importance of scaffolding *does* indicate is a main reason for the invalidity of direct inferences from the stability of an economic regularity to the existence of any matching cognitive regularity. In the case of any given regularity, institutions may be carrying more or less of the load. The issue is always purely empirical and contingent.

There is an orthogonal sense in which appeal by economists to institutional scaffolds implies indirect reference to cognitive elements. Institutions are not merely self-reproducing social regularities; they are, in addition, partly sustained by norms. Conte and Castelfranchi (1995) argue that norms are, in turn, more than just the focal points for coordination recognized (though still not formally axiomatized) by game theorists. Norms must be represented by agents *as* norms in order for agents who internalize them to expect to suffer the distinctive costs that accompany norm violation. Thus norms imply cognitive architectures that can support the relevant representations. Note, however, that this does not necessarily imply the inclusion of the implicated architectural variables in economic models. A model might simply distinguish between normatively regulated and other behavior, in ways that are econometrically detectible, without specifying the cognitive elements that must underlie the normative regulation; these could, in principle, remain black-boxed. This strategy is exemplified, for example, by the well-known work of Bicchieri (2005), who incorporates norms into individual utility functions for application in game-theoretic models.

Over the past few decades, some economists have joined other social scientists in becoming interested in multi-agent models aimed at simulating social dynamics. Representative selections of this work can be found in Anderson, Arrow, and Pines (1988), Arthur, Durlauf, and Lane (1997), and Blume and Durlauf (2005), all based on work emanating from the Santa Fe Institute in New Mexico. Its theoretical basis lies in evolutionary game theory, for which foundational mathematical relationships with classical equilibrium solution concepts have been extensively though incompletely studied (Weibull, 1995; Cressman, 2003). Almost all of this work to date involves cognitively simple agents, and thereby black-boxes both the influence of cognitive variables on social interactions, and feedback effects of the latter on the former. The program urged by Sun (2006) for combining cognitive with social modeling has thus not yet been taken up by economists, except in a few isolated instances focused on restricted topics (e.g., Conte & Paolucci, 2002). This might be regarded as surprising because, as will be discussed in detail in the next section, economists have devoted considerable attention to models of individual people as disunified agents. The explanation probably lies in the strong interdisciplinary preference in economics for models that can be represented by closed-form equations. There is little or no stability of state variables from one complex-system model of the economy to the next. As long as this is the case, economists are likely to doubt that this methodology promises to deliver accumulation of theoretical knowledge. Under such circumstances, it is perhaps less puzzling that few have ventured to combine complexity at the social scale with yet more complexity at the intra-agent scale.

At this point, therefore, there is little to be reported on the possibility that alternative models of cognitive architecture might support systematic differences in economic outcomes and dynamics. On the other hand, we can view the recent proliferation of economic models of cooperation and competition among intra-personal agents as preliminary steps toward the introduction of computational architectures that will allow for simulation of processes of mediation between microeconomics and macroeconomics along the lines suggested fifteen years ago by Conte and Castelfranchi (1995). Therefore, the next section reviews the current state of theoretical modeling of the disunified economic agent at a finer level of detail than has characterized the discussion so far. To provide additional focus, the comparison of models will concentrate on a specific target phenomenon, intertemporal discounting of utility as the basis for regretted consumption, procrastination, and addiction. The investigation has a specific objective. Models of cognitive architecture, as opposed to models from cognitive neuroscience, have so far had little impact on or role in economic modeling. The history of efforts to model the sources of preference inconsistency shed some revealing light on this.

11.2   Theories of Intertemporal Discounting and Impulsive Consumption in Psychologically Complex Economic Agents

In economics, rationality is generally understood as comprising two elements: consistency of choice, and full use of such information that isn't too costly to be worth gathering. Behavioral economists typically motivate inclusion of psychological variables in economic models to the extent that (1) people are systematically inconsistent in their choices and (2) are prone to act on scanty information even when more could be obtained with relatively little effort.

It is a common observation that people are prone to impulsive consumption, that is, patterns of choice that are subsequently regretted. Many people knowingly choose to consume substances they know to be addictive, then pay later costs to try to overcome their dependence on these substances. Others accumulate debts to credit card companies that they know cannot be rationalized by comparison with their levels of income and wealth. In almost all rich countries, the majority of retired people eventually regret their prior allocations of income between savings and consumption. Ross (2010) argues that all of these phenomena can be understood as instances of procrastination, that is, as cases in which people postpone expenditure of effort aimed at optimizing their wealth in favor of enjoying their current income. In the case of addiction, procrastination involves a particularly interesting phenomenological twist. Being intoxicated is typically incompatible with effortful investment. Therefore, the person who procrastinates by getting high *commits* to her choice for a period of time, during which she is relieved of the anxiety of knowing she could "get back to work" at any moment given some level of effort. Anxiety caused by a behavioral choice is a cost associated with that behavior, so people can be expected to be attracted to mechanisms that reduce this cost. This is part of the pleasure of intoxication. The other main source of its pleasure lies in the direct action of neurochemical mechanisms that stimulate pleasure centers and, more importantly, heighten stimulus salience by producing dopamine surges in the midbrain reward circuit (Ross et al., 2008). A property of substance addiction that is common to procrastination processes in general is that the longer a person puts off her investments in the future, the higher becomes the cost of switching from procrastination to investment at any particular moment. Thus procrastination generally, and addiction specifically, have the form of behavioral traps.

It is in principle possible to theoretically model procrastination and addiction in a way that is consistent with the hypothesis that a person is a unified rational agent. In general, it is rational to discount future consumption relative to present consumption, simply because uncertainty increases with distance into the future; reward prospects may disappear, or the consumer may die or become incapacitated. Because people have idiosyncratic levels of tolerance for risk, rationality alone recommends no specific rate of intertemporal reward discounting. Under the idealization of a linear relationship between uncertainty and the passage of time, a rational agent should discount according to the formula

$$v_i = A_i e^{-kD_i}, \tag{11.1}$$

where v_i, A_i, and D_i represent, respectively, the present value of a delayed reward, the amount of a delayed reward, and the delay of the reward; e is the base of the natural logarithms; and the parameter $0 > k > 1$ is a constant that represents the influence of uncertainty and the agent's idiosyncratic attitude to risk. (Note that k is a psychological variable.) Becker and Murphy (1988) showed that there are values for these variables and parameters such that the agent will rationally choose to consume in a way that causes her welfare to steadily decline *over* time, because *at* any given moment of choice she is best off choosing to consume an addictive drug or other good that improves the payoff from future procrastination relative to future prudence, while lowering the value of both relative to what would have been available had prudence been chosen in the first place.

This so-called *rational addiction* model, using either an average value of k or some motivated distribution of k-values, is widely applied by economists engaged in such enterprises as discovering tax rates on addictive goods that are optimal with respect to revenue and public health, or to both. However, most economists now regard the model as inadequate for purposes of representing and predicting the consumption pattern of a specific individual agent. Addicts, in particular, seldom exhibit life-cycle consumption patterns that accord with it. According to the model, *if* an addict successfully overcomes withdrawal *and* intended at the time of entering withdrawal to quit for good, then either she should never relapse or she should never permanently quit. But the overwhelming majority of addicts repeatedly undergo the rigors of withdrawal, planning never to relapse, and then return to addictive consumption—before eventually quitting successfully. Thus the rational addiction model is empirically refuted by a basic aspect of the standard addictive pattern. (For the detailed logic underlying this argument, see Ross et al., 2008, chapter 3.)

The roots of this misprediction by the rational addiction model lie in the fact that it has no device for allowing an individual's preferences to change over time while preserving her identity as the same agent. The economist trying to cope with this problem can do so by one of three strategies for trading off the two aspects of rational agency (stability of preference and full use of information) against one another. Her first general option is to suppose that people procrastinate (consume impulsively) in full awareness that they are doing so, but exhibit intertemporal preference inconsistency. That is, they choose courses of behavior that are rationalized by the payoff that would accrue in completion of various specific investments, then choose not to complete the investments in question, then subsequently reverse preferences a second time and suffer regret over their irresoluteness. This strategy *diachronically* divides the agent into a *sequence* of sub-agents. A second general possibility is to model procrastinators as having intertemporally consistent utility functions, but as lacking accurate information about their probable future behavior when they choose schedules of activities. Since what is mainly of interest is *recurrent* procrastination, the poverty of sound expectations here must be comparatively radical: the procrastinator fails to learn to predict future procrastination from her own history of past procrastination. This approach further divides into two strategies. One is to *synchronically* divide the agent into a *community* of sub-agents, in which different sub-agents have different information, and communication among them is imperfect. The alternative is to preserve agent unity by "shrinking" the agent and moving aspects of cognitive architecture or of the brain onto the environment side of the agent-environment boundary. In this framework, parts of the agent's mind or brain that are not "parts" of the agent generate exogenous and unpredicted costs and benefits for the agent at particular points in time.

This choice is not the only dimension of modeling discretion concerning the procrastination–impulsive consumption cycle. The modeler must also decide which discounting function to use for fitting the intertemporal choice behavior of the whole agent. Musau (2009) surveys the full range of mathematically distinct options, but it is standard at the level of conceptual description to sort them into two families: hyperbolic functions and quasi-hyperbolic, or β-δ, functions. The most common version of the hyperbolic function is Mazur's formula:

$$v_i = \frac{A_i}{1 + kD_i}.$$
(11.2)

This produces discount curves that decline steeply from an origin at the temporal point of choice, then become (much) flatter between points further into the future. The quasi-hyperbolic function is a step function, borrowed from Phelps's and Pollack's (1968) model of the motives in intergenerational wealth transfers, and often referred to as "β-δ." This class of functions is expressed by

$$v_i = A_i \beta \delta^D$$
(11.3)

where β is a constant discount factor for all delayed rewards and δ is a per-period exponential discount factor. Where β = 1, the equation reduces to standard exponential discounting as in equation (11.1). Where β < 1, discounting is initially steeper up to some inflection point, then flattens. β-δ discounting predicts that value drops precipitously from no delay to a one-period delay, but then declines more gradually, and exponentially over all periods thereafter. In early applications to intertemporal inconsistency (e.g., Laibson, 1994, 1997), β-δ discounting is defended as merely an idealized approximation to hyperbolic discounting, to be preferred for the sake of mathematical tractability (compatibility with standard identifications of unique optima). However, more recently β-δ discounting has been promoted over hyperbolic discounting, because the former lends itself to a particular interpretation in terms of cognitive neuroscience.

To explain this last point, we must first introduce the third dimension of discretion in modeling intertemporally inconsistent choice. This is whether one intends a *molar* or a *molecular* interpretation of the model in respect to cognitive architecture or neuroscience.

The contrast between molar and molecular scales of description and explanation is a well-established one in psychology, and was carried over into cognitive science as the "bottom-up/top-down" distinction. Molar-scale descriptions situate behavioral systems in environmental contexts, sorting their dispositions and properties by reference to equivalence classes of problems they face. Molecular-scale descriptions begin by distinguishing parts of the processing system on the basis of neuroanatomy or biochemistry, or, in an artificial intelligence (AI) setting, on the basis of hardware or encapsulated architectural modules. In AI, designers often deliberately simplify systems by ensuring that molecular-scale and molar-scale boundaries do not cross-cut one another. In general, however, this is only possible in contexts where problems that the system must solve are artificially restricted (that is, for systems functioning in artificial or "toy" environments). In systems that co-evolve with complex environments, molar-scale equivalence classes tend to be highly heterogeneous from the molecular point of view while remaining stable objects for scientific generalization due to external environmental pressures that "capture" different molecular processes within distinctive patterns. The logic here is the same as that which explains convergence in evolution by adaptation to niches. At the level of phylogeny, the relevant external pressures are ecological; in the case of people, they are mainly social, and frequently institutional. Philosophers have discussed this under the rubric of "the multiple realization of the mental" (Putnam, 1975). In cognitive science it is reflected in Marr's (1982) highly influential methodology. According to this approach, cognitive architectures at the *algorithmic level* link molar-scale *computational level* descriptions with independently discovered molecular-scale *implementation level* descriptions of hardware or brain anatomy.

In the case of the models of intertemporal individual choice that interest economists, the most accurate verdict at the present time would be that the methodological sophistication of Marr's framework is not yet manifest, though one of the pioneers of neuroeconomics, Paul Glimcher (2003), has specifically urged its adoption. The four families of existing models will now be described, with a view to substantiating this verdict. We will see that specifications of cognitive architectures are largely missing. In light of this, it is hardly surprising that no beginning at all has yet been made on the further step of embedding economic models of inconsistent individual choice in social settings, as called for by, among others, Ross (2005, 2008) and Castelfranchi (2006).

A first family of models emerges from the students and intellectual descendents of psychologist Richard Herrnstein, and has been dubbed *picoeconomics* by its most systematic exponent, George Ainslie (1992, 2001). It denotes applications of game theory to model procrastination–impulsive consumption phenomena as occasional equilibrium outcomes of games played among sub-personal interests. These interests are not derived from either molar-scale descriptions or models of cognitive architecture. Instead, they are constructed directly as abstract manifestations of the pursuit of goals attributed at the personal scale and then hyperbolically discounted from points of choice. Thus, for example, a person trying to quit smoking has a short-range interest in having a cigarette and a long-range interest in not having one. These interests interact strategically. Thus, the short-range interest might strengthen its prospects by allying with and promoting an interest in going to the bar, where a smoking lapse is more likely, while the longer-range interest might advance its cause by teaming up with an interest in going jogging. Hyperbolic discounting emerges to the extent that the short-range interest is not suppressed by defense of "personal rules" that reframe choices as between temporally extended *sequences* of alternatives. The successful quitter, on Ainslie's account, recognizes that a triumph of the short-run interest today predicts future such triumphs in similar future circumstances. This can lead her to notice that her effective choice is not between a cigarette now and one fewer lifetime cigarette, but between a series of indulgent choices and a series of prudent choices. The value of the latter may outweigh the value of the former from the current perspective *despite* hyperbolic discounting, because more valuable future prudent choices are summed by the reframing, a perception that Ainslie calls *reward bundling*. The bundler then chooses to pay a cost in the form of voluntary suffering caused by restraint, which would be pointless if relapse were sure. In economic terms, such behavior is a form of investment.

Note that on Ainslie's model, willpower is a matter of becoming aware of, and maintaining salience of, certain *information*—specifically, the information that present choices carry about future choices. In light of this, the picoeconomic model has the same structural character as Prelec and Bodner's (2003) *self-signaling* model. The variation introduced by Prelec and Bodner is that the information that triggers bundling is not so much about the choice situation as about the agent herself: the bundler's prudent choice informs her that she is the sort of agent who makes prudent choices and thus will likely make more such choices in the future, whereas impulsive choice signals the opposite sort of character.

Another framework that is logically akin to picoeconomics, though it does not use the label, has been introduced by the late behavioral economist Michael Bacharach (2006). This understands prudent choice as involving *team reasoning*, in which diachronically distinct sub-selves involved in games with one another re-frame their very agency so as to identify with one another. Formally, this changes the relevant games and so may change equilibrium outcomes. A present self that is in a prisoner's dilemma (PD) with future selves, and therefore defects by choosing impulsively, transforms the PD into an assurance game if it frames itself as part of a team with future selves and aims to maximize team utility. The assurance game model, unlike the PD, has Nash equilibria in which present selves cooperate with future selves by making prudent choices.

A cognitive scientist or AI researcher would be likely to ask for, or try to develop, a cognitive architecture that implemented Ainslie's or Bacharach's molecular-scale accounts. To date, however, such work has not been forthcoming. We may speculate that this reflects the intellectual origins of picoeconomics in the moderate behaviorism associated with Richard Herrnstein, who taught Ainslie and many of his leading interlocutors. There is no reason to suppose that Ainslie, Prelec, or other currently active students of Herrnstein would be *opposed to* their models being set into specific cognitive architectures; the point is merely that this is not the kind of modeling that has an established place in the research tradition of which they are members.

The second family of models of impulsive choice and control, and the family that is most popular among economists, is based on β-δ discounting. These models also parse the agent into sub-agents, with the difference that, in any given choice problem, one sub-agent can be identified with the β discounter and another sub-agent with the δ discounter. This means that each agent is an exponential discounter as in equation (11.1) above, so standard economic techniques for identifying unique optimization equilibria can be applied. Examples of models that apply this approach are O'Donoghue and Rabin (2001); Bénabou and Tirole (2004); Benhabib and Bisin (2004); Bernheim and Rangel (2004); and Fudenberg and Levine (2006). Recently, some neuroeconomists and behavioral economists have exploited β-δ discounting to interpret neuroimaging data in support of a molecular-scale account of intertemporal preference reversal. McClure, Ericson, Laibson, Loewenstein et al. (2007) and McClure, Laibson, Loewenstein, and Cohen (2004) obtained fMRI evidence they interpret as suggesting that evolutionarily older orbitofrontal brain areas discount more steeply than later-evolving prefrontal areas. They then propose that hyperbolic discounting at the molar scale be understood as an aggregation of the tug of war between neurally localized β-discounting orbitofrontal and δ-discounting frontal sub-agents. This interpretation has lately been very widely embraced and promoted by behavioral economists.

The McClure et al. (2005, 2007) hypothesis does not include, in Marr's terms, an algorithmic-level account. Its focus is entirely at the implementation level, where it suggests that firing rates in different groups of neurons respectively implement β and δ discounting. The hypothesized "tug of war," which can presumably be influenced by cognitive structures and social influences, would need to be modeled at the level of cognitive architecture. Since these processes are black-boxed in the current model, it offers no immediate purchase on the motivations and conditions for reward bundling, as pointed out by Ross, Ainslie, and Hofmeyr (2010).

The dual brain center interpretation of the McClure et al. (2005, 2007) findings has encountered direct empirical difficulties reported by Glimcher, Kable, and Louie (2007), which are set in a wider context by Ross et al. (2010), who conclude that the weight of evidence is against the hypothesis. However, what is most relevant in the present context is the clear preference among behavioral economists for an explanation of the molar-scale behavioral pattern that appeals to the implementation level and black-boxes the algorithmic level.

The third family of models conceptually shrinks the processes identified with the agent, treating parts of a person's brain as generating exogenous environmental impacts on the agent. Allowing for important variations in details, this modeling approach is shared by Loewenstein (1996, 1999), Read (2001, 2003), and Gul and Pesendorfer (2001). These models (of which only Gul and Pesendorfer's are fully explicit in economic terms) all explain personal-scale violations of thin economic rationality as resulting from "visceral" temptations to immediately consume certain sorts of rewards, which the agent may or may not successfully resist. In these models, resisting temptation is expensive for agents (paid for in short-range suffering), but so is succumbing (paid for in lower longer-range utility). Thus the appearance of a temptation constitutes a negative shock along the agent's optimizing path. How agents respond to such shocks is a function of relative costs, which agents minimize subject to an exponential discount function. The resulting behavioral pattern, if graphed as though it were all just discounting behavior, yields a β-δ curve. This family of models implicitly denies any mediating role for cognitive architecture between the economic and molecular accounts, because the former is a purely abstract description of a relationship between price changes and behavior. It is not, in general, evident that Marr's picture of cognitive explanation applies to Gul and Pesendorfer's model. This would, of course, be consistent with their denial, in their methodological work, that economists and cognitive or brain scientists are interested in the same phenomena as one another.

Finally, a fourth style of modeling is urged by Glimcher (2009). He defines a new, distinctively neuroeconomic concept he calls *subjective value* (SV), which resembles the traditional economist's idea of utility in respects *other than* those relevant to its social and welfare properties. The units of SV are action potentials per second, defined as the mean firing rates of specific populations of neurons. SVs are hypothesized as being always stochastically consistent with choice, even when expected utilities are not. Candidate axioms for specifying SV are proposed, though Glimcher is explicit that he expects these to be modified under empirical pressure. The hypothesis is rendered into a specifically neuroscientific one by Glimcher's suggestion that the weight of fMRI and other evidence indicates that the SV of an action or good is encoded by the mean activity in the medial prefrontal cortex and the ventral striatum. He speculates that the former encodes goods valuation and the latter encodes action valuation.

The suggestion that SV reliably predicts molar-scale choice gives Glimcher's hypothesis a reductionist character. Ross et al. (2008) criticize this aspect of it. Molar-scale utility, they argue in deference to models of ecological rationality discussed above, may be partly calculated by the person in conjunction with external systems in the environment. Thus it may systematically diverge from value-in-the-brain. The picoeconomist grants that SV or something much like it may be a crucial *input* to personal choice, but also allows for other sorts of input, many of which are simply "choice-governing tracks" laid out in the subject's cultural, social, or market environment and which the brain may never explicitly evaluate. Of course, brains must always do *something* to produce behavior that implements choices; but this may not generally, let alone always, be direct neural computation of comparative reward values.

All of the modeling approaches to intertemporal discounting reviewed so far share the assumption that there is a single best functional form for use in modeling. This assumption is strongly called into question by recent experimental work reported in Andersen, Harrison, Lau, and Rutström (2008, 2010). They estimate a so-called mixture model that allows sufficiently rich data to empirically indicate proportions of intertemporal choices best captured by exponential, hyperbolic, and β-δ functions. Using a battery of experiments with a representative sample of over 400 adult Danes, they acquired data of sufficient richness to permit structural estimation of the theoretical models, allowing for joint identification of utility functions and discount functions. They found much shallower discount rates than are typically reported by behavioral economists, who do not control for attitudes to risk and the convexity of utility functions. This should not be regarded as an unexpected discovery, but rather as empirical confirmation of a methodological shortcoming in most behavioral economics. At least as strikingly, they found no evidence to support β-δ models, only modest evidence that any choices were best modeled by hyperbolic functions, and no evidence that such hyperbolic discounting as was supported made a substantial difference in economic magnitudes.

The survey of efforts to settle on models of intertemporal reward discounting for incorporation in larger economic models indicates general and representative failure to map a readily navigable frontier between cognitive science and economics. The final section of the chapter will offer reflections on this failure.

11.3 General Discussion

Ross (2005) and Ross et al. (2008) argue that the main factors that assist people in avoiding impulsive consumption—that is, in approximating traditional, intertemporally consistent economic agency—are social structures and the judgments and expectations of other people. This account of the sample phenomenon from section 2 is broadly in accord with Smith's (2007) emphasis on ecological rationality. It is also congruent with Castelfranchi's (2006) remark that "models [of architectures of mind] should not only 'compete' but should also be compared to and integrated with the models of the other sciences more 'entitled' to model mind or society" (p. 357). Castelfranchi follows this with comments that pertain more specifically to economics:

[The] AI approach to cognitive architectures—as not identical to psychological models—will for example … avoid the bad alternative currently emphasized in economic studies between (Olympic or bounded) formal, simple models of rationality and too empirically oriented, data-driven, non-formal and non-general models of human economic choices based on experiments (behavioral economics). (p. 357)

Part of the claim here is that cognitive science cannot be expected to integrate usefully with any economics that restricts its attention entirely to axiomatic theories and reduced-form models. It is not necessary to endorse currently popular, Luddite hostility to such models (as shown by, e.g., Moss, 2006) in order to agree with this. As noted earlier, Gul and Pesendorfer (2008) make the same point from the other side. With respect to the other alternative mentioned by Castelfranchi, the material reviewed in this chapter demonstrates the limitations of standard behavioral economics: notwithstanding its devotion to socially mediated preferences, it cleaves too tightly to methodological individualism by trying to locate economic valuation and choice entirely "within the agent's head." The same point applies to the rival neuroeconomic approaches of McClure et al. (2004, 2007) and Glimcher (2009).

So Castelfranchi's judgment is sound. However, he doesn't mention what has recently come to be the most common methodology in the parts of economics that closely border cognitive science and psychology with respect to target explanatory phenomena. This is to estimate and econometrically test models with varying, but typically nonlinear, structures featuring statistical choice patterns as dependent variables. In this framework, the only natural path to integration with cognitive science would be identification by cognitive scientists of independent variables for inclusion on the right-hand sides of econometric models.

This immediately raises problems. Models of cognitive architectures are *processing* models. They are typically evaluated by qualitatively or quantitatively comparing their outputs to human or animal behavior. They seldom come attached with indicator variables, of the sort that could be embedded in econometric models. Furthermore, economic models that avoid the charge of implicit methodological individualism, for example the Andersen et al. (2008, 2010) model discussed at the end of section 2, generally do so by aggregating the choices of many individuals. The Andersen et al. criticism of non-standard discounting functions is mute on the question of whether the heterogeneity they observe is between individuals or within individuals; thus, it is unclear what kind of cognitive model one might best compare with it.

In general, economists other than behavioral economists are more interested in aggregated than in individual behavior. Thus one might suppose that the most fruitful interanimations of economic and cognitive studies will involve multi-agent simulations. As pointed out in section 1, however, to date such simulations of economic phenomena have involved only simple agents. Interesting dynamics in such models are social rather than cognitive.

It must thus be concluded that cognitive science has not yet become a significant supplier of variables or parameters to constrain economic models. Future, more productive interdisciplinary collaboration likely depends on progress in greater integration of cognitive models and multi-agent models of social interaction more generally (Sun, 2006). However, we must sound a note of caution due to the fact that the trend among economists who are interested in ecological rationality is to model agents as driven more by institutional structures and less by inboard computations. The parallel rise of neuroeconomics has not incorporated interest in Marr's algorithmic level—precisely the ground on which peoples' strategic problems would be considered in tandem with biological restrictions on the computations they perform.

Perhaps in decades ahead a survey of the role of cognitive variables in economics will report a very different set of contributions and conclusions. There may be considerable space open for methodologically innovative researchers. They must begin, however, from recognition of the reasons that have kept the space open. It is hoped that the present chapter points toward progress by helping to consolidate these reasons.

References

Ainslie, G. (1992). *Picoeconomics*. Cambridge: Cambridge University Press.

Ainslie, G. (2001). *Breakdown of will*. Cambridge: Cambridge University Press.

Andersen, S., Harrison, G., Lau, M., & Rutström, E. (2008). Eliciting risk and time preferences. *Econometrica, 76*, 583–619.

Andersen, S., Harrison, G., Lau, M., & Rutström, E. (2010). Discounting behavior: A reconsideration (working paper 2011–03). Retrieved from the Georgia State University Center for the Economic Analysis of Risk website: http://cear.gsu.edu/papers/index.html

Anderson, P., Arrow, K., & Pines, D. (Eds.). (1988). *The economy as an evolving complex system*. Boston, MA: Addison-Wesley.

Angner, E., & Loewenstein, G. (in press). Behavioral economics. In U. Mäki (Ed.), *Handbook of the philosophy of science: Vol. 13. Philosophy of economics*. Amsterdam, The Netherlands: Elsevier.

Ariely, D. (2008). *Predictably irrational*. New York: Harper Collins.

Arrow, K., & Debreu, G. (1954). Existence of equilibrium for a competitive economy. *Econometrica, 22*, 265–290.

Arthur, W. B., Durlauf, S., & Lane, D. (Eds.). (1997). *The economy as an evolving complex system II*. Boston: Addison-Wesley.

Bacharach, M. (2006). *Beyond individual choice*. Princeton, NJ: Princeton University Press.

Becker, G. (1962). Irrational behavior and economic theory. *Journal of Political Economy, 70*, 1–13.

Becker, G., & Murphy, K. (1988). A theory of rational addiction. *Journal of Political Economy, 96*, 675–700.

Bénabou, R., & Tirole, J. (2004). Willpower and personal rules. *Journal of Political Economy, 112,* 848–886.

Benhabib, J., & Bisin, A. (2004). Modeling internal commitment mechanisms and self-control: A neuroeconomics approach to consumption-saving decisions. *Games and Economic Behavior, 52,* 460–492.

Bernheim, B. D., & Rangel, A. (2004). Addiction and cue-triggered decision processes. *American Economic Review, 94,* 1558–1590.

Bernheim, B. D., & Rangel, A. (2008). Choice-theoretic foundations for behavioral welfare economics. In A. Caplin & A. Schotter (Eds.), *The foundations of positive and normative economics: A handbook* (pp. 155–192). Oxford: Oxford University Press.

Bicchieri, C. (2005). *The grammar of society.* Cambridge: Cambridge University Press.

Binmore, K. (2009). *Rational decisions.* Princeton, NJ: Princeton University Press.

Blume, L., & Durlauf, S. (Eds.). (2005). *The economy as an evolving complex system III.* Oxford: Oxford University Press.

Camerer, C., Loewenstein, G., & Prelec, D. (2005). Neuroeconomics: How neuroscience can inform economics. *Journal of Economic Literature, 43,* 9–64.

Castelfranchi, C. (2006). Cognitive architecture and contents for social structure and interactions. In R. Sun (Ed.), *Cognition and multi-agent interaction* (pp. 355–390). Cambridge: Cambridge University Press.

Clark, A. (1997). *Being there.* Cambridge, MA: MIT Press.

Conte, R., & Castelfranchi, C. (1995). *Cognitive and social action.* London: UCL Press.

Conte, R., & Paolucci, M. (2002). *Reputation in artificial societies.* Dordrecht, The Netherlands: Springer.

Cressman, R. (2003). *Extensive form games and evolutionary dynamics.* Cambridge, MA: MIT Press.

Debreu, G. (1959). *Theory of value.* New York: Wiley.

Debreu, G. (1974). Excess demand functions. *Journal of Mathematical Economics, 1,* 15–23.

Fudenberg, D., & Levine, J. (2006). A dual-self model of impulse control. *American Economic Review, 96,* 1449–1476.

Gigerenzer, G., Todd, P., & the ABC Research Group. (1999). *Simple heuristics that make us smart.* Oxford: Oxford University Press.

Glimcher, P. (2003). *Decisions, uncertainty and the brain.* Cambridge, MA: MIT Press.

Glimcher, P. (2009). Choice: Towards a standard back-pocket model. In P. Glimcher, C. Camerer, E. Fehr, & R. Poldrack (Eds.), *Neuroeconomics: Decision making and the brain* (pp. 503–521). London: Elsevier.

Glimcher, P., Camerer, C., Fehr, E., & Poldrack, R. (Eds.). (2009). *Neuroeconomics: Decision making and the brain* (pp. 323–329). London: Elsevier.

Glimcher, P., Kable, J., & Louie, K. (2007). Neuroeconomic studies of impulsivity: Now or just as soon as possible? *American Economic Review, 97*, 142–147.

Gode, D., & Sunder, S. (1993). Allocative efficiency of markets with zero-intelligence traders: Market as a partial substitute for individual rationality. *Journal of Political Economy, 101*(1), 119–137.

Grassé, P. (1959). La reconstruction du nid et les coordinations individuelles chez *Bellicositermes natalensis* et *Cubitermes* sp. La théorie de la stigmergie: Essai d'interprétation du comportement des termites constructeurs. *Insectes Sociaux, 6*, 41–83.

Gul, F., & Pesendorfer, W. (2001). Temptation and self-control. *Econometrica, 69*, 1403–1436.

Gul, F., & Pesendorfer, W. (2008). The case for mindless economics. In A. Caplin & A. Schotter (Eds.), *The foundations of positive and normative economics: A handbook* (pp. 3–39). Oxford: Oxford University Press.

Harrison, G. (2008). Neuroeconomics: A critical reconsideration. *Economics and Philosophy, 24*, 303–344.

Harsanyi, J. (1967). Games with incomplete information played by "Bayesian" players I–III. *Management Science, 14*, 159–182.

Heyman, G. (2009). *Addiction: A disorder of choice*. Cambridge, MA: Harvard University Press.

Hutchins, E. (1995). *Cognition in the wild*. Cambridge, MA: MIT Press.

Jevons, W. S. (1871). *The theory of political economy*. London: Macmillan.

Kahneman, D., Slovic, P., & Tversky, A. (1982). *Judgment under uncertainty: Heuristics and biases*. Cambridge, MA: Cambridge University Press.

Laibson, D. (1994). *Hyperbolic Discounting and Consumption*. (Unpublished doctoral dissertation). MIT, Cambridge, MA.

Laibson, D. (1997). Golden eggs and hyperbolic discounting. *Quarterly Journal of Economics, 6*, 443–479.

Lichtenstein, S., & Slovic, P. (2006). *The construction of preference*. Cambridge: Cambridge University Press.

Lipsey, R., & Lancaster, K. (1956). The general theory of second best. *Review of Economic Studies, 24*, 11–32.

Loewenstein, G. (1996). Out of control: Visceral influences on behavior. *Organizational Behavior and Human Decision Processes, 65*, 272–292.

Loewenstein, G. (1999). A visceral account of addiction. In J. Elster & O.-J. Skog (Eds.), *Getting hooked: Rationality and addiction* (pp. 235–264). Cambridge, MA: Cambridge University Press.

Mantel, R. (1974). On the characterization of aggregate excess demand. *Journal of Economic Theory, 7*, 348–353.

Mantel, R. (1976). Homothetic preferences and community excess demand functions. *Journal of Economic Theory, 12*, 197–201.

Marr, D. (1982). *Vision*. San Francisco: W.H. Freeman.

McClure, S., Ericson, K., Laibson, D., Loewenstein, G., & Cohen, J. (2007). Time discounting for primary rewards. *Journal of Neuroscience, 27*, 5796–5804.

McClure, S., Laibson, D., Loewenstein, G., & Cohen, J. (2004). The grasshopper and the ant: Separate neural systems value immediate and delayed monetary rewards. *Science, 306*, 503–507.

Menger, C. (1981). *Principles of economics* (J. Dingwall & B. Hoselitz, Trans.). New York: New York University Press. (Original work published 1871)

Moss, S. (2006). Cognitive science and good social science. In R. Sun (Ed.), *Cognition and multi-agent interaction* (pp. 393–400). Cambridge, MA: Cambridge University Press.

Musau, A. (2009). Modeling alternatives to exponential discounting (MPRA Paper No. 16416). Retrieved from http://mpra.ub.uni-muenchen.de/16416.

O'Donoghue, T., & Rabin, M. (2001). Choice and procrastination. *Quarterly Journal of Economics, 116*, 121–160.

Pareto, V. (1971). *Manual of political economy*. New York: Augustus Kelley. (Original work published 1909)

Phelps, E., & Pollack, R. (1968). On second-best national saving and game-equilibrium growth. *Review of Economic Studies, 35*, 185–199.

Prelec, D., & Bodner, R. (2003). Self-signaling and self-control. In G. Loewenstein, D. Read, & R. Baumeister (Eds.), *Time and decision: Economic and psychological perspectives on intertemporal choice* (pp. 277–298). New York: Russell Sage.

Putnam, H. (1975). *Mind, language and reality*. Cambridge: Cambridge University Press.

Read, D. (2001). Is time-discounting hyperbolic or subadditive? *Journal of Risk and Uncertainty*, *23*, 5–32.

Read, D. (2003). Subadditive intertemporal choice. In G. Loewenstein, D. Read, & R. Baumeister (Eds.), *Time and decision: Economic and psychological perspectives on intertemporal choice* (pp. 301–322). New York: Russell Sage.

Robbins, L. (1935). *An essay on the nature and significance of economic science* (2nd ed.). London: Macmillan.

Ross, D. (2005). *Economic theory and cognitive science: Microexplanation*. Cambridge, MA: MIT Press.

Ross, D. (2008). Economics, cognitive science and social cognition. *Cognitive Systems Research*, *9*, 125–135.

Ross, D. (2010). Economic models of procrastination. In C. Andreou & M. White (Eds.), *The thief of time* (pp. 28–50). Oxford: Oxford University Press.

Ross, D., Ainslie, G., & Hofmeyr, A. (2010). Self-control, discounting and reward: Why picoeconomics is economics (working paper). Retrieved from the Georgia State University Center for the Economic Analysis of Risk website: http://www.cear.gsu.edu/files/Ross.pdf.

Ross, D., Sharp, C., Vuchinich, R., & Spurrett, D. (2008). *Midbrain mutiny*. Cambridge, MA: MIT Press.

Samuelson, P. (1938). A note on the pure theory of consumer's behavior. *Economica*, *5*, 61–72.

Samuelson, P. (1983). *Foundations of economic analysis* (Enlarged ed.). Cambridge, MA: Harvard University Press. (Original work published 1947)

Savage, L. (1954). *The foundations of statistics*. New York: Wiley.

Simon, H. (1969). *The sciences of the artificial*. Cambridge, MA: MIT Press.

Simon, H. (1972). Theories of bounded rationality. In C. B. McGuire & R. Radner (Eds.), *Decision and organization* (pp. 161–176). Amsterdam, The Netherlands: North-Holland.

Smith, V. (2007). *Rationality in economics*. Cambridge: Cambridge University Press.

Sonnenschein, H. (1972). Market excess demand functions. *Econometrica*, *40*, 549–563.

Sonnenschein, H. (1973). Do Walras identity and continuity characterize the class of excess demand functions? *Journal of Economic Theory*, *6*, 345–354.

Sun, R. (2006). Prolegomena to integrating cognitive modeling and social simulation. In R. Sun (Ed.), *Cognition and multi-agent interaction* (pp. 3–26). Cambridge: Cambridge University Press.

Sunder, S. (2003). Market as artifact: aggregate efficiency from zero intelligence traders. In M. Augier & J. March (Eds.), *Models of a man: Essays in memory of Herbert A. Simon* (pp. 501–519). Cambridge, MA: MIT Press.

Von Mises, L. (1949). *Human action: A treatise on economics.* New Haven, CT: Yale University Press.

Von Neumann, J., & Morgenstern, O. (1944). *The theory of games and economic behavior.* Princeton, NJ: Princeton University Press.

Weibull, J. (1995). *Evolutionary game theory.* Cambridge, MA: MIT Press.

Zak, P. (2008). The brains behind economics. *Journal of Economic Methodology, 15,* 301–302.

12 Neuroeconomics: How Neuroscience Can Inform the Social Sciences

Joseph W. Kable

12.1 Introduction

Over the past decade, there has been a concerted research effort at the intersection of economics and neuroscience. The new field of neuroeconomics attempts to understand decision making through an integrative mapping between the social, psychological, and biological levels. This chapter argues for the utility of multilevel neuroeconomic research and discusses some early work in this field. More broadly, neuroeconomics provides examples of how the social sciences can be grounded not just in cognitive psychology, but also, at yet another level of analysis, in cognitive neuroscience.

Neuroeconomics is not the only attempt to ground the social sciences in neurobiology. Both evolutionary psychology (Barkow, Cosmides, & Tooby, 1992; Kurzban, 2010) and social neuroscience (Cacioppo, 2002; Ochsner & Lieberman, 2001) have brought biological facts to bear on the understanding of social phenomena, although the social scientists involved in these fields are mostly centered in psychology departments. Neuroscience has also been applied in the domains of cultural psychology (Kitayama & Uskul, 2011); political psychology (Amodio, Jost, Master, & Yee, 2007); legal studies (Chorvat & McCabe, 2004; Garland & Glimcher, 2006; Greene & Cohen, 2004); business (Ariely & Berns, 2010; Plassmann, Ambler, Braeutigam, & Kenning, 2007; Shiv, Bechara, Levin, Alba et al., 2005); and art (Hasson, Landsman, Knappmeyer, Vallines et al., 2008).

However, neuroeconomics stands out as one of the more sustained and developed of these interdisciplinary efforts. The annual meeting of the Society for Neuroeconomics draws over 500 scientists, and a recent social network analysis shows that the field represents a true integration and collaboration between biological and social scientists (Levallois, Smidts, & Wouters, 2010). Work in neuroeconomics has been featured in the highest-profile neuroscience and economics journals, including special review issues in both fields (Cohen & Blum, 2002; Rustichini, 2005). The field has generated several books (Glimcher, 2003; Montague, 2006), and recently the first textbook of neuroeconomics was published, which summarizes the methodological approaches and early results in the field (Glimcher, Camerer, Fehr, & Poldrack, 2009).

The field of neuroeconomics, therefore, can provide valuable lessons regarding the promises and pitfalls of drawing links across the biological and social sciences. This chapter highlights some of those lessons in the course of providing a general overview of neuroeconomics, as well as a closer look at neuroeconomics studies of intertemporal choice.

The goal of neuroeconomics is to build theories of decision making that account for both individuals' choices and the psychological and neural mechanisms that generate those choices. Neuroeconomics aims for causal explanations of decision making—what some have called *because* models (Kable & Glimcher, 2009)—which incorporate constraints from multiple levels of analysis, from social to psychological to neural. Neuroeconomics is thus situated at a logical contact point between the social and biological sciences—that of specifying models of individual behavior.

Current work in neuroeconomics focuses on questions in several broad domains, including (1) decisions where a single agent is involved, including choices between different goods, decisions under risk or uncertainty, and choices involving intertemporal tradeoffs; (2) decisions where multiple agents are involved, so that concerns such as social norms, altruism, and strategic considerations come into play; and (3) dynamic changes that occur in both kinds of decisions during learning. These three domains will be familiar to social scientists as ones that have been modeled with the mathematical formalisms of, respectively, (1) utility theory, prospect theory, and the like, (2) game theory and behavioral game theory, and (3) dynamic optimization and reinforcement learning.

The development of neuroeconomics has been shaped by two major motivations (Glimcher et al., 2009). The first motivation for neuroeconomics comes from within the field of economics, and concerns the "bottom-up" influence of neuroscience. Economists are interested in how neuroscience can inform economics, including how neuroscience methods could be used in economics and how neuroscience data could be incorporated into economic models. Many economists involved in neuroeconomics are influential in the "psychology and economics" movement and are interested in expanding the economic approach beyond the "neoclassical" paradigm. The neoclassical paradigm, described in detail below, is exclusively concerned with a theory's implications for observable choices, and not with the (putatively unobservable) process that generates those choices. The second motivation for neuroeconomics comes from within the field of neuroscience, and concerns the "top-down" influence of economics. Neuroscientists are interested in how economics can inform neuroscience, including how decision theory could provide a theoretical framework to guide investigations into the neural mechanisms of choice. A very similar goal has motivated the incorporation of computational models from computer science and mathematical psychology into neuroscientific studies.[1]

There seems to be little disagreement that economics might inform neuroscience by providing tools and ideas that are useful in studying the neural mechanisms of choice.[2] However, whether influence might flow in the opposite direction as well—whether neuroscience can inform economics—has proven to be extremely controversial (Gul & Pesendorfer, 2008; Harrison, 2008). Many economists question the relevance of psychological or neural factors to economics. Of course, even if these economists are correct, there would still be many good reasons to do neuroeconomic research, and many neuroeconomists would continue their research programs unfazed.[3] Nevertheless, readers of this chapter are likely most interested in how neuroscience could inform the social sciences, and this chapter therefore aims to show several (albeit small) ways neuroscience already has informed economics.

Before turning to specific ways that neuroscience can prove useful, two general objections to neuroeconomics need to be considered. One of these has been advanced by economists, the other by psychologists. Both question the utility of neuroscience for the study of decision making, in principle. Answering these objections demonstrates the potential of neuroeconomics. This chapter then turns to specific ways the field is fulfilling this potential, providing examples from early neuroeconomic research on intertemporal choice.

12.2 Neurobiology Can Inform the Social Sciences: Countering the Standard Objections to Neuroeconomics

12.2.1 The Objection from Economics: Economic Theory Should Be "Mindless"

The first objection to neuroeconomics has been made primarily by economists.[4] The claim is that psychology, and even more so neuroscience, are fundamentally irrelevant to economics. Gul and Pesendorfer (2008) made the argument most pointedly in their "case for mindless economics" (see chapter 11 for another take on this objection):

Economics and psychology address different questions, utilize different abstractions, and address different types of empirical evidence. Neuroscience evidence cannot refute economic models because the latter make no assumptions and draw no conclusions about the physiology of the brain. Conversely, brain science cannot revolutionize economics because the latter has no vehicle for addressing the concerns of economics. (p. 4)

The following sections unpack three different facets of this argument.

Economics Uses Different Abstractions: The "As If" Argument

One facet of Gul and Pesendorfer's (2008) argument is that psychology and economics "utilize different abstractions." A central example is the typical agent in an economic model, who chooses as if all the costs and benefits of each available option can be summarized by a single number, or *utility*, and the option with the maximal utility is selected. The notion that agents maximize utility is often criticized by non-economists as being an unrealistic assumption, but the "as if" part of this conception is often misunderstood. Binmore (2007) clarifies exactly what modern utility theory does and does not claim:

To speak of utility is to raise the ghost of a dead theory. Victorian economists thought of utility as measuring how much pleasure or pain a person feels. Nobody doubts that our feelings influence the decisions we make, but the time has long gone when anybody thought that a simple model of a mental utility generator is capable of capturing the complex mental process that swings into action when a human being makes a choice. The modern theory of utility has therefore abandoned the idea that a util can be interpreted as one unit more or less of pleasure or pain. One of these days psychologists will doubtless come up with a workable theory of what goes on in our brains when we decide something. In the interim, economists get by with *no theory at all* of why people choose one thing rather than another. The modern theory of utility makes no attempt to explain choice behavior. It assumes that we already know what people choose in some situations and uses this data to deduce what they will choose in others—on the assumption that their behavior is consistent. (p. 111, as quoted in Harrison, 2008, p. 308)

Here Binmore is describing neoclassical utility theory as developed by Samuelson (1937). Samuelson showed that the empirical consequence of utility theory for choice data was that a decision maker's choices would be consistent. This pioneered an approach that focuses exclusively on the predictions of economic theories for observable choice data, and omits any concern whatsoever with the decision process. In this respect, economics diverges from psychology, where the decision process is a central concern.[5,6] Friedman (1953) provided perhaps the most famous defense of the neoclassical approach pioneered by Samuelson, arguing that the realism of its assumptions were irrelevant:

> One confusion that has been particularly rife and has done much damage is confusion about the role of "assumptions" in economic analysis. A meaningful scientific hypothesis or theory typically asserts that certain forces are, and other forces are not, important in understanding a particular class of phenomena. It is frequently convenient to present such a hypothesis by stating that the phenomena it is desired to predict behave in the world of observation as if they occurred in a hypothetical and highly simplified world containing only the forces that the hypothesis asserts to be important. ...
>
> Such a theory cannot be tested by comparing its "assumptions" directly with "reality." Indeed, there is no meaningful way in which this can be done. Complete "realism" is clearly unattainable, and the question of whether a theory is realistic "enough" can be settled only by seeing whether it yields predictions that are good enough for the purposes in hand or that are better than predictions from alternative theories. Yet the belief that a theory can be tested by the realism of its assumptions independently of the accuracy of its predictions is widespread and the source of much of the perennial criticism of economic theory as unrealistic. Such criticism is largely irrelevant, and, in consequence, most attempts to reform economic theory that it has stimulated have been unsuccessful. (pp. 40–41)

The neoclassical approach is therefore *behaviorist* and *instrumentalist*. This approach is behaviorist[7] because the neoclassical economist, like the behaviorist psychologist, develops theories that deal exclusively with the relationship between observable variables. Unobservable constructs may be called upon in the mathematics of making this link, but these unobservables are *not* considered part of the content of the theory.[8] This approach is instrumentalist (Cacioppo, Semin, & Berntson, 2004; Craver & Alexandrova, 2008; Quartz, 2008) because the neoclassical economist, like the physicist who can treat light as a particle or wave, chooses modeling conventions largely because they prove useful for making predictions in certain domains, not because of any belief that these conventions ultimately describe the true state of the world. This instrumentalist streak runs from Pareto's (1897/1973) warning against concern with "the essence of things" to Friedman's (1953) admonition to ignore the realism of a model's assumptions to Gul and Pesendorfer's (2008) case for mindless economics.

There are two responses to the argument that neoclassical economics and psychology utilize different abstractions. First, many economists have argued that, rather than stick to the same old abstractions, economists should be open to new ones that might be inspired by work in psychology (Harrison, 2008; Rubinstein, 2008; Spiegler, 2008). For example, Spiegler (2008) comments:

[A] standard model always ends up representing choice behaviour with some kind of utility maximization. Thus, the implied decision process is invariably a cost-benefit analysis. Gul and Pesendorfer (2005) argue that the implied decision process is only a metaphor, a rhetorical device. The utility-maximization model is not meant to serve as a "model of the brain." Still, why should we restrict ourselves to the cost-benefit metaphor? As Salant and Rubinstein argue, there is no reason why we cannot conduct similar decision-theoretic exercises, in which observable behavior ... is shown to be the possible result of some decision process other than a cost-benefit analysis. (p. 517)

Similarly, Rubinstein (2008) argues, "once we have enriched economic models ... it would make sense not to simply invent procedures off the top of our heads but to use models based on our understanding of the mind" (p. 493). Although one might characterize psychology's role in such cases as mere inspiration for economics, Spiegler (2008) warns against doing so dismissively:

On the contrary; inspiration is a scarce resource, and if neuroeconomics ends up being successful in giving economic theorists the motivation, focus and minimal empirical knowledge they need to come up with new, interesting and insightful models, that will be a wonderful achievement. (p. 520)

Second, if economists are willing to be inspired by psychology and neuroscience, then why not admit the abstractions are the same across fields and consider this link a strength? Why are theories that make predictions about both observable choices *and* the psychological and neural mechanisms that generate those choices *not* economic ones? There is reason to favor such theories because they explain a wider range of facts, and in doing so might be expected to be more robust in the long run. Camerer (2007), for example, argues: "Theories that can explain neural facts *and* choices should have some advantage over theories which explain *only* choices, if they are comparably tractable" (p. C39). Craver and Alexandrova (2008) go further, tracing Friedman's exclusive focus on prediction to the then-prominent *covering law model* of explanation, in which explanation is synonymous with prediction. Craver and Alexandrova (2008) argue this approach "cedes the explanatory ambitions of economics"[9] (p. 388), since the covering law model has almost universally been rejected as a philosophy of explanation. Most biological and social sciences have adopted the (largely successful) strategy of seeking explanations that are causal and mechanistic, rather than aiming for mere prediction—why should economics be any different?

Thus, skeptics and proponents of neuroeconomics agree that mechanistic theories about how individuals make decisions are a potentially fruitful contact point for economics, psychology, and neuroscience—whether those theories are treated as mere metaphors in economic modeling or instead as causal explanations (Camerer, 2007; Camerer, Loewenstein, & Prelec, 2004; Camerer, Loewenstein, & Prelec, 2005; Clithero, Tankersley, & Huettel, 2008; Harrison, 2008; Kable & Glimcher, 2009; McCabe, 2008; Rubinstein, 2008; Rustichini, 2009; Spiegler, 2008).

Economics Addresses Different Data: The "Behavioral Sufficiency" Argument

Another facet of Gul and Pesendorfer's (2008) argument is that psychology and economics "address different types of empirical evidence." Most importantly, economics is traditionally interested in observed choices, and not in verbal self-reports, reaction times, neural activities, and the like. This issue is related to but distinct from the "as if" argument. Even if one permits that economics can be concerned with mechanistic theories of decision making, there is still the question of how to best test such theories. Since mechanistic theories have implications for behavior, behavioral data alone are sufficient to test these theories (what Clithero et al., 2008, call the "behavioral sufficiency" argument). Since economists (and psychologists) are interested in mechanistic theories because of their implications for behavior, then it would make sense to test these theories using behavioral data, especially when neural data are much more expensive to collect. And if two mechanistic theories do not have different implications for behavior, then distinguishing between the two might not be of any interest. Even if economists are concerned with the ideas in psychology and neuroscience, then, they need not necessarily be concerned with the data. While the "as if" argument posits that psychological and neural data are irrelevant, the behavioral sufficiency argument merely posits that these data are unnecessary.

However, the behavioral sufficiency argument presents a false choice. There is no reason to limit ourselves to behavioral data alone. To argue otherwise assumes that we are, or could be, awash in all the behavioral data necessary to test any possible theory. Yet often the most critical question in behavioral research is what data to collect next. In this regard, findings in neuroscience may suggest novel and important directions for future behavioral research.

Through such a "behavior to brain to behavior" approach (Clithero et al., 2008), considering multiple levels of analysis can lead to a more efficient development of theory.

There is also a clear reason to be interested in distinguishing between mechanistic theories that do not currently make different predictions about extant behavioral data. Fundamentally, it is the same reason that economists try to provide micro-foundations for macro-level models, which is that such models are more likely to correctly predict "out-of-sample" under novel or changing conditions (Rustichini, 2009). Camerer (2008) summarizes the argument nicely:

One might wonder why we care about separating theories that have the same observable implications about field data. The answer is that theories will often make different predictions about responses to changes in the economic environment. Recall Friedman and Savage's justly celebrated example of the pool players who "act as if" they understand the laws of physics (probably from learning to play by trial-and-error). Whether they act as if, or truly understand physical laws will make a difference when the friction of the surface is changed or the pool table is warped and uneven. Act-as-ifers will make mistakes in the new environments but physical-law-knowers won't. Since economic environments are also undergoing constant change, the "as if" vs. "really do know" distinction is important in economics too. (p. 372)

Economics Addresses Different Questions: The "Emergent Phenomenon" Argument

The third and final facet of Gul and Pesendorfer's (2008) argument is that psychology and economics "address different questions." Economists are often interested in analyzing how institutions (e.g., markets, firms) aggregate the decisions of individual agents, rather than the decisions of the individual agents per se (Gul & Pesendorfer, 2008; McCabe, 2008). This analysis may not depend on a particularly accurate characterization of the individual agent. Further, phenomena at the aggregate level may not be reducible at all to phenomena at the individual level, but rather may be emergent—what Clithero and colleagues (2008) call the "emergent phenomenon" argument (see also Ross, 2008; Wilcox, 2008). The potential for independence between the two levels of analysis is discussed by Sunder (2006):

The marriage of economics and computers led to a serendipitous discovery: there is no internal contradiction in suboptimal behaviour of individuals yielding aggregate-level outcomes derivable from assuming individual optimization. Individual behavior and aggregate outcomes are related but distinct phenomena. Science does not require integration of adjacent disciplines into a single logical structure. As the early twentieth century unity of science movement discovered, if we insist on reducing all sciences to a single integrated structure, we may have no science at all. In Herbert Simon's words: "This skyhook-skyscraper construction of science from the roof down to yet unconstructed foundations was possible because the behavior of the system at each level depended on only a very approximate, simplified, abstracted characterization of the system at the level next beneath. This is lucky; else the safety of bridges and airplanes might depend on the correctness of the 'Eightfold Way of looking at elementary particles.'" (p. 322, as quoted in Harrison, 2008, p. 338)

One can admit the potential emergence of some economic phenomena, though, while simultaneously maintaining the potential benefits of neuroeconomics. In fact, one prominent proponent of neuroeconomics suspects that at least some economic phenomena will turn out to be emergent and not reducible to lower-level foundations (Glimcher, 2011). Correspondingly, though, there will also be some phenomena that are reducible, some predictions at the institution level that depend on assumptions about individual agents. To discover these correspondences, we will need to look for them, which in the process will require good theories of the individual decision maker. Craver and Alexandrova (2008) succinctly summarize the benefit of a multilevel approach:

If there is one privileged level of explanation for decision-making phenomena, the best way to find it quickly is to embrace research at multiple levels. If there are multiple levels of explanation for decision-making, the best way to find them quickly is again to do research at multiple levels. (p. 383)

The aim of neuroeconomics is not to abandon approaches that have served economics well, but to supplement them. The aim is not to assume reductionism in all cases, but to test potential mappings through a coordinated investigation of decision phenomena at multiple levels of analysis.

12.2.2 The Objection from Psychology: Hardware Isn't Software
The second objection to neuroeconomics has been made predominantly by psychologists. The concern here is not that psychology is irrelevant to economics, but rather that neuroscience is irrelevant to psychology (and thus presumably to economics as well). The objection is exemplified by a recurring question along the lines of, "How does knowing *where* a decision occurs in the brain tell us anything about *how* that decision is made?"

This argument predates neuroeconomics. The same basic objection has been raised for many years by cognitive psychologists about cognitive neuroscience. Coltheart (2004) put it pointedly in describing his ("ultra-cognitive-neuropsychological") position:

The other possible aim of cognitive neuroimaging is to use imaging data for testing or adjudicating between cognitive models. Here the ultra-cognitive-neuropsychological position is a particularly extreme one: The assertion is that this aim is impossible to achieve in principle, because facts about the brain do not constrain the possible nature of mental information-processing systems. No amount of knowledge about the hardware of a computer will tell you anything serious about the nature of the software that the computer runs. In the same way, no facts about the activity of the brain could be used to confirm or refute some information-processing model of cognition. (p. 22)

Coltheart's computer analogy is rooted in a long tradition of functionalist thinking in cognitive science; namely, that an understanding of the mind could and should be sought independently of its physical implementation (Quartz, 2008). In his treatise *Vision*, David Marr provided one description of this distinction (1982). Marr argued that any information processing system had to be understood at three levels of analysis: (1) computational theory, which describes the goal the system is trying to achieve or the problem it is trying to solve; (2) algorithm and representation, which describes how the problem is solved in terms of the specific representations used and the processes that act on those representations; and (3) hardware implementation, which describes how the algorithm used is realized in a physical device, such as a brain or computer.[10] Marr also argued that these levels were formally independent of each other—that the same problem could be solved using different algorithms and that the same algorithm could be implemented in different kinds of hardware. Seen within Marr's framework, the psychologist's objection, first to cognitive neuroscience and now to neuroeconomics, is that psychological theories are concerned with the algorithmic level, while neuroscience data only informs us about the (formally independent) implementation level.

One response to this objection is to reject the assumption that different levels of analysis are independent. This argument has been made many times in defense of cognitive neuroscience (e.g., Wager, 2006) and multilevel cognitive modeling (Sun, Coward, & Zenzen, 2005). Most psychologists are interested in not just any minds, but specifically with those minds implemented in neural tissue. Marr himself even noted that the hardware available limits the space of possible algorithms. Facts about the brain therefore constrain the space of otherwise plausible information-processing models of cognition, and can interact with phenomena at other levels of analysis (Sun et al., 2005).

However, there is another response to this objection, which is that it fundamentally mischaracterizes much of the research in neuroscience.

Many, if not most, studies in cognitive, computational, or systems neuroscience are aimed at understanding the mind and brain *at the algorithmic level of analysis*. This includes many, if not most, neuroeconomic studies. Single-cell neurophysiological studies provide paradigmatic examples. Dopamine neurons are shown to encode a reward prediction error signal (Schultz, Dayan, & Montague, 1997); striatal neurons are shown to encode the value of a specific motor action (Lau & Glimcher, 2008; Samejima, Ueda, Doya, & Kimura, 2005); lateral intraparietal neurons are shown to encode first the value of a particular eye movement, and later a categorical signal if that movement is selected (Louie & Glimcher, 2010). These results are all explicitly couched in terms of the *representations* discovered and *how* those representations are transformed. Many neuroimaging studies have similar goals. It is true that these studies all make assumptions about implementation (for example, that quantities can be represented in the nervous system by the spike rate of single neurons), and that any conclusions made regarding the algorithmic level hinge on the correctness of those assumptions. But behavioral studies are absolutely no different in this regard—to draw conclusions about the algorithmic level, assumptions have to be made to link the algorithmic level to the observed data (choices, reaction times, etc.).

These responses to the objections commonly raised by economists and psychologists illustrate the scope and promise of neuroeconomics. Neuroeconomists argue that neuroscience can and should influence psychology, and that psychology in turn can and should influence economics (Padoa-Schioppa, 2008). Recognizing that "the major challenge facing theory formation in the neural and behavioral sciences is that of being under-constrained by data" (Quartz, 2008, p. 470), neuroeconomics aims to build theories that incorporate constraints from each of its three parent disciplines—neuroscience, psychology, and economics. The next section discusses five specific ways in which neurobiology can inform economics, providing an example of each from early research on the neuroeconomics of intertemporal choice. These examples show that neuroeconomics is not just an approach full of promise, but rather a field that is already making progress.

12.3 How Neurobiology Can Inform the Social Sciences: Examples from the Neuroeconomics of Intertemporal Choice

12.3.1 Testing What Representations Are Used in the Brain
The first way that neuroscience can be informative is by testing what representations are used in the brain—what variables and quantities are

encoded when individuals make decisions (Willingham & Dunn, 2003). This evidence can inform judgments about algorithmic theories. If an algorithmic theory predicts that a certain representation should be used in a decision, and we observe that the representation is encoded in the brain during the decision, then we should infer greater confidence in that theory relative to ones that cannot martial such evidence.

Single-neuron electrophysiology is perhaps the technique best suited for these kinds of tests, but this technique is largely restricted to animal models (though see, for example, Zaghloul, Blanco, Weidemann, McGill et al., 2009). In humans, neuroimaging techniques like functional magnetic resonance imaging (fMRI) can also test what representations are used in the brain. Here, though, it is important to take care in making such inferences. It is not sufficient to simply compare one task that is believed to require a certain class of representations to a control task that is not believed to depend on these representations. Aside from the usual inferential issues with such "cognitive subtraction" designs (Friston, Price, Fletcher, Moore et al., 1996), observing different brain activity across the two tasks would not be surprising. If a person behaves differently in the two tasks, then presumably that difference was generated somewhere in the person's brain.[11] To paraphrase Cacioppo and colleagues (2003), we already know that decision making involves the brain. Interpreting the completely expected activity differences as a result of certain representations being activated depends on an a priori specification of the representations used in the two tasks. This specification is a maintained hypothesis that is not tested in the experiment—some interpretation would be given to the activity differences regardless.[12]

In neuroeconomics, a specific kind of analysis is often used in neuroimaging experiments, which strengthens the inference that certain representations are being encoded in the brain. Usually referred to as "model-based fMRI," the first step in the analysis is to fit a model to each subject's behavioral data (O'Doherty, Hampton, & Kim, 2007). This model relates the stimuli the subject experienced (choice sets, choice outcomes) to the actions (choices) they made via a set of hidden functions and parameters. "Fitting" involves estimating the hidden parameters. Often multiple models are compared in terms of their ability to account for the subject's choices.

The next step is to look for a parametric relationship, in the neural data, between fMRI activity and one or more of the hidden values in the behavioral model. This differs from typical fMRI analyses, which relate activity to observable stimuli (presented to the subject) or observable actions (taken by the subject). Observing a relationship between activity in a region and hidden values from a behavioral model provides evidence that activity in that region is associated with the hidden variable. Observing a parametric relationship provides stronger inference than a categorical one (i.e., a comparison between high and low values of the variable), as does seeing that activity is more strongly correlated with the chosen hidden variable than with observable stimuli, observable choices, or hidden variables from alternative behavioral models. In model-based fMRI, both facts about psychology and facts about the brain are jointly tested. How a brain region responds—that its activity is more strongly correlated with a hidden variable from one model than either observable variables or a hidden variable from another model—provides one kind of test of an algorithmic theory. Which brain regions demonstrate this response, or where in the brain this activity is, tests facts about the brain rather than psychology.

Kable and Glimcher (2007) provide one example of model-based fMRI. They measured neural activity with fMRI while individuals made intertemporal choices, or choices between immediate and delayed outcomes. Specifically, individuals chose between an immediate monetary reward, which remained constant from trial to trial, and a delayed monetary reward, the magnitude and receipt time of which changed from trial to trial. These choices had real monetary consequences—subjects were paid according to their decisions on a few randomly selected trials, with payments delivered through a novel debit-card mechanism. The first step in their analysis was fitting a decision model to subjects' behavior. Kable and Glimcher (2007) fit a model that assumed the subjective value of a monetary reward declined hyperbolically as a function of delay:

$$SV(A,D) = \frac{A}{1+kD},$$

where SV is the subjective value, A the amount, and D the delay to the receipt of the monetary reward, and k is the one free parameter called the discount rate. This model provided a good fit to the behavioral data and outperformed a model assuming an exponential decline in subjective value. They found that discount rates varied widely across individuals, from those who are very patient (e.g., choosing $21 in 60 days over $20 now) to those who are very impatient (e.g., choosing $20 now over $150 in 60 days).

Next, they tested for brain regions where the fMRI signal was correlated with the subjective value of the delayed reward, estimated by the behavioral model, as this varied from trial to trial. This analysis is greatly simplified by the fact that only the delayed reward was changing. They found that activity in three distinct brain regions tracked the subjective value of the delayed reward: the medial aspects of prefrontal cortex, including the ventral anterior cingulate cortex; the posterior aspects of the cingulate cortex; and the ventral aspects of the sub-cortical striatum. All three of these regions are logical candidates for encoding subjective value, given that they have previously been implicated in reward evaluation (Bowman, Aigner, & Richmond, 1996; Breiter, Aharon, Kahneman, Dale et al., 2001; Cromwell & Schultz, 2003; Delgado, Nystrom, Fissell, Noll et al., 2000; McCoy, Crowley, Haghighian, Dean et al., 2003; O'Doherty, Kringelbach, Rolls, Hornak et al., 2001; Tremblay & Schultz, 1999).

Finally, they showed that the subjective value from the behavioral model accounted for activity in these regions better than alternative possibilities. The fMRI signal in these regions was more strongly associated with the subjective value from the behavioral model than with the objective characteristics of the options (amounts, delays). The fMRI signal was also more strongly correlated with the single subjective value from the hyperbolic model than with either of the value components of an alternative dual-systems model.

These results provide evidence that a neural representation of subjective value is present at the time of a decision. These data thus support mechanistic models that include this important decision variable. In such models, the various attributes of each option are integrated into a single summary measure of subjective value, and comparisons between options rely on this abstract scale, a kind of "common currency" for choice. Such mechanistic models are broadly compatible with standard economic thinking. There is abundant data consistent with individuals choosing "as if" they are maximizing a value function, and this modeling convention is standard in economics (beginning with Samuelson, 1937) and widespread in behavioral economics (e.g., Kahneman & Tversky, 1979). Kable and Glimcher's (2007) results suggest that these conventions are as successful as they are because the choice mechanism incorporates value maximization.

One important aspect of this example is how it shows that neuroeconomics (and social neuroscience more broadly) is not fundamentally about looking for "centers" (Cacioppo et al., 2003; Harrison, 2008). When using model-based fMRI, one does not have to assume that the computations occur in only one "center"—in Kable and Glimcher (2007), *several* brain regions were found to track subjective value. Similarly, one does not have to assume that the same brain regions perform the same computations under all conditions. This approach is thus compatible with non-localist assumptions, because it can focus on what computations are being performed rather than where these computations are being performed.

This example also emphasizes the critical role of alternative hypotheses in neuroimaging studies. Neuroscience can only inform the social sciences if competing theories are being tested. If a single social science theory is assumed to hold, then only implementation hypotheses are being tested. Neuroscience evidence may then be most informative in situations where the explanations for behavior are under debate, situations where social scientists have not yet converged on considering a single theory as correct.[13]

There are several other examples of model-based fMRI studies within neuroeconomics. One particularly rich area of inquiry has used this approach to demonstrate neural signals associated with dynamical learning models, and to further test competing models of the learning mechanism (Daw, O'Doherty, Dayan, Seymour et al., 2006; Hampton, Bossaerts, & O'Doherty, 2006; O'Doherty, Dayan, Friston, Critchley et al., 2003; O'Doherty et al., 2007; Preuschoff, Quartz, & Bossaerts, 2008). Others have used this technique to study the inputs to a subjective value computation, for example in the contexts of decision making under risk (Preuschoff, Bossaerts, & Quartz, 2006) or charitable giving (Hare, Camerer, Knoepfle, & Rangel, 2010).

12.3.2 Testing What Processes Are Instantiated in the Brain

The second way that neuroscience can be informative is by providing information about the cognitive processes that occur in the brain during decision making (Shiv et al., 2005; Willingham & Dunn, 2003). Again, this evidence can inform judgments about algorithmic theories. If an algorithmic theory predicts that certain cognitive processes are used in making a decision, and we observe brain activation consistent with that process, then we should infer greater confidence in that theory relative to ones that cannot martial such evidence.

Here again, though, extreme care needs to be taken in drawing such inferences from functional imaging data. One style of reasoning that has sometimes been employed is (1) finding that task Z is associated with activity in brain area Y; (2) noting that in previous studies cognitive process X has been associated with activity in brain area Y; and (3) concluding that task Z involved cognitive process X. For example, one might find that a certain kind of decision is associated with activity in the insula, note that the insula has previously been implicated in the emotion of disgust, and then conclude that the decision under consideration involved disgust. This style of reasoning is called "reverse inference,"[14] and, in general, reverse inference is problematic (Poldrack, 2006). First, it is not deductively valid, but rather exemplifies the logical error of "affirming the consequent" ("If X, then Y"; "Y"; "therefore X"). To be valid, one would need to assume that there is a one-to-one mapping between the presence of a cognitive process and activity in a brain region (i.e., "If *and only if X*, then Y"), and such a one-to-one mapping is unlikely. The inference can be restated in probabilistic terms (so that now what is of interest is $P(Y|X)$ and $P(Y|{\sim}X)$). Poldrack (2006) works through an example of this to show that reverse inferences are unlikely to provide strong evidence, at least at the spatial scale of brain activity and coarseness of the cognitive ontology that are typically used.

While reverse inference is generally problematic, there may be specific cases where the technique proves useful. For example, using the same approach as Poldrack (2006), Ariely and Berns (2010) show that activation of the nucleus accumbens seems to be fairly selective to the presence of a reward. There are many caveats to this conclusion (e.g., the database of studies probably does not sample activations or cognitive processes randomly), but it does suggest that the strength of evidence can differ for different reverse inferences. Reverse inferences could also prove useful when viewed as hypotheses that should then be tested, rather than as conclusions. In this way, researchers can bootstrap across several imaging or behavioral studies to refine their hypothesis about the cognitive processes involved in a task and the brain activity associated with those processes. This approach would benefit from expansion of the databases of functional imaging results to allow more sophisticated comparisons across studies, and from efforts to refine the ontology of cognitive processes used by such databases (Poldrack, 2006).

The ability of neuroimaging to provide evidence regarding cognitive processes does not hinge entirely on reverse inference, however. If one assumes that similarities and differences in neural activation imply similarities and differences in cognitive processing, then one can compare neural activation in two tasks or conditions to see if they involve overlapping cognitive processes (Camerer et al., 2004; Willingham & Dunn, 2003). This approach uses neural activations to "lump" or "split" tasks according to whether they involve the same cognitive processes, rather than using neural activation to infer directly what cognitive processes are involved in a task.

Kable and Glimcher (2010) used this approach to test whether immediate and delayed rewards are valued using fundamentally different mechanisms. Some theories propose that anomalies in intertemporal choice arise

because immediate and delayed rewards are treated in categorically different ways, and two previous fMRI studies argued that specific neural structures were engaged only during decisions involving immediate rewards
(McClure, Ericson, Laibson, Loewenstein et al., 2007; McClure, Laibson,
Loewenstein, & Cohen, 2004). Interestingly, the areas proposed to exclusively process immediate rewards—medial prefrontal cortex, posterior cingulate cortex, ventral striatum—were the same ones in which Kable and
Glimcher (2007) had observed subjective value signals. Thus, an alternative
possibility is that these regions encode a subjective value signal for both
immediate and delayed rewards during intertemporal choice, and the
greater activity observed for immediate rewards in previous studies
(McClure et al., 2007; McClure et al., 2004) occurred simply because these
rewards were more valuable.

Kable and Glimcher (2010) directly tested this idea by measuring neural
activity with fMRI while individuals made decisions in two conditions: the
first condition involved choices between an immediate and a delayed
monetary reward; the second condition was based on the first, but with a
fixed delay of 60 days added to both options, so that now the choice was
between two delayed rewards. Choices in both conditions were intermixed
randomly, and both sets of choices were designed so that the effects of
subjective value on neural activity could be estimated. Kable and Glimcher
(2010) again found that activity in the medial prefrontal cortex, posterior
cingulate cortex, and ventral striatum was modulated by subjective value.
Furthermore, after accounting for these effects of subjective value, these
brain regions did not exhibit differential activity in the two conditions.
Neither these brain regions, nor any others, exhibited a categorical difference in activity when an immediate reward was available. Kable and Glimcher (2010) concluded that choices involving immediate rewards and those
involving only delayed rewards do not recruit different neural, or different
cognitive, processes.

These data support mechanistic models in which immediate and
delayed monetary rewards are not processed in categorically different
ways. Such mechanistic models would depart from the current convention
in behavioral economics, β-δ discounting, which models individuals as if
they treat immediate rewards in a categorically different manner (Laibson,
1997). In line with their conclusions based on the neural data, Kable and
Glimcher (2010) present behavioral data that β-δ discounting fails to
account for choices under conditions similar to those used in their fMRI
experiment. Individuals were not more patient when choosing between
delayed monetary rewards, compared to choosing between immediate and

delayed rewards, as β-δ discounting would predict. These behavioral data are consistent with other empirical results that have been largely ignored by theoreticians (Baron, 2000; Green, Myerson, & Macaux, 2005; Holcomb & Nelson, 1992; Read, 2001; Read & Roelofsma, 2003; Scholten & Read, 2006). These results suggest that current efforts to account for anomalies in intertemporal choice as arising out of processes that work on both immediate and delayed rewards, such as temporal perception (Zauberman, Kim, Malkoc, & Bettman, 2009), may meet with greater empirical success.

This example deals with one of the early debates within neuroeconomics, and some might think such debates would make it difficult for neuroscience to influence the social sciences. How can a social scientist be confident of a neural finding if there is a chance it will be refuted in a few short years? However, such debates are, on balance, a positive sign. They show that empirical facts are at stake, and that neuroeconomists revise their theories in light of new evidence. Such debates clearly refute the unfair characterization of some critics that the field merely uses brain pictures as marketing tools for already established ideas.

Several other neuroeconomic studies have used the approach of inferring similarities and differences in cognitive processes from similarities and differences in neural activation. Similar to the debate regarding intertemporal choice, different groups have argued that decision making under risk and ambiguity involve similar (Levy, Snell, Nelson, Rustichini et al., 2010) versus different (Hsu, Bhatt, Adolphs, Tranel et al., 2005) neural and cognitive mechanisms. Regarding valuation, Chib and colleagues (2009) used overlapping neural activations to argue for a shared valuation process across several different domains (money, goods, and food). In social decision making, Singer and colleagues (2004) demonstrated that watching someone else's pain and experiencing pain yourself evince overlapping neural activations. And in marketing, Yoon and colleagues (2006) showed that thinking about brands involved different neural regions than thinking about people, suggesting that it is mistaken to assume that brands have a "personality" that is processed in the same way as a person's.

12.3.3 Revealing New Representations and Processes

In addition to testing between existing theories on the basis of the representations and processes observed in the brain, neuroscience can also inspire fundamentally new theories based on novel constructs and concepts. Several scholars have identified this exciting possibility as key to the enormous potential of neuroeconomics (Camerer, 2007, 2008; Camerer et al., 2005; Shiv et al., 2005; Spiegler, 2008). For example, Camerer (2007) argues:

The largest payoff from neuroeconomics will not come from finding rational-choice processes in the brain for complex economic decisions, or from supporting ideas in behavioural economics derived from experimental and field data. ... The largest innovation may come from pointing to biological variables which have a large influence on behavior and are underweighted or ignored in standard theory. (p. C35)

Several reviews (Camerer, 2008; Camerer et al., 2004; Camerer et al., 2005) have provided examples of the new constructs and concepts that are being incorporated into neuroeconomic models, generating a fairly long list: emotions (e.g., anger, disgust, fear); other visceral states like stress, fatigue, hunger, lust, and pain; cognitive processes like attention and memory; and other constructs like willpower and self-regulation. One thing that is striking about this list, however, is that all of these constructs originally emanated from psychology, rather than from neuroscience. Of course, neuroscience may contribute to these new models by influencing psychology, providing support for or against certain constructs. But a stronger argument for *neuro*economics would be showing that neuroscience also generates novel ideas.

The prevalence of such examples is likely to lag behind examples where neuroscience leads to refinements of existing theories. As Smith (2008) notes, new tools are usually first applied to old questions. One example of neuroscience generating a novel hypothesis, though, concerns an issue raised by Harrison (2008), which is how to model the stochasticity widely observed in agents' choices. In linking economic models to actual data, this is usually done by specifying that the probability of observing a choice follows a probit or logit function of parameters in the core model (e.g., expected utility theory, discounted utility theory, etc.). Different specifications of the probability function arise from assuming stochasticity occurs in the computation of value, the comparison of values, or the execution of a choice. Recently Glimcher and colleagues (2007), motivated by the results of neuroscience experiments, proposed that choice probability is a function of a particular form of *relative subjective value* representation. This form of representation arises from a computational principle, called *cortical normalization*, which has been observed across several different cortical regions.

This idea was generated from several studies, which all involved recording action potentials from single neurons in the lateral intraparietal region

(LIP) of macaque monkeys. In these experiments, thirsty subjects had to decide what visual targets to look at in order to obtain juice rewards. Single neurons in LIP have spatial response fields. These neurons exhibit increased activity when potential targets for eye movements are present within these response fields, as well as preceding movements that bring the eyes' focus into the response field. Louie and Glimcher (2010) examined the responses of these neurons during a choice between two targets, one associated with an immediate reward and a second with a delayed reward. Under these conditions, neurons in LIP responded early in the trial in proportion to the discounted value of a movement into the response field—firing rates were maximal when this movement was associated with an immediate reward, and decreased monotonically as delay to reward increased. This signal in LIP evolved into a binary representation later in the trial, immediately preceding the choice—firing rates were high preceding movements into the response field and low preceding other movements. This study shows how activity in LIP reflects the dynamics of the choice process during intertemporal decision making, including both inputs (the discounted value of each movement) and outputs (what movement is selected).

Louie, Grattan, and Glimcher (2011) further investigated value-related activity in LIP. In Louie and Glimcher (2010), the value of an eye movement into the response field changed, while the value of the other movement remained constant over the course of the experiment. Louie et al. (2011) showed that the firing rate of LIP neurons also decreased (increased) when the movement away from the response field increased (decreased) in value, even though the value of the movement into the response field did not change. The firing rate of LIP neurons also decreased as more potential movements were added to the choice set, even though the value of the movement into the response field was again held constant. Both of these results suggest that, rather than encoding the value of a movement into the response field in isolation, LIP neurons encode the value of a movement into the response field, *relative* to that of all possible movements.

This representation of relative subjective value is consistent with the general principle of cortical normalization. Although there are some subtle differences, different cortical areas share the same general anatomical structure in terms of layer organization, basic types of neurons, and connectivity between layers. Thus, different cortical areas may perform the same types of computations on different types of inputs. Computational models of different sensory cortical areas have shown how the firing of neurons in these regions can be explained by a general normalization process, whereby the specific input to a particular neuron is normalized by a weighted sum of the inputs to an entire region. For example, Schwartz and Simoncelli (2001) proposed that the neuronal response scaled according to:

$$\frac{V_i}{C + \sum_j w_{ji} V_j},$$

where V's are the inputs, w's are weights, and C is a constant.

Glimcher and colleagues (2007) proposed that a similar computation might occur in LIP, where the inputs (V's) in this case would be the subjective values of the different possible eye movements. Since LIP reflects the dynamics of the choice process, then choice probability might be a function of relative subjective values, calculated according to the above Schwartz-Simoncelli cortical normalization equation. One possibility is that choice probabilities are simply proportional to relative subjective values. This would result in a choice probability function similar but not identical to Luce's choice rule (the probability of choosing an option equals the value of that option divided by the sum of all option values). Other possibilities could assume that neural representation of relative subjective values is stochastic, across a population of neurons or across time, and that the option with the highest (stochastic) relative subjective value is chosen. Depending on the assumptions, this would result in choice probability functions that differ from the standard probit and logit forms. It remains to be seen whether these proposals can better describe choice stochasticity compared to currently used forms, but the hypothesis is one that only arose from considering the results from neuroscience experiments.

12.3.4   Identifying and Characterizing Individual Differences

The preceding three sections all show how neuroscience can inform the social sciences by improving models of the individual decision maker. However, the potential influence of neuroscience goes beyond building better algorithmic theories of the decision process. A different way that neuroscience can be informative is by identifying and characterizing individual differences in decision making (Kosslyn, 1999). In the same way that brain imaging can be used to see whether the same individual performs different decisions using similar or different processes, it can also be used to see if *different* individuals perform the *same* decision using similar or different processes.

Several economists have identified the ability to detect and characterize heterogeneity as a particular promise of neuroeconomics. For example, Rubinstein identified this as one way neuroeconomics could prove useful: "One potentially important task for the neuroeconomics approach is to identify 'types' of economic agents, namely to determine characteristics of agents that predict their behavior in different choice problems" (p. 490). Spiegler (2008) similarly identifies this as a potential contribution:

As neuroeconomists gather non-choice data in experiments, they may stumble upon correlations between personal characteristics that are independent a priori. ... The implications of such correlations for economic theory is that we impose them as a constraint on models that incorporate several personal characteristics. In particular, such restrictions on the domain of personal characteristics may be relevant for mechanism design problems. (pp. 516–517)

With respect to mechanism design, Houser and colleagues (2007) demonstrate exactly how and when the kinds of information that could be gained from imaging experiments—the existence of differences in decision strategy among observationally identical individuals, and the conditional probabilities relating those differences to observable characteristics—could improve the predictions made by econometric analyses.

Neuroeconomics has already made some progress in determining correlations between the personal characteristics of individual decision makers. Within the domain of intertemporal choice, for example, the Kable and Glimcher (2007) study discussed above shows that differences in degree of impatience are related to how certain brain regions (medial prefrontal cortex, posterior cingulate cortex, ventral striatum) respond to immediate and delayed rewards. More patient individuals have a greater neural response in these regions to delayed rewards. The involvement of the striatum, in particular, suggests that these differences could be related to differences in the dopaminergic system, and indeed there is preliminary evidence that impatience differs as a function of genetic differences in the dopaminergic system (Boettiger, Mitchell, Tavares, Robertson et al., 2007). Other studies have identified regions of the lateral prefrontal cortex where there is greater activity in more patient individuals (Monterosso, Ainslie, Xu, Cordova et al., 2007; Wittmann, Leland, & Paulus, 2007), and this has motivated investigations into the correlation between patience and fluid intelligence (Shamosh, DeYoung, Green, Reis et al., 2008). Obviously much work remains to solidify these links, but current evidence points toward an association between patience, dopaminergic genes, and intelligence.

As a next step, ongoing work in the author's and others' labs is seeking "proof of concept" for how the identification of individual differences based on neural data could aid predictions. A basic demonstration would have two steps. First, it would show how two (or more) groups of individuals, who cannot be distinguished easily on the basis of their choices alone, can be distinguished based on their neural activity and imputed cognitive processes during these choices. Second, it would then demonstrate that the observed differences can be used to predict how the two groups respond differently to changes in the choice situation.

12.3.5 Using Neurobiological Variables to Predict Economic Data

A fifth way that neuroscience could inform the social sciences is by using neurobiological variables to predict social science data. The previous section provides one example of this, using neural data regarding individual differences to parameterize the types of agents in economic models. In addition to this example, several scholars have suggested that neural measures could prove to be independent predictors of economic behavior, that they might be used to add precision to behavioral estimates of standard predictors (e.g., discount rates or degrees of risk aversion), or that they might prove useful in cases where behavioral measures are unavailable or unreliable (Camerer et al., 2004; Camerer et al., 2005).

Work in the author's laboratory provides one example of using neuroscience data in this way. Recall that in previous studies of intertemporal choice, subjects were actually paid on a debit card according to one or more of their choices. One feature of this mechanism is that an individual's expenditures from the debit card are also available. Ongoing research is examining whether neural responses to intertemporal choices in the scanner predict a person's expenditures outside the scanner, over and above the concurrent behavioral measures (e.g., discount rates) that are gathered. Preliminary evidence suggests that they do. This finding provides "proof of concept" that neural measures can serve as independent predictors of economic data—in this case, consumer spending.

This work builds on foundational demonstrations that neural activity can be used to predict simultaneously measured choices (Knutson, Rick, Wimmer, Prelec et al., 2007; Kuhnen & Knutson, 2005). While predicting simultaneously measured choices is unlikely to impress economists (after all, very few economic decisions are made in brain scanners), it is an essential first step and provides a critical testing ground for improving statistical techniques for aggregating activity measures across different brain regions (Grosenick, Greer, & Knutson, 2008). Subsequent work has shown that neural activity when passively viewing goods can predict preferences between those goods (Levy, Lazzaro, Rutledge, & Glimcher, 2011), a finding that could be developed into practical uses in situations where behavioral measures are unavailable or unreliable (e.g., contingent valuation). Ariely and Berns (2010) suggest another potential use of neural predictions: designing and selecting products for development. Berns and Moore (in press) provide one example where neural data predicts product success—neural activity when adolescents listen to music from unknown bands predicts the later gross earnings of those bands, and does so better than behavioral ratings of the song's likability.

12.4   Conclusion

Several reviews have summarized what neuroeconomic research has uncovered about the neural mechanisms of decision making (Kable & Glimcher, 2009; Rangel, Camerer, & Montague, 2008). This chapter has covered similar ground, but with a different focus on how these neurobiological discoveries can inform economics. Some have argued that such influence is impossible, in principle (Gul & Pesendorfer, 2008). However, neuroscience can inform economics through psychology (Padoa-Schioppa, 2008): neuroscience informs psychology by testing hypotheses regarding cognitive representations and processes, and psychology in turn informs economics by providing mechanistic models of individual decision making. There are already (small) ways in which this promise of neuroeconomics is being fulfilled. In the domain of intertemporal choice, neuroeconomic studies have shown that (1) the brain encodes a subjective value signal during choice, in a manner consistent with standard value-maximization models; (2) immediate monetary rewards are not processed in a categorically different manner from delayed rewards, which is inconsistent with what some prominent models of intertemporal choice might suggest; and (3) the rescaling of subjective value signals in motor preparation regions suggests that a novel, neuroscience-inspired choice function might best account for the pattern of stochasticity observed in choices. Ongoing work is investigating how neuroscience methods can identify and characterize individual differences in intertemporal choice, and how neural measures can predict intertemporal choices in the field. The ultimate weight of these implications in economics will only be known with time, but clearly the field of neuroeconomics is already making progress toward multilevel mechanistic explanations of decision making. Such explanations of individual decision making should provide a natural contact point for the social and biological sciences.

In the future, one improvement to the neuroeconomic approach would be a less exclusive reliance on functional neuroimaging, and an increased use of multiple methods to gather converging evidence. While introductions to neuroeconomics have explained the benefits of a multimethod approach (Camerer, 2007; Camerer et al., 2004; Camerer et al., 2005), the empirical data suggest that most neuroeconomic research relies exclusively on fMRI (Kable, 2011). Of course, functional MRI has several clear advantages. It is unparalleled in its ability to image the entire human brain at high spatial resolution, and advanced analysis techniques like model-based fMRI allow fine-grained questions to be answered. However, other neuroscience methods have important, and complementary, inferential strengths. Both neuropsychology and non-invasive brain stimulation can uncover key associations and dissociations between tasks and processes, based on the deficits that occur with the disruption of different brain systems. Genetics can identify critical markers of individual differences. Psychophysiological techniques (e.g., electroencephalography, electromyography, skin conductance, eye tracking) will be easier to use in the field, and thus might prove more practical as independent predictors of the behavior that most interests economists (Lo & Repin, 2002; Wang, Spezio, & Camerer, 2010). As neuroeconomics matures, these techniques should be used to their full potential.

This chapter is organized to make it apparent how the effort in neuroeconomics—to ground a social science not just in cognitive psychology, but also in cognitive neuroscience—could be generalized to other social sciences. Competing theories in the domains of culture, politics, or religion may make competing claims about the cognitive representations and processes that occur in the human mind (sections 3.1–3.3) or competing claims about the categories of individuals that should be observed (section 3.4). Neuroscience can help test these claims, and may also provide additional predictors of important data in the social sciences (section 3.5). Given this promise, investigators have already started to explore the use of neuroscience in cultural psychology (Kitayama & Uskul, 2011); political psychology (Amodio et al., 2007); the law (Chorvat & McCabe, 2004; Garland & Glimcher, 2006; Greene & Cohen, 2004); business (Ariely & Berns, 2010; Plassmann et al., 2007; Shiv et al., 2005); and art (Hasson et al., 2008). Given the integration between cognitive psychology and cognitive neuroscience over the past decade, such explorations are a natural extension of efforts to ground the social sciences in cognition. This does not mean, of course, that all questions in the social sciences will benefit from turning to cognitive neuroscience. However, when possible, grounding in neurobiology will certainly enhance the explanatory richness, explanatory reach, and explanatory strength of theories in the social sciences.

Notes

1. These two motivations recur in Ross's division of neuroeconomics into *behavioral economics* "in the scanner" and *neurocellular economics* (Ross, 2008). Interestingly, Quartz proposes an analogous split guiding the melding of cognitive science with neuroscience, resulting in cognitive neuroscience and computational neuroscience (Quartz, 2008).

2. Why neuroscientists seem much less disturbed than economists about this interdisciplinary effort is an interesting question. One possibility is that the field of neuroscience is already characterized by methodological pluralism, and is also so diverse that there is no singular core tradition. Another possibility is that neuroscientists simply spend less time writing critical theoretical pieces about such issues.

3. As one prominent economist said to me regarding their research in neuroeconomics, "Is it economics? I don't know. But it is just so *interesting*!"

4. Though not exclusively: some aspects of the argument, particularly the notion that choice behavior should be the primary object of study, are also made by many psychologists.

5. This divorce of economic theory from psychology has even earlier roots in the emergence of neoclassical economics at the beginning of the twentieth century. For example, consider Pareto (1897/1973): "It is an empirical fact that the natural sciences have progressed only when they have taken secondary principles as their point of departure, instead of trying to discover the essence of things. ... Pure political economy has therefore a great interest in relying as little as possible on the domain of psychology" (as quoted in Glimcher et al., 2009, p. 3). Or Pareto's contemporary Fisher, commenting here on the idea of linking the notion of utility to psychophysical laws proposed by Weber and Fechner: "This foisting of Psychology on Economics seems to be *in*appropriate and vicious" (Fisher, 1925, p. vi, as quoted in Colander, 2007, p. 220). Fisher (1925) continues: "To fix the idea of utility the economist should go no farther than is serviceable in explaining *economic* facts. It is not his province to build a theory of psychology" (p. 11, as quoted in Colander, 2007, p. 220). Or, in a similar vein, von Wieser and Hinrichs (1927): "Economic theory would be benefited, had scientific psychology advanced further beyond its beginnings; but our discipline does not seek and could not find direct aid from this source. The tasks of the two branches of knowledge are entirely distinct" (p. 2, as quoted in Colander, 2007, p. 221).

6. Of course, there were also several prominent early economists who did seek out connections with psychology. Neumarker (2007) notes three prominent examples: "Smith's *The Theory of Moral Sentiments* (1759) rested upon the emotion of sympathy. In his *Mathematical Psychics*, Edgeworth (1881, p. 104) accepted that 'man is for the most part an impure egoist,' and Hayek, the Nobel laureate always being interested [*sic*] in constitutional conditions of the individual, political and social life, published with *The Sensory Order* (1952) an investigation into the nature of human mental events, the role of the nervous system in determining principles and systems of order inside the brain and the consequences of these human factors for the social order" (p. 60). Camerer (2008) is even more forceful on this point: "It is useful to point out, by the way, that the foundation of choice was not always considered only a matter of linking unobserved constructs to observed choice and ignoring mechanism. Ramsey, Fisher and Edgeworth were among the classical economists who fantasized about measuring hedonic bases of utility directly (Ramsey spoke of a 'psychogalvanometer' and Edgeworth of a 'hedonimeter'; see Colander, 2008). Admittedly, they were mostly interested in being able to establish a cardinal measure of utility, which is far from the focus of modern neuroeconomics. Given how interested these early economists were in direct measurement—when they could not readily do so—one might think that their intellectual descendants would at least be interested when tools like those they fantasized about are actually available" (p. 370).

7. It is interesting to note that the neoclassical approach was developed in economics around the same time that the behaviorist approach was developed in psychology. So thinkers in the two disciplines may have been influenced by each other. Consider, for example, this statement by Fisher (1925): "So economists cannot afford to be too academic and shirk the great practical problems pressing upon them merely because these happen to touch on unsolved, perhaps insoluble, philosophical problems. The psychologist has set the example by becoming a 'behaviorist.' He can thereby deal practically with phenomena the essential nature of which he confesses he cannot fathom" (p. 180, as quoted in Colander, 2007, p. 220).

8. As Gul and Pesendorfer (2008) say, "In standard economics, the testable implications of a theory are its content; once they are identified, the non-choice evidence that motivated a novel theory becomes irrelevant" (p. 8). Since unobservable constructs are not directly testable, they cannot be part of the content of a theory.

9. In a similar vein, Camerer (2007) argues against Friedman: "If assumptions A are false but lead to an accurate prediction, they presumably do so because of a hidden 'repair' condition R (that is, (not-A and R) $\rightarrow P$ is a more complete theory at both ends than $A \rightarrow P$). Then the proper focus of progressive research should be specifying the repair assumption R and exploring its implications, in conjunction with more accurate assumptions" (p. C27).

10. Newell and Simon (1976) proposed a related division between the *knowledge* level, *symbol* level, and *physical* level. Sun and colleagues (2005) argue that the focus in these two proposals on computational constructs is misguided, and that it would be more productive to focus on levels of analysis defined by phenomena to be modeled (inter-agent processes, agents, intra-agent processes, or substrates).

11. And if you cannot detect a difference, this just says something about *your technique's ability* to detect the relevant difference.

12. In other words, this approach to neuroimaging involves making some assumptions about the mind and then using these to draw conclusions about the brain (Aguirre & D'Esposito, 1999). Such an approach has been referred to as "forward inference," because it contrasts with the "reverse inference" discussed below (Henson, 2006).

13. Drazen Prelec, in particular, has emphasized this point (personal communication).

14. This contrasts with the "forward inference" approach described above. Here an assumption is made about the brain, and this is used to draw conclusions about the mind (Aguirre & D'Esposito, 1999).

References

Aguirre, G. K., & D'Esposito, M. (1999). Experimental design for brain fMRI. In C. T. W. Moonen & P. A. Bandettini (Eds.), *Functional MRI* (pp. 369–380). Berlin: Springer.

Amodio, D. M., Jost, J. T., Master, S. L., & Yee, C. M. (2007). Neurocognitive correlates of liberalism and conservatism. *Nature Neuroscience, 10*(10), 1246–1247.

Ariely, D., & Berns, G. S. (2010). Neuromarketing: The hope and hype of neuroimaging in business. *Nature Reviews Neuroscience, 11*(4), 284–292.

Barkow, J. H., Cosmides, L., & Tooby, J. (1992). *The adapted mind: Evolutionary psychology and the generation of culture.* Oxford: Oxford University Press.

Baron, J. (2000). Can we use human judgments to determine the discount rate? *Risk Analysis, 20*(6), 861–868.

Berns, G. S., & Moore, S. E. (in press). A neural predictor of cultural popularity. *Journal of Consumer Psychology.* Retrieved from doi:10.1016/j.jcps.2011.05.001.

Binmore, K. G. (2007). *Playing for real: A text on game theory.* Oxford: Oxford University Press.

Boettiger, C. A., Mitchell, J. M., Tavares, V. C., Robertson, M., Joslyn, G., D'Esposito, M., et al. (2007). Immediate reward bias in humans: Fronto-parietal networks and a role for the catechol-O-methyltransferase 158(Val/Val) genotype. *Journal of Neuroscience, 27*(52), 14383–14391.

Bowman, E. M., Aigner, T. G., & Richmond, B. J. (1996). Neural signals in the monkey ventral striatum related to motivation for juice and cocaine rewards. *Journal of Neurophysiology, 75*(3), 1061–1073.

Breiter, H. C., Aharon, I., Kahneman, D., Dale, A., & Shizgal, P. (2001). Functional imaging of neural responses to expectancy and experience of monetary gains and losses. *Neuron, 30*(2), 619–639.

Cacioppo, J. T. (2002). Social neuroscience: Understanding the pieces fosters understanding the whole and vice versa. *American Psychologist, 57*(11), 819–831.

Cacioppo, J. T., Berntson, G. G., Lorig, T. S., Norris, C. J., Rickett, E., & Nusbaum, H. (2003). Just because you're imaging the brain doesn't mean you can stop using your head: A primer and set of first principles. *Journal of Personality and Social Psychology, 85*(4), 650–661.

Cacioppo, J. T., Semin, G. R., & Berntson, G. G. (2004). Realism, instrumentalism, and scientific symbiosis: Psychological theory as a search for truth and the discovery of solutions. *American Psychologist, 59*(4), 214–223.

Camerer, C. F. (2007). Neuroeconomics: Using neuroscience to make economic predictions. *Economic Journal, 117*(519), C26–C42.

Camerer, C. F. (2008). The potential of neuroeconomics. *Economics and Philosophy, 24*(3), 369–379.

Camerer, C. F., Loewenstein, G., & Prelec, D. (2004). Neuroeconomics: Why economics needs brains. *Scandinavian Journal of Economics, 106*(3), 555–579.

Camerer, C. F., Loewenstein, G., & Prelec, D. (2005). Neuroeconomics: How neuroscience can inform economics. *Journal of Economic Literature, 43*(1), 9–64.

Chib, V. S., Rangel, A., Shimojo, S., & O'Doherty, J. P. (2009). Evidence for a common representation of decision values for dissimilar goods in human ventromedial prefrontal cortex. *Journal of Neuroscience, 29*(39), 12315–12320.

Chorvat, T., & McCabe, K. (2004). The brain and the law. *Philosophical Transactions of the Royal Society B: Biological Sciences, 359*(1451), 1727–1736.

Clithero, J. A., Tankersley, D., & Huettel, S. A. (2008). Foundations of neuroeconomics: From philosophy to practice. *PLoS Biology, 6*(11), e298.

Cohen, J. D., & Blum, K. I. (2002). Reward and decision. *Neuron, 36*(2), 193–198.

Colander, D. (2007). Edgeworth's hedonimeter and the quest to measure utility. *Journal of Economic Perspectives, 21*(2), 215–226.

Coltheart, M. (2004). Brain imaging, connectionism, and cognitive neuropsychology. *Cognitive Neuropsychology, 21*(1), 21–25.

Craver, C. F., & Alexandrova, A. (2008). No revolution necessary: Neural mechanisms for economics. *Economics and Philosophy, 24*, 381–406.

Cromwell, H. C., & Schultz, W. (2003). Effects of expectations for different reward magnitudes on neuronal activity in primate striatum. *Journal of Neurophysiology, 89*(5), 2823–2838.

Daw, N. D., O'Doherty, J. P., Dayan, P., Seymour, B., & Dolan, R. J. (2006). Cortical substrates for exploratory decisions in humans. *Nature, 441*(7095), 876–879.

Delgado, M. R., Nystrom, L. E., Fissell, C., Noll, D. C., & Fiez, J. A. (2000). Tracking the hemodynamic responses to reward and punishment in the striatum. *Journal of Neurophysiology, 84*(6), 3072–3077.

Fisher, I. (1925). *Mathematical investigations in the theory of value and prices*. New Haven, CT: Yale University Press.

Friedman, M. (1953). *Essays in positive economics* (pp. 3–43). Chicago: University of Chicago Press.

Friston, K. J., Price, C. J., Fletcher, P., Moore, C., Frackowiak, R. S., & Dolan, R. J. (1996). The trouble with cognitive subtraction. *Neuroimage, 4*(2), 97–104.

Garland, B., & Glimcher, P. W. (2006). Cognitive neuroscience and the law. *Current Opinion in Neurobiology, 16*(2), 130–134.

Glimcher, P. W. (2003). *Decisions, uncertainty and the brain: The science of neuroeconomics*. Cambridge, MA: The MIT Press.

Glimcher, P. W. (2011). *Foundations of neuroeconomic analysis*. Oxford: Oxford University Press.

Glimcher, P. W., Camerer, C. F., Fehr, E., & Poldrack, R. A. (Eds.). (2009). *Neuroeconomics: Decision making and the brain*. New York: Academic Press.

Glimcher, P. W., Kable, J. W., & Louie, K. (2007). Neuroeconomic studies of impulsivity: Now or just as soon as possible? *American Economic Review, 97*(2), 142–147.

Green, L., Myerson, J., & Macaux, E. W. (2005). Temporal discounting when the choice is between two delayed rewards. *Journal of Experimental Psychology. Learning, Memory, and Cognition, 31*(5), 1121–1133.

Greene, J., & Cohen, J. (2004). For the law, neuroscience changes nothing and everything. *Philosophical Transactions of the Royal Society B: Biological Sciences, 359*(1451), 1775–1785.

Grosenick, L., Greer, S., & Knutson, B. (2008). Interpretable classifiers for fMRI improve prediction of purchases. *IIEEE Transactions on Neural Systems and Rehabilitation Engineering, 16,* 539–548.

Gul, F., & Pesendorfer, W. (2008). The case for mindless economics. In A. Caplin & A. Schotter (Eds.), *The foundations of positive and normative economics: A handbook* (pp. 3–39). Oxford: Oxford University Press.

Hampton, A. N., Bossaerts, P., & O'Doherty, J. P. (2006). The role of the ventromedial prefrontal cortex in abstract state-based inference during decision making in humans. *Journal of Neuroscience, 26*(32), 8360–8367.

Hare, T. A., Camerer, C. F., Knoepfle, D. T., & Rangel, A. (2010). Value computations in ventral medial prefrontal cortex during charitable decision making incorporate input from regions involved in social cognition. *Journal of Neuroscience, 30*(2), 583–590.

Harrison, G. W. (2008). Neuroeconomics: A critical reconsideration. *Economics and Philosophy, 24*(3), 303–344.

Hasson, U., Landesman, O., Knappmeyer, B., Vallines, I., Rubin, N., & Heeger, D. J. (2008). Neurocinematics: The neuroscience of making movies. *Projections: The Journal for Movies and Mind, 1*(2), 2–26.

Henson, R. (2006). Forward inference using functional neuroimaging: Dissociations versus associations. *Trends in Cognitive Sciences, 10*(2), 64–69.

Holcomb, J. H., & Nelson, P. S. (1992). Another experimental look at individual time preference. *Rationality and Society, 4*(2), 199–220.

Houser, D., Schunk, D., & Xiao, E. (2007). Combining brain and behavioral data to improve econometric policy analysis. *Analyse & Kritik, 29,* 86–96.

Hsu, M., Bhatt, M., Adolphs, R., Tranel, D., & Camerer, C. F. (2005). Neural systems responding to degrees of uncertainty in human decision-making. *Science, 310*(5754), 1680–1683.

Kable, J. W. (2011). The cognitive neuroscience toolkit for the neuroeconomist: A functional overview. *Journal of Neuroscience, Psychology, and Economics, 4*(2), 63–84.

Kable, J. W., & Glimcher, P. W. (2007). The neural correlates of subjective value during intertemporal choice. *Nature Neuroscience, 10*(12), 1625–1633.

Kable, J. W., & Glimcher, P. W. (2009). The neurobiology of decision: Consensus and controversy. *Neuron, 63*(6), 733–745.

Kable, J. W., & Glimcher, P. W. (2010). An "as soon as possible" effect in human inter-temporal decision making: Behavioral evidence and neural mechanisms. *Journal of Neurophysiology, 103,* 2513–2531.

Kahneman, D., & Tversky, A. (1979). Prospect theory: An analysis of decision under risk. *Econometrica, 47*(2), 263–291.

Kitayama, S., & Uskul, A. K. (2011). Culture, mind, and the brain: Current evidence and future directions. *Annual Review of Psychology, 62*, 419–449.

Knutson, B., Rick, S., Wimmer, G. E., Prelec, D., & Loewenstein, G. (2007). Neural predictors of purchases. *Neuron, 53*(1), 147–156.

Kosslyn, S. M. (1999). If neuroimaging is the answer, what is the question? *Philosophical Transactions of the Royal Society B: Biological Sciences, 354*(1387), 1283–1294.

Kuhnen, C. M., & Knutson, B. (2005). The neural basis of financial risk taking. *Neuron, 47*(5), 763–770.

Kurzban, R. (2010). *Why everyone (else) is a hypocrite: Evolution and the modular mind.* Princeton, NJ: Princeton University Press.

Laibson, D. (1997). Golden eggs and hyperbolic discounting. *Quarterly Journal of Economics, 112*(2), 443–477.

Lau, B., & Glimcher, P. W. (2008). Value representations in the primate striatum during matching behavior. *Neuron, 58*(3), 451–463.

Levallois, C., Smidts, A., & Wouters, P. (2010). *Whose field is it? Disciplinary interactions in neuroeconomics.* Paper presented at the Society for Neuroeconomics 8th Annual Meeting, Evanston, IL.

Levy, I., Lazzaro, S. C., Rutledge, R. B., & Glimcher, P. W. (2011). Choice from non-choice: Predicting consumer preferences from blood oxygenation level-dependent signals obtained during passive viewing. *Journal of Neuroscience, 31*(1), 118–125.

Levy, I., Snell, J., Nelson, A. J., Rustichini, A., & Glimcher, P. W. (2010). Neural representation of subjective value under risk and ambiguity. *Journal of Neurophysiology, 103*(2), 1036–1047.

Lo, A. W., & Repin, D. V. (2002). The psychophysiology of real-time financial risk processing. *Journal of Cognitive Neuroscience, 14*(3), 323–339.

Louie, K., & Glimcher, P. W. (2010). Separating value from choice: Delay discounting activity in the lateral intraparietal area. *Journal of Neuroscience, 30*(16), 5498–5507.

Louie, K., Grattan, L., & Glimcher, P. W. (2011). Reward value-based gain control: Divisive normalization in parietal cortex. *Journal of Neuroscience, 30*(29), 10627–10639.

Marr, D. (1982). *Vision: A computational investigation into the human representation and processing of visual information.* New York: Henry Holt.

McCabe, K. A. (2008). Neuroeconomics and the economic sciences. *Economics and Philosophy*, *24*(3), 345–368.

McClure, S. M., Ericson, K. M., Laibson, D. I., Loewenstein, G., & Cohen, J. D. (2007). Time discounting for primary rewards. *Journal of Neuroscience*, *27*(21), 5796–5804.

McClure, S. M., Laibson, D. I., Loewenstein, G., & Cohen, J. D. (2004). Separate neural systems value immediate and delayed monetary rewards. *Science*, *306*(5695), 503–507.

McCoy, A. N., Crowley, J. C., Haghighian, G., Dean, H. L., & Platt, M. L. (2003). Saccade reward signals in posterior cingulate cortex. *Neuron*, *40*(5), 1031–1040.

Montague, R. (2006). *Why choose this book? How we make decisions*. New York: Dutton.

Monterosso, J. R., Ainslie, G., Xu, J., Cordova, X., Domier, C. P., & London, E. D. (2007). Frontoparietal cortical activity of methamphetamine-dependent and comparison subjects performing a delay discounting task. *Human Brain Mapping*, *28*(5), 383–393.

Neumarker, B. (2007). Neuroeconomics and the economic logic of behavior. *Analyse & Kritik*, *29*, 60–85.

Newell, A., & Simon, H. (1976). Computer science as empirical inquiry: Symbols and search. *Communications of the ACM*, *19*, 113–126.

O'Doherty, J. P., Dayan, P., Friston, K., Critchley, H., & Dolan, R. J. (2003). Temporal difference models and reward-related learning in the human brain. *Neuron*, *38*(2), 329–337.

O'Doherty, J. P., Hampton, A., & Kim, H. (2007). Model-based fMRI and its application to reward learning and decision making. *Annals of the New York Academy of Sciences*, *1104*, 35–53.

O'Doherty, J. P., Kringelbach, M. L., Rolls, E. T., Hornak, J., & Andrews, C. (2001). Abstract reward and punishment representations in the human orbitofrontal cortex. *Nature Neuroscience*, *4*(1), 95–102.

Ochsner, K. N., & Lieberman, M. D. (2001). The emergence of social cognitive neuroscience. *American Psychologist*, *56*(9), 717–734.

Padoa-Schioppa, C. (2008). The syllogism of neuro-economics. *Economics and Philosophy*, *24*(3), 449–457.

Pareto, V. (1973). *Epistolario, 1890–1923*. G. Busino (Ed.) Rome: Accademia nazionale dei Lincei.

Plassmann, H., Ambler, T., Braeutigam, S., & Kenning, P. (2007). What can advertisers learn from neuroscience? *International Journal of Advertising*, *26*(2), 151–175.

Poldrack, R. A. (2006). Can cognitive processes be inferred from neuroimaging data? *Trends in Cognitive Sciences, 10*(2), 59–63.

Preuschoff, K., Bossaerts, P., & Quartz, S. R. (2006). Neural differentiation of expected reward and risk in human subcortical structures. *Neuron, 51*(3), 381–390.

Preuschoff, K., Quartz, S. R., & Bossaerts, P. (2008). Human insula activation reflects risk prediction errors as well as risk. *Journal of Neuroscience, 28*(11), 2745–2752.

Quartz, S. R. (2008). From cognitive science to cognitive neuroscience to neuroeconomics. *Economics and Philosophy, 24*, 459–471.

Rangel, A., Camerer, C., & Montague, P. R. (2008). A framework for studying the neurobiology of value-based decision making. *Nature Reviews Neuroscience, 9*(7), 545–556.

Read, D. (2001). Is time-discounting hyperbolic or subadditive? *Journal of Risk and Uncertainty, 23*(1), 5–32.

Read, D., & Roelofsma, P. H. M. P. (2003). Subadditive versus hyperbolic discounting: A comparison of choice and matching. *Organizational Behavior and Human Decision Processes, 91*(2), 140–153.

Ross, D. (2008). Two styles of neuroeconomics. *Economics and Philosophy, 24*, 473–483.

Rubinstein, A. (2008). Comments on neuroeconomics. *Economics and Philosophy, 24*, 485–494.

Rustichini, A. (2005). Neuroeconomics: Present and future. *Games and Economic Behavior, 52*(2), 201–212.

Rustichini, A. (2009). Is there a method of neuroeconomics? *American Economic Journal: Microeconomics, 1*(2), 48–59.

Samejima, K., Ueda, Y., Doya, K., & Kimura, M. (2005). Representation of action-specific reward values in the striatum. *Science, 310*(5752), 1337–1340.

Samuelson, P. (1937). A note on the measurement of utility. *Review of Economic Studies, 4*, 155–161.

Scholten, M., & Read, D. (2006). Discounting by intervals: A generalized model of intertemporal choice. *Management Science, 52*(9), 1424–1436.

Schultz, W., Dayan, P., & Montague, P. R. (1997). A neural substrate of prediction and reward. *Science, 275*(5306), 1593–1599.

Schwartz, O., & Simoncelli, E. P. (2001). Natural signal statistics and sensory gain control. *Nature Neuroscience, 4*(8), 819–825.

Shamosh, N. A., DeYoung, C. G., Green, A. E., Reis, D. L., Johnson, M. R., Conway, A. R. A., et al. (2008). Individual differences in delay discounting: Relation to intelligence, working memory, and anterior prefrontal cortex. *Psychological Science, 19*(9), 904–911.

Shiv, B., Bechara, A., Levin, I., Alba, J. W., Bettman, J. R., Dube, L., et al. (2005). Decision neuroscience. *Marketing Letters, 16*(3–4), 375–386.

Singer, T., Seymour, B., O'Doherty, J. P., Kaube, H., Dolan, R. J., & Frith, C. D. (2004). Empathy for pain involves the affective but not sensory components of pain. *Science, 303*(5661), 1157–1162.

Smith, V. L. (2008). *Rationality in economics: Constructivist and ecological forms.* New York: Cambridge University Press.

Spiegler, R. (2008). Comments on the potential significance of neuroeconomics for economic theory. *Economics and Philosophy, 24,* 515–521.

Sun, R. L., Coward, A., & Zenzen, M. J. (2005). On levels of cognitive modeling. *Philosophical Psychology, 18*(5), 613–637.

Sunder, S. (2006). Economic theory: Structural abstraction or behavioral reduction? *History of Political Economy, 38,* 322–342.

Tremblay, L., & Schultz, W. (1999). Relative reward preference in primate orbitofrontal cortex. *Nature, 398*(6729), 704–708.

von Wieser, F., & Hinrichs, A. F. (1927). *Social economics.* New York: Greenberg.

Wager, T. D. (2006). Do we need to study the brain to understand the mind? *APS Observer, 19*(9), 25–27.

Wang, J. T.-y., Spezio, M., & Camerer, C. F. (2010). Pinocchio's pupil: Using eyetracking and pupil dilation to understand truth telling and deception in sender-receiver games. *American Economic Review, 100*(3), 984–1007.

Wilcox, N. T. (2008). Against simplicity and cognitive individualism. *Economics and Philosophy, 24,* 523–532.

Willingham, D. T., & Dunn, E. W. (2003). What neuroimaging and brain localization can do, cannot do and should not do for social psychology. *Journal of Personality and Social Psychology, 85*(4), 662–671.

Wittmann, M., Leland, D. S., & Paulus, M. P. (2007). Time and decision making: Differential contribution of the posterior insular cortex and the striatum during a delay discounting task. *Experimental Brain Research, 179*(4), 643–653.

Yoon, C., Gutchess, A. H., Feinberg, F., & Polk, T. A. (2006). A functional magnetic resonance imaging study of neural dissociations between brand and person judgments. *Journal of Consumer Research, 33*(1), 31–40.

Zaghloul, K. A., Blanco, J. A., Weidemann, C. T., McGill, K., Jaggi, J. L., Baltuch, G. H., et al. (2009). Human substantia nigra neurons encode unexpected financial rewards. *Science, 323*(5920), 1496–1499.

Zauberman, G., Kim, B., Malkoc, S. A., & Bettman, J. R. (2009). Discounting time and time discounting: Subjective time perception and intertemporal preferences. *Journal of Marketing Research, 46*(8), 543–556.

13 Cognitive Aging and Human Capital

John J. McArdle and Robert J. Willis

13.1 Introduction

Aging citizens around the world are required to make complex decisions as they prepare for and then enjoy increasingly long periods of retirement. They need to make complex choices about financing retirement and insuring long-term care and health risk. Financial, health, and care risks compound as lifetimes lengthen. Moreover, as people age, there is seems to be a mismatch between the complexity of the decisions they face and the often gradual and sometimes sudden cognitive declines that accompany aging. Meanwhile, societies all over the world are getting older, making these problems more prevalent.

The complexity of later-life decisions has grown over time as defined benefit pensions, which provide a fixed retirement income determined by salary and length of service, have been replaced by defined contribution pensions such as 401(k) plans. The planning problems faced by workers with such plans have grown much more complicated, and their well-being following retirement is far more dependent on the decisions they make many years before retirement about participation, contribution rates, portfolio contributions, rollovers, and so on. Once people retire, the complications continue. Rather than learning how to live on a fixed income, retirees must manage their wealth in the face of uncertainty from many different sources—financial markets, personal health, and family circumstances.

People vary greatly in their ability to think and reason about which among a set of complex alternatives would be best for them (i.e., maximize their preferences). They also vary greatly in their knowledge of the specifics of each alternative, such as, for example, institutional arrangements, contractual specifications, relevant aspects of pricing, and understanding of risks and benefits. Traditional economic theory assumes either that all economic actors can deal with such matters effortlessly and accurately or, following Friedman's famous pool player example, behave "as if" they can do so (Friedman, 1953). Obviously, the existence of population heterogeneity in cognitive ability would be of no consequence for economic behavior under these assumptions.

Conversely, this chapter treats cognitive ability as form of human capital that is useful in improving the quality of economic decisions people make in preparing for retirement, and then while managing their wealth and health during retirement. While there is an extensive literature, largely based on human capital theory, that relates cognitive ability to economic performance, most of it focuses on the first part of the life cycle spent in school and the labor market and devotes relatively little attention to the latter part of the life cycle. The logic of human capital theory, however, applies equally well to the issues that confront older people. A person's reasoning ability and stock of relevant knowledge is valuable to the extent that it leads to better decisions and planning. Over time, the stock of knowledge may be augmented by investment. The demand for investment depends on the marginal value of new knowledge, and the supply depends on the individual's learning capability and willingness to devote effort to learning.

There is a remarkable parallel between human capital theory and the theory of fluid and crystallized intelligence (*Gf-Gc* theory), the leading theoretical framework used by psychologists to study the development of ability, knowledge, and skills over the life cycle. Just as remarkable, with a few notable exceptions, until recently economists were largely unaware of this theory, while psychologists were equally ignorant of the theory of human capital in economics, despite the fact that both theories began to be developed more than fifty years ago, have generated vast literatures, and continue to be highly active fields of research. In this chapter, key features of these two theories are described, showing the ways in which they are parallel and, more importantly, the ways in which they complement one another and ultimately may be merged into a genuinely interdisciplinary theory. The chapter then describes the creation of new data, inspired by this theory, that links detailed assessment of the cognitive abilities of a sample of older Americans with a rich set of measures of their wealth, portfolio choices, financial knowledge, and risk preferences. It concludes with some illustrative findings from this data.

13.2   The Life-Span Cognitive Theory of Fluid and Crystallized Intelligence

It is clear that the creation of a score for a person's cognitive ability is a main objective of much cognitive psychological research. In terms of cognitive ability, the first such theory of unified cognitive ability was developed by Spearman (1904). This intelligence concept identified that each person had an ability within them that is measurable and differentiable, which he termed the g-factor. By administering a set of items, psychologists could measure and compare persons on their intelligence score. This revolutionary idea was based in the theory that such items and abilities have related variances and will load together to form a factor. Factor analysis became a staple for analyzing the relationship of a set of related items. In this set of rules, a vocabulary score would be indicative of intelligence in a similar way as math ability or spatial ability. The first set of such tests of general intelligence were developed by Binet and Simon (1905). This set of intelligence tests, based largely on the single intelligence factor, was a way to broadly identify scholastic achievement.

The focus of a single intelligence construct has been questioned since the inception of Thurstone's primary mental abilities (PMA) model (Thurstone, 1935, 1938). The PMA were theorized as a set of seven identifiable factors of cognitive ability in human behavior. The seven mental abilities measured by the PMA are verbal comprehension, word fluency, number facility, spatial visualization, associative memory, perceptual speed, and reasoning. Thurstone separated specific abilities into groups of related processes and believed there was something special about each process; examining only a single, overall intelligence factor, he believed, would hide individual variability in each category. One can note that perceptual speed has been one of the defining factors of human ability since multiple ability theories have existed. In subsequent studies the factor structure of the PMA has been tested for stability (Smith, 1958; Kaiser, 1960; Zimmerman, 1953).

Recent work has clearly rejected the original notion of a g-factor and instead focused on measuring multiple abilities. For example, the original form of Gf-Gc theory (Cattell, 1941, 1987) considered only two broad factors, *crystallized knowledge* (Gc) and *fluid reasoning* (Gf). Gf-Gc theory made three key predictions about the complex nature of human intellectual abilities. The first predictions are *structural*: a single general factor (i.e., Spearman's "g") will not account for the patterns of variation seen among

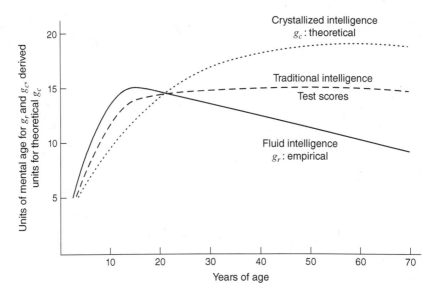

Figure 13.1
Cattell (1941) and Horn's (1967) theory of cognitive changes.

multiple abilities, and at least two broad factors are required for a reasonable level of fit to observations. One broad factor, *Gc*, is thought to represent *acculturated knowledge*, and the other broad factor, *Gf*, is thought to represent *reasoning and thinking* in novel situations. A second set of predictions of *Gf-Gc* theory are kinematic—over the early phases of the life-span, there is an expected rise of *Gc* together with an expected rise of *Gf*, but in early adulthood there is a divergence of these two processes. In theory, there is continued growth of *Gc*, but *Gf* levels off and exhibits rapid declines that increase in older ages. These differences are depicted in figure 13.1 (from Cattell, 1998). A third set of predictions are kinetic—the "investment" in *Gf*, coupled with other factors in a context of educationally relevant settings, leads to individual differences in the development of *Gc*. Only recently have these dynamic propositions been put to the test empirically (e.g., McArdle, 2001; McArdle, Hamagami, Meredith, & Bradway, 2001; McArdle et al., 2002; Ferrer & McArdle, 2004). But, and in general, the structural, kinematic, and kinetic aspects of adult cognitive development now have broad empirical support.

The current form of *Gf-Gc* theory (Horn, 1988, 2003) is an integration of results from studies of adult aging using both experimental and differential paradigms, and contains eight broad cognitive functions, as depicted

Table 13.1

Broad cognitive functions contained in *Gf-Gc* theory

---

1. *Fluid reasoning* (*Gf*) is measured in tasks requiring inductive, deductive, conjunctive, and disjunctive reasoning to arrive at understanding of relations among stimuli, comprehend implications, and draw inferences.

2. *Acculturation knowledge* (*Gc*) is measured in tasks indicating breadth and depth of knowledge of the dominant culture. A prime exemplar of *Gc* is acquired knowledge of the vocabulary of a culture.

3. *Short-term memory* (*Gsm*), also called *short-term apprehension-retention* (SAR), is measured in tasks that require one to apprehend, maintain awareness of, and recall elements of immediate stimulation.

4. *Processing speed* (*Gs*) is measured most purely in rapid scanning and responding to intellectually simple tasks for which almost all people would get the right answer given unlimited amounts of time.

5. *Long-term retrieval* (*Glr*), also termed *fluency of retrieval*, is measured in tasks requiring consolidation for storage, and retrieval through association, of information stored from minutes to years before.

6. *Visual processing* (*Gv*) is measured in tasks involving visual closure, constancy, and fluency in viewing the way objects appear in space as they are rotated and reorganized in various ways.

7. *Auditory processing* (*Ga*) is measured in tasks that involve perception of sound patterns under distraction or distortion, maintaining awareness of order and rhythm, and discerning the elements of complex sounds.

8. *Quantitative knowledge* (*Gq*) is measured in tasks requiring understanding and application of the concepts and skills of mathematics. *Gq* may be regarded as a component of *Gc*.

---

in table 13.1. This form of *Gf-Gc* theory is used here as a set of guiding hypotheses and as a classification system for the collection of key cognitive abilities.

Repeated examinations using independent sets of data have led to broad support for *Gf-Gc* theory in both cross-sectional (e.g., Lindenberger & Baltes, 1997; Horn & Cattell, 1967; Horn & Noll, 1997; McArdle & Prescott, 1992) and longitudinal studies (e.g., Baltes & Mayer, 1999; Donaldson & Horn, 1992; McArdle, Prescott, Hamagami, & Horn, 1998; Schaie, 1996; Salthouse et al., 2003). The empirical synthesis of over 400 research studies (Carroll, 1993, 1998) showed a remarkable set of similarities among the theories used by most researchers in this area, and identified *Gf* as the central *G* (see McArdle & Woodcock, 1998).

The current eight-factor model has been related to many widely used test batteries (e.g., the WAIS; McGrew & Flanagan, 1998; McArdle et al., 2007) and used as the theoretical foundation of the widely used *Woodcock-Johnson* tests (WJ-R, WJ III; Woodcock et al., 2001). These eight factors are important in aging research because they are expected to change in different ways as aging occurs, and expected to have different relations to health, decision making, dementia, and other non-cognitive outcomes. Most importantly, we are not now sure about which of these cognitive functions may be required for dealing with complex decisions.

These functional changes in *Gf-Gc* even match some recent ideas about "brain plasticity"—that is, the possibility that brain functions actually change as a result of experiences and learning (see the work of Baltes & Kliegel, 1992; Li & Lindenberger, 2002; Lövdén, Bäckman, Lindenberger, Schaefer, & Schmiedek, 2010; Lustig, Shah, Seidler, & Reuter-Lorenz, 2009). In one study of London cab drivers, for example, researchers found that cab drivers have larger hippocampi than matched controls and that the longer an individual had worked as a cab driver, the larger the hippocampus (Maguire et al., 2000).[1] According to this theory, areas of the brain may shrink or expand—become more or less functional—based on experience. In other words, the brain, like the rest of the body, builds the "muscles" it uses most, sometimes at the expense of other abilities. In a recent study using cross-national data and country-specific retirement policy variables as instrumental variables to control for endogeneity, Rohwedder and Willis (2010) found that early retirement appears to have a substantial causal negative effect on episodic memory scores. While they speculate about mechanisms that might underlie this result, they have no evidence regarding changes in brain structure or the psychological mechanisms that might be involved. More generally, although most of this work is at the early stage of speculation, there is obviously a great interest in understanding the neurological changes that accrue with age and life experience.

13.3 Human Capital Theory

Education provides the canonical example of human capital investment on which Becker (1962, 1964) focused his pioneering work on human capital. Willis (1986, p. 527) summarized the "theory in a nutshell" as follows:

Additional schooling entails opportunity costs in the form of forgone earnings plus direct expenses such as tuition. To induce a worker to undertake additional schooling, he must be compensated by sufficiently higher lifetime earnings. To command higher earnings, more schooled workers must be sufficiently more productive than their less schooled fellow workers. In long-run competitive equilibrium, the relationship between lifetime earnings and schooling is such that (a) the supply and demand for workers of each schooling level is equated and (b) no worker wishes to alter his schooling level.

Human capital theory can be extended to explain the life-cycle pattern of earnings growth by replacing the term "schooling" with "on-the-job training" in the preceding nutshell summary. In addition, it was recognized early on that the concept of human capital can be extended to other forms of productive investments, including migration (Sjaastad, 1962), health (Grossman, 1972), information (Stigler, 1962), and many others. While the effect of human capital investment on labor market earnings has been the most prominent application of the theory, it is important to note that the payoff to human capital investment may occur both in the form of monetary returns and non-market returns. For example, in addition to increasing a person's market wage, the benefits of education may include increased utility from non-market activities, such as enjoyment from music or reading novels. Likewise, improved health is something valued for itself in addition to whatever value it may have in improving a person's market earnings.

Similarly, the cost of investments in human capital may include direct expenditures like tuition or a doctor's bill, but often the opportunity cost of the investor's time is the most important component of cost (Becker, 1965). For example, the opportunity cost of a mother's time in augmenting the human capital of her child may be determined by her market wage if she works, or by the (unobservable) shadow price implied by alternative non-market uses of her time if she is not in the labor market (Willis, 1973; Heckman, 1974). Other, usually unobservable, costs of investment include things like the pleasure or disutility that students experience in effort devoted to studying or that workers experience in doing their jobs well.

Early studies of the returns to education and on-the-job training found that wage differentials by education and age or experience generated rates of return that were commensurate with returns on physical capital (see Willis, 1986, for a survey). However, one of the most notable trends in the US economy is the substantial widening of the wage structure that began during the 1980s (Bound & Johnson, 1992; Katz & Murphy, 1992; Murphy & Welch, 1992; Juhn, Murphy, & Pierce, 1993). Wage differentials by education, by occupation, and by age and experience group all rose substantially during the 1980s and have continued to be large. Although other factors such as changes in minimum wages and the erosion of unions played a role in this widening of differentials, in a recent review of the evidence, Autor, Katz, and Kearney (2008, p. 300), summarizing findings from a number of influential studies, conclude that the growth of inequality reflected a secular rise in the demand for skill, together with a slowdown in the relative supply of college-equivalent workers.

An important question is why the supply of more skilled workers has not increased sufficiently to reduce these wage differentials and drive down the increasing rate of return to investment in human capital to previous levels. Part of the explanation for the persistence of large skill premiums is the heterogeneity in the ability distribution, which reduces the elasticity of the aggregate supply of skill. In the next section, parallels between the role of ability in human capital theory and the theory of fluid and crystallized intelligence are examined.

13.4   Parallels between *Gc-Gf* Theory and Human Capital Theory

The early statements of human capital theory by Becker (1964) and others did not address the key question of the mechanism by which investments in education or on-the-job training increased a worker's labor market productivity and, thus, increased the wage he or she could obtain in a competitive labor market. In a seminal paper, Ben-Porath (1967) filled this gap by introducing the concept of a human capital production function. He assumed this function is of the Cobb-Douglas form,

$$Q_t = \beta_0 (s_t K_t)^{\beta_1} D_t^{\beta_2} \qquad \beta_1 > 0; \beta_2 > 0; \beta_1 + \beta_2 < 0; 0 \le s_t \le 1. \tag{13.1}$$

In words, this function says that during period t individuals combine their ability (β_0), effort (s_t), stock of human capital (K_t), and other inputs (D_t) in order to produce additional human capital (Q_t) or, equivalently, to produce Q_t units of gross investment in human capital.

The stock of human capital represents the amount of labor market skill that an individual possesses at time t.[2] The stock grows according to the amount of net investment, so that

$$K_{t+1} = Q_t - \delta K_t , \tag{13.2}$$

where the parameter δ is the rate of depreciation.

The parallels between the human capital production function in equation (13.1) and Gf-Gc theory are clear. In the language of the psychological theory, an individual combines her fluid intelligence or reasoning ability, β_0; her crystallized intelligence or knowledge, K_t; her effort or time devoted to learning, s_t, and other inputs, D_t, in order to acquire additional crystallized knowledge through learning. D_t may be considered to be a vector of variables that includes other factors such as, for example, parents, teachers, supervisors, co-workers, books, and computers that exist in the individual's home, school, and work environment which influence the productivity of time spent learning. D_t may also contain environmental inputs that influence a person's ability to learn, such as the state of technology, culture, language, or political system, which vary over time and across countries.

Equation (13.2) also has an obvious psychological interpretation: the capacity of a person to remember what she has learned is represented by $(1 - \delta)$, where δ is the rate of forgetting. Note that technological change may erode the economic value of person's effective stock of knowledge through obsolescence, in much the same way forgetting does.

As discussed above in Section 2, Gf-Gc theory is dynamic. It describes the time path of the development of fluid and crystallized intelligence over the entire life cycle. As stated there, the kinetic aspect of cognitive development is explicitly capital-theoretic: the "investment" in Gf, coupled with other factors in a context of educationally relevant settings, leads to individual differences in the development of Gc. In effect, the psychological theory says that Gf, a perishable resource that reaches its peak in early adulthood, is used to create a durable cognitive resource, Gc, in the form of knowledge that is productive in both market and non-market activities. Whether these changes in Gc are associated with changes in brain structure as suggested by the theory of brain plasticity, also discussed in section 2, remains an open question.

The Ben-Porath model also predicts a dynamic pattern of cognitive development. While the psychological theory allows for individual variation in the rate of development and informally discusses the idea that Gc is the product of an investment process that involves Gf, the Ben-Porath model (and more recent variants of it to be discussed shortly) provides explicit predictions about the dynamics of cognitive development for an individual with given initial values of Gf, Gc, and other variables. These models also generate "comparative dynamic" predictions of the time path of the cognitive variables and key economic variables, such as the duration of schooling and the life-cycle pattern of earnings for individuals who differ in their ability endowment, in market wages, earned interest rates, and environmental variables. In the rest of this section, a few implications of these models are briefly discussed in order to give the reader a flavor of the kinds of results they generate. Readers interested in a more complete analysis are referred to the cited papers.

Beginning with the Ben-Porath model, assume the time path of investment in human capital is chosen so as to maximize the present discounted value of labor market earnings from the time the individual leaves school and enters the labor market until he retires. The only choice variable in the model is the trade-off between earning and learning implied by the decision in each period about the fraction of time, s_t, devoted to investment and the proportion, $1 - s_t$, devoted to working in the market. Assuming that p is the market-determined "skill price" per unit of human capital, the earning capacity of a person at time t is proportional to her stock of human capital: $E_t = pK_t$. However, actual earnings are equal to $Y_t = (1 - s_t)pK_t$, reflecting the assumption that the individual pays for investment in on-the-job training, whose opportunity cost is $I_t = s_t pK_t$.[3] When a person attends school, it is assumed that $s_t = 0$, so that earnings are zero and the opportunity cost of schooling is equal to earnings capacity.

Ben-Porath shows that the optimal pattern of life-cycle investment in human capital follows three phases: a person engages in full-time schooling for the first S years of life, then enters the labor market and works continuously to an exogenously fixed retirement age, at which time work ceases. Schooling is optimal at the beginning of life for two reasons. First, because a unit of human capital is durable, the value of the flow of benefits it produces is worth more the earlier in life that it is created. Put differently, there is a "use it or lose it" aspect to fluid intelligence that is used as an input into the production of the more durable crystallized intelligence. The potential productivity of Gf in adding to knowledge in period t is lost forever if it is not devoted to learning.

The second reason for specialization in school during the first part of life is that the marginal value of adding to K_t is especially high, because human capital is self-productive and because K_0, the initial stock of capital, is at its lowest level at the beginning of life. Human capital is self-productive in the sense that the multiplicative input, $s_t K_t$, in equation (13.1) implies that the marginal product of time devoted to learning is greater in proportion to the existing stock of knowledge, and that a unit of knowledge gained today will enhance the productivity of future learning. Early in life, the value of the marginal product of time spent in full-time learning exceeds the market wage, and the person remains in school. As schooling continues, the stock of human capital grows rapidly and at some time, $t = S$, the stock of capital reaches a threshold at which the marginal value of time for a full-time student is equal to the market wage. At that point, the individual leaves school and enters the labor force.

During working life, the individual continues to accumulate additional human capital through on-the-job training. However, as the worker ages and retirement approaches, the marginal value of additional capital falls, causing the rate of investment to decrease. At some point late in the career, new investment may fail to offset the loss of capital due to depreciation, and the person's earnings capacity will decline prior to retirement. Recall that an individual's observed market earnings are equal to earnings capacity minus the opportunity cost of investment. The decline in the rate of investment implies that the rate of growth of observed earnings will be higher than the rate of growth of earnings capacity and that earnings will peak later than capacity.

The dynamics of the Ben-Porath model bear many similarities to the dynamics of the Gf-Gc model portrayed in figure 13.1. The parallel between the models is closest if β_0 in equation (13.1) is interpreted to be an endowment of fluid intelligence, which is determined by genes and the uterine environment that remains constant for the rest of life, and if we think of fluid intelligence at later ages during childhood and adulthood as indicated by the rate at which an individual can learn new things. Note that the output of the human capital production function, Q_t, is the amount of new knowledge that an individual learns during a given period of time.

The sign and magnitude of the percentage rate of change of Q_t is a good indicator of the shape of the time path of Gf. Taking the log of both sides of equation (13.1) and differentiating with respect to time, the following expression for the percentage rate of change of Q_t as a function of the percentage rates of change of inputs to the production of human capital is obtained:

$$\dot{q}_t = \beta_1 \dot{s}_t + \beta_1 \dot{k}_t + \beta_2 \dot{d}_t, \tag{13.3}$$

where lower-case variables are used to indicate the natural log of the corresponding upper-case variable, and the dot indicates a time derivative. First, consider the shape of the Gf curve during the schooling phase. Since all time is allocated to investment during this phase, $\dot{s}_t = 0$. The second term in (13.3) involving the percentage rate of growth of human capital can be rewritten as

$$\dot{k}_t = \frac{d\ln(K_t)}{dt} \frac{dK_t}{dt} \frac{1}{K_t} = \frac{Q_t}{K_t} - \delta. \tag{13.4}$$

This term is certainly positive during the schooling phase and may well be accelerating as the ratio of gross investment to the capital stock grows. The final term involving other inputs, such as teachers, books and so on, is also almost surely positive and perhaps accelerating as, say, a student moves from nursery school through medical school. In all, the Ben-Porath model suggests rapid growth in Gf during the schooling phase.

In contrast, the slope of the Gf curve tends to become negative during the working phase of life. Time spent learning decreases as a worker becomes trained and devotes an ever-larger fraction of time to earning.

The slowing pace of investment, combined with growth in the total stock of capital, mean that the ratio of gross investment tends to fall, eventually falling below the rate of depreciation. The final term is also likely to be negative as the pace of investment slows and reduces the need for other inputs to on-the-job training. Thus, the Ben-Porath model suggests that Gf is likely to become negatively sloped during the working phase of life. In all, therefore, the dynamics of the Ben-Porath model seem quite consistent with age trajectories of Gf and Gc in figure 13.1.

In a series of path-breaking papers, James Heckman and colleagues have revised and extended the concept of the human capital production function, created new econometric methods to estimate it, and applied these models to a variety of data sets in order to study the role of cognitive and non-cognitive skills in determining educational and labor market success (Borghans, Duckworth, Heckman, & ter Weel, 2008; Cunha & Heckman, 2008; Cunha, Heckman, & Schennach, 2010). Although this line of research has primarily focused on the first phases of the life cycle, it contains several features that are highly relevant in thinking about applications of human capital theory to older individuals.

First, unlike Ben-Porath, who assumes that human capital is homogeneous stuff that is equally productive in all applications, Heckman and colleagues assume that skills are heterogeneous. Second, dynamic complementarities between skills mean that failure to acquire key skills early may have long-lasting negative impacts on the ability to produce new skills. Finally, Heckman and colleagues emphasize the role of non-cognitive skills such as conscientiousness, patience, and trustworthiness. These skills may be complementary to cognitive skills in the production of new skills by influencing how hard a person is willing to study; in the labor market by increasing the degree to which a worker can be trusted to carry out a task without close supervision; and in valuing investment by increasing the weight given to future benefits relative to current costs.

13.5   Ability and Economic Behavior

There is a vast body of evidence that cognitive ability is strongly related to educational attainment and labor market earnings, implying that heterogeneity in cognition is importantly related, often causally, to performance in school and at work. Behavioral economics has called attention to a variety of departures of economic behavior from the predictions of conventional economic theory (Camerer, Loewenstein, & Rabin, 2003). Until recently, however, little attention was given in this literature to population heterogeneity in behavior, perhaps because it is inconvenient to acknowledge heterogeneity when attempting to generalize findings from lab experiments conducted with undergraduates to the broad population. For example, Frederick (2005, p. 25) wrote, "Studies on time preference, risk preference, probability weighting, ambiguity aversion, endowment effects, anchoring and other widely researched topics rarely make any reference to the possible effects of cognitive abilities (or cognitive *traits*)." Frederick goes on to state: "In the domain of risk preferences, there is no widely shared presumption about the influences of cognitive ability and almost no research on the topic."

The situation has changed dramatically since 2005. Noting a precedent in the writings of John Rae (1834/1905), Frederick suggests that cognitive ability is somehow related both to rates of time preference and to risk tolerance. He produces some evidence for this hypothesis based on surveys of a variety of non-random samples, mostly consisting of students from different universities, finding that more cognitively able persons tend to be both more risk tolerant and more patient. Since then, the connection between cognitive ability, risk tolerance, and time preference has been studied systematically in probability samples of the population, with results strongly confirming this pattern (e.g., Benjamin, Brown, & Shapiro, 2006; Dohmen, Falk, Huffman, & Sunde, 2010; Shapiro, 2010).

Time preference and risk tolerance are key preference parameters in economic models of saving, wealth accumulation, and portfolio choice. Survey-based measures of these parameters have been shown to be correlated with such outcomes by Dohmen et al. (2010), among others. Focusing on older individuals in the United States and England, respectively, McArdle, Smith, and Willis (2011) and Banks, O'Dea, and Oldfield (2010) found that wealth is correlated with several measures of ability, particularly one that measures basic numeracy. In addition, Smith, McArdle, and Willis (2010) and Hsu (2011) found that cognitive ability measures predict the division of labor between husbands and wives in determining the responsibility for household financial decisions.

Why is cognitive ability related to time preference and risk tolerance? Commenting in passing on Frederick's findings, Heckman (2007) conjectures, "Lower rates of time preference are associated with greater cognitive skills. Those with higher IQs are more farsighted (have lower time preference) because they envision future scenarios more clearly." In a similar vein, Lillard and Willis (2001) suggest that cognitively more able people who have more clearly defined probability beliefs tend to behave less risk aversely because, to them, the world looks less risky.

Kézdi and Willis (2008) find that people with higher ability and greater knowledge of the stock market have subjective probability beliefs of higher expected returns and lower risk than persons of lower ability or less knowledge and, consistent with standard theory, find that their share of risky assets in total wealth is an increasing function of expected returns and a decreasing function of the variance of returns. Hudomiet, Kézdi, and Willis (2011) find that the subjective stock market beliefs of more able and knowledgeable individuals reacted more sharply to the 2008 financial crisis than did the less able and knowledgeable. However, their most notable result is that there was a dramatic increase in "disagreement" about the direction of future stock returns as a consequence of the crisis, especially among the most able and knowledgeable. Despite their presumably clearer vision of what was going on in financial markets, the broader economy, policy, and politics, more able people apparently do not share a common "mental model" of how news from these spheres during the crisis would influence future stock returns.

Among all measured aspects of ability, recent research has convincingly demonstrated that mathematical or numerical ability is the most strongly related to economic success. Much of the research leading to this conclusion focused on the role of ability in determining educational attainment and the determinants of labor market earnings. As an early example, Willis and Rosen (1979) found that mathematical ability had a strong effect on the lifetime earnings of those who choose to go to college and little effect on those who stop at college. This implies that increased mathematical ability raises the rate of return to college, leading to self-selection of more able individuals vis-à-vis college. Using more recent and more representative data, Carneiro, Heckman, and Vytlacil (forthcoming) confirm the Willis-Rosen finding that students self-select into college on the basis of the returns to education, which in turn are related to a cognitive score in which mathematical ability plays a large role.[4] Obviously, these findings suggest that quantitative skills are complementary to the kinds of tasks that college-trained workers perform, while such skills are not particularly productive for the tasks that less educated workers perform.

13.6 New Data for the Analysis of Cognitive Aging and Economic Behavior

The Health and Retirement Study (HRS) was designed to provide academic researchers, policy analysts, and program managers with reliable, current data on the economic and physical well-being of men and women 50 years of age and older in America. As this population ages, assessing their cognitive skills is becoming increasingly important. The HRS has contained cognitive measures since it began in 1992.[5] Initially, the HRS focused on measuring aspects of cognition related to cognitive decline and early signs of dementia, an effort that was intensified with the ADAMS sub-study, which conducted an in-home clinical assessment of dementia of a stratified random sample of HRS participants age 70 and over beginning in 2001 (Langa, Plassman, Wallace, Herzog et al., 2005).

By the early 2000s, the HRS designers—an interdisciplinary team of economists, demographers, health researchers, and psychologists—had become convinced that developments in both policy and the private economy placed important demands on the cognitive resources of older Americans and believed that it would be important to both the scientific and policy communities to add measures of higher-order cognitive abilities to the HRS. There were three major challenges to implementing this idea: (1) it was unclear whether it is possible to obtain valid measures of cognition in the HRS because, unlike most psychological tests which are conducted in person, the HRS conducts a personal interview at baseline, followed by telephone interviews in subsequent longitudinal waves; (2) conventional psychological tests of any given type of ability are typically far too long to administer on the HRS, which covers many domains related to economics, health, and the family; and (3) the HRS team did not know which aspects of ability were most important to measure in order to help understand how well or badly respondents respond to complex decisions, fluctuations in the economy, and changing policies in the public and private sectors.

The first two of these challenges were addressed by analyses of existing HRS cognitive data and of data from a new survey called Cognition in the USA (CogUSA) designed by John McArdle. The third challenge stimulated the collection of a new Cognitive Economics (CogEcon) survey, designed by a team of economists led by Robert Willis, using CogUSA respondents as a sample frame. When merged, the two surveys allow an empirical investigation of which components of cognition are most closely related to economic behavior and outcomes that are of most interest. New cognitive measures were introduced into the 2010 HRS, based on recommendations that were heavily influenced by analyses of the CogUSA/CogEcon data. In the next section, the psychological measures and analysis in the CogUSA study are described, followed in section 13.8 by a description of the CogEcon measures and some illustrative findings about the relationship between cognitive abilities and economic behavior and outcomes.

13.7 The New CogUSA Study

The CogUSA study was designed to look at many measures across different modalities, with two goals in mind: (1) to identify measures for possible inclusion in future waves of the HRS project; and (2) to collect information that can be used to better understand how various cognitive abilities change with age. The CogUSA participants were selected based on the general overall principles of the HRS, but these participants had no prior HRS experiences, and they were chosen to test effectiveness of cognitive tasks in different modalities. Because measures might vary across different modalities used in the HRS (i.e., in-person and over the telephone), CogUSA specifically tested a telephone and in-person version of cognitive scores and examined cross-mode reliability.

The CogUSA study consisted of three cognitive survey components. Initial contact was established by telephone calls to selected households, and the study began with a 40-minute telephone interview that largely replicated the sections of the HRS questionnaire on demography, health, and cognition. This telephone survey was followed (usually within a few days) by a three-hour face-to-face cognitive assessment measuring components of HRS and *Gf-Gc*–based intelligence (i.e., the Woodcock-Johnson III cognitive battery, or WJ-III, and the Wechsler Abbreviated Scale of Intelligence, or WASI). A second telephone survey took place at a randomized interval of one to twenty-four months following the face-to-face interview. In addition to retesting several components of ability using a telephone administration, this second telephone interview included many of the same survey questions administered in the HRS 2008 wave.

The initial sample of respondents for the CogUSA Study consisted of $N = 3,224$ individuals obtained from a two-stage list to target households with individuals located in twenty-eight primary sampling units (PSUs) across the nation. Eligible sample members included individuals born in 1956 or earlier (age 51+) and their spouses or partners (regardless of age). Respondents were required to 1) be cognitively able to do the interview, and 2) be able to perform the interview in English. Among the individuals asked to participate in the telephone interview, a total of $N = 1,514$ individuals completed the telephone interview. The 47% response rate obtained was slightly higher than anticipated. Among the $N = 1,514$ individuals who completed a telephone interview in wave 1, $N = 1,230$ completed a face-to-face interview, for a multiple testing response rate of 81%, which is slightly lower than the original goal of 85%. Sample weights to post-stratify the CogUSA sample back to the HRS were developed to help adjust for non-response bias in the attrition between the telephone and face-to-face interviews (but see McArdle, 2010a).

The *Gf* reasoning factor described earlier was measured in this sample with several tasks requiring thinking and solving novel problems. The best measure of *Gf* was thought to be the Number Series test (NS) based on the WJ-III, as this requires a participant to solve a small numerical puzzle. Each item is considered correct or incorrect, and the final score is based on the WJ score scaling. The mean of the W-scale is 500 (for a person with a fifth-grade education) and each increase (or decrease) of each 10 points in the scale supposedly indicates the score reached was twice as difficult (or easy).

In order to be able to examine the multivariate characteristics of this construct, *Gf* was also measured in several other ways. A second WJ-III indicator of *Gf* is likely to be the Concept Formation (CF) test, because this is a task that asks participants to suggest a changing rule of inclusion with unfamiliar objects (i.e., is it the same color, shape, size, etc.). Each item is scored as correct or incorrect, again on the WJ scoring W-scale. The *Gf* concept is based on a very broad factor, so it can be represented by the Matrix Reasoning test (MS from the WASI), because this is a task that asks participants to provide the next best answer for a specific spatial item. Calculated scores are based on how far the participant was able to get and the number of items answered correctly.

The *Gc* knowledge factor was also measured in this sample, but with several verbal tasks of knowledge. The best measure of *Gc* was thought to be Picture Vocabulary (PV), based on the WJ-III, as this requires a participant to correctly provide the name of pictures or elements of pictures, using the same W-scale. A second WJ-III indicator of *Gc* is likely to be Word Attack (WA), a task that asks participants to correctly pronounce a written word. This is scored as correct or incorrect, again on the WJ scoring W-scale. The *Gc* concept is based on a very broad factor, so it can be represented by the Verbal Comprehension test (VC from the WASI), because this is a task that asks participants to provide verbal descriptions of specific items. In this case, the transcript of a participant's response was read after the fact, and answers were judged to be right or wrong based on how well the description fit an a priori rubric. Perfect answers were given 2 points credit, while close answers were given 1 point in credit. Calculated scores were based on how far the participant was able to get and the number of items answered correctly.

In addition, three demographic variables were used here—chronological age, educational attainment level, and gender. These variables were rescaled for easy interpretation. (Tables of simple statistics—means, deviations, and correlations—about the measures on N = 1,230 persons measured in our in-person tests are available from the authors). Most importantly here, the guiding Gf-Gc theory indicates that the non-verbal measures chosen will be separable from the verbal measures, and these factors will be related to the demographics in different ways—i.e., Gf will have a much larger age-related decline, and Gc will be more positively related to education. Due to the relatively large sample size, the significance of all effects were evaluated at a relatively high level of stringency (a< = 001).

The analyses of the CogUSA data have been carried out in a sequence of increasing complexity (see McArdle, Fisher, Kadlec, Prindle, & Woodcock, 2010), and only a brief summary will be presented here. First results from a standard multiple regression model were examined. In one such model, multiple indicators are used to provide clearer age differences in performance and seem to theoretically be related to the Gf factor, and less to the Gc factor. For example, in the prediction of Picture Vocabulary (PV) scores, the explained variance is approximately R^2 = 0.11, with significant independent differences from education (β = +10.9 W scores per each 4 years of education) and gender (β = –6.8 favoring males), and no significant age effect. In contrast, the prediction of Number Series (NS) scores was stronger, at approximately R^2 = 0.35, with significant independent differences from education (β = +17.8 W scores per each 4 years of education), age (β = –6.7 loses per decade), and gender (β = –5.2 favoring males). The pattern seems similar, but the effects on the Gf indicator are larger.

In the multivariate analysis of these data, the data were used to carry out a very simple latent variable path analysis (following McArdle & Prescott, 1996, 2010) using only two indicators of cognition, PV and NS. Basically, this kind of model tests the hypothesis that only one general function (g; see Spearman, 1904; Horn & McArdle, 2007) is indicated by these two variables. That is, factor loadings are calculated and used to describe the parallel effects from the demographic variables. The first variable (PV) has a fixed loading (λ_1 = 1) to identify the scale of the single latent variable. The second loading (λ_2 = 1.9) indicates that the second variable, NS, is much closer to the common factor, and both unique variances are estimated. The explained variance for the demographics related to the factor is much larger than seen before (R^2 = 0.466) because this is essentially a multiple regression with a disattenuated outcome score (see Fleiss & Shrout, 1977). The assumption that one factor fits the data is tested using the overall fit of the model to the data, as indicated by the likelihood ratio (i.e., χ^2 = 24 on df = 2, ε_a = 0.095, with lower ε_a = 0.064 and upper

$\varepsilon_a = 0.131$). This five-variable model has very few variables, so there are few restricted correlations and very few ways to go wrong.

In a more advanced version of this multivariate model, six indicators (NS, CF, MR, PV, WA, and PV) are used to form a single common factor (g) that will be related to the demographic variables. The first variable (PV) was assigned a fixed loading ($\lambda_1 = 1$) to identify the scale of the single latent variable (g), but the other factor loadings are allowed to be free ($\Lambda = \{ = 1$, 1.4, 0.7, 1.4, 0.7, 1.5 \}). This model indicates that the final variable (NS) is much closer to the common factor (h), and all six unique variances are estimated. The explained variance for the demographics related to the factor is larger ($R^2 = 0.488$) and, again, this is a multiple regression on the disattenuated scores. The assumption that one factor fits the data is tested both internally (i.e., does one factor account for all covariation within the six variables?) and externally (i.e., do the loadings λ_j multiplied by the β_k-weights account for the covariation between the demographics and the six variables?). Using the overall fit of the model to the data as indicated by the likelihood ratio (i.e., $\chi^2 = 613$ on $df = 24$, $\varepsilon_a = 0.141$, with lower $\varepsilon_a = 0.132$ and upper $\varepsilon_a = 0.151$), this does not seem to be very good in contrast to other models fitted here.

A final latent model is depicted here as figure 13.2. In this model, the latent Gc factor is indicated by PV, WA, and VC, while the latent Gf factor is indicated by CF, MR, and NS. The identification of each common factor was ensured by forcing the first loading to be fixed (i.e., $\lambda_1 = 1$) and the other two indicators to be freely estimated. This structural model is restrictive and does not fit perfectly (i.e., $\chi^2 = 223$ on $df = 20$, $\varepsilon_a = 0.09$, with lower $\varepsilon_a = 0.08$ and upper $\varepsilon_a = 0.10$). The relative estimates suggest that VC is probably the best index of Gc (where $\lambda_3/\lambda_1 = 0.97$), and MR is the best index of Gf (where $\lambda_2/\lambda_1 = 0.82$). In this model of common factor scores, in terms of the demographics, the explained variance of the disattenuated Gc score is approximately $R^2 = 0.31$, and there is only a significant independent difference from education ($\beta = +13.3$ W scores per each four years of education), and no significant age or gender effects. In contrast, the explained variance of the disattenuated Gf score is approximately $R^2 = 0.54$, and there is only a significant independent difference in education ($\beta = +15.6$ W scores per each four years of education), and a significant age effect ($\beta = -8.3$ loses per decade), with no significant gender effect.

In general, overall loss of fit of this two latent factor model to the data of the eight variables may be accepted. This final eight-variable model has many more measured variables, so there are more restricted correlations and very few ways to go wrong. When using the two factors, Gf and Gc

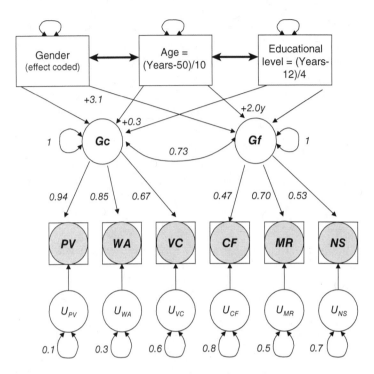

Figure 13.2
Final model of *Gf-Gc* results for CogUSA measures related to demographics ($\chi^2 = 228$, $df = 20$, $\varepsilon_a = .090$).

are estimated to have a high positive correlation ($\rho = 0.73$), but it may be questioned whether or not this is a perfect correlation. The third and fourth models fitted here differ in this respect (i.e., model 4 assumes $\rho = 1.00$). A test of the unit valued correlation from the difference between these models (i.e., $d\chi^2 = 613–223 = 390$ on $ddf = 4$). The previous one-factor model does not fit as well as this final two-factor model (figure 13.2).

As illustrated by this discussion, the possibility of these novel kinds of analyses are the main reasons why the CogUSA data were collected, and the results are useful. For example, the CogUSA results suggest that the concept of a single *g* factor is not a very good idea when considering data collected over the later part of the life-span (see McArdle & Woodcock, 1997; Horn & McArdle, 2007; McArdle, 2011). Indeed, theoretical statements such as *Gf-Gc* provide much better empirical fit to real data. The initial use of multiple indicators of each of the two common factors has enhanced effects for the education and age relationships. The *Gc* factor is positively related to educational attainment, but does not decline much with age, while the Gf factor is also positively related to educational attainment but declines dramatically with age. The empirical findings, to the degree cross-sectional data beyond age 50 can be used, are strongly consistent with the predictions of figure 13.1.

13.8 The New CogEcon Survey

The primary aims of the CogEcon survey were to evaluate the predictive validity of cognitive abilities for economic outcomes, and to increase understanding of portfolio allocation and financial sophistication, both in general and in relation to cognitive abilities. To this end, in addition to extensive measures of financial literacy and financial beliefs (which are referred to by the shorthand term "financial knowledge"), the survey gathered data on risk aversion, self-assessed financial knowledge and math abilities, economics and financial education background, assets, debt, portfolio allocation, household spending, credit-worthiness, and preferences.

The survey was fielded as a mixed-mode, self-administered mail and Internet survey of persons who had participated in the CogUSA baseline telephone and in-person surveys. These modes were chosen in order to reduce costs below those entailed for the interviewer-administered CogUSA survey. The survey was fielded in two waves (3/2008 and 8/2008) to avoid conflict with CogUSA study follow-up telephone interviews. The first wave was fielded as an Internet survey to those respondents who indicated in the baseline CogUSA interview that they used the Internet, and was fielded as a mail survey to the remaining respondents. Because analysis of CogUSA data showed that Internet access is strongly correlated with age and cognitive ability in this sample of persons over age 50, it was important to offer a mail survey in order to represent the older and less able portions of this population. (See Hsu, Fisher, & Willis, 2011, for details.) The second wave was fielded solely as a mail survey. Because respondents were randomly assigned to the first or second wave, this created a mode experiment. In the end, roughly half of all responses were submitted via the Internet, and half via mail. In total, 985 of the 1222 eligible Cognition Study participants completed CogEcon surveys, for about an 80% response rate. In addition, there is little evidence of selectivity by age, education, or ability among CogUSA respondents who accepted or did not accept the invitation to participate in CogEcon. These results bolster confidence that these data are nationally representative and applicable to the older adult population of the United States.

The success of the CogEcon-2008 data collection effort, coupled with its timing shortly before the financial crisis and stock market crash in the fall of 2008, ideally positioned this survey to study the effects of this economic crisis via a "post-crash" wave of the CogEcon survey. This CogEcon-2009 wave was fielded in May-June 2009.[6] Early analysis of these data by Shapiro (2010a) describes some of the short-run effects of the economic crisis on the older adult population. In continuing collaboration with the CogUSA project, CogEcon has plans to collect two more longitudinal waves in 2011 and 2013, and CogUSA plans to field new waves in 2012 and 2014.

13.9   How Fluid and Crystallized Intelligence Are Related to Wealth and Portfolio Choice

The CogEcon survey has several very distinctive features. In addition to being linked to the cognitive assessments contained in the CogUSA data, it includes high-quality data on wealth and portfolio allocation and a unique 25-item "financial sophistication" battery that allows for more comprehensive and detailed assessment and scoring of financial literacy, attitudes, and knowledge than other existing data sets. Delavande, Rohwedder, and Willis (2008) present a theoretical model in which they argue that the acquisition of financial knowledge is a form of human capital that permits an individual to obtain a higher risk-adjusted rate of return on their savings through a variety of mechanisms such as choosing a more efficiently diversified portfolio, shopping for financial products with lower fees, and obtaining and understanding higher-quality advice about financial matters.

In this model, cognitive ability affects wealth and portfolio choice through a number of channels. A person with high fluid intelligence may be able to acquire financial knowledge more rapidly and easily than a less able person. Put differently, Gf enters the human capital production function for financial knowledge, which is a form of Gc. In addition—and probably more importantly—the value of financial knowledge is (roughly) proportional to the number of dollars to which it is applied. Thus, as discussed earlier, the incentives to acquire financial knowledge strongly increase with the volume of saving, which in turn is an increasing function of lifetime earnings. Savings and portfolio choice are also affected by "non-cognitive" factors such as patience, risk tolerance, conscientiousness, leisure preference, and so on. Given the volume of savings, portfolio choice depends on risk preference and beliefs about stock returns.

Given the complexity of the various pathways by which cognitive ability operates over the life cycle and the early stage of research on this topic, it is not possible to make strong statements about causal relationships between cognitive aging and economic behavior and outcomes during the latter part of the life cycle. However, it is useful to consider a couple of basic descriptive relationships from the CogUSA/CogEcon data, which illustrate that further research into these relationships is likely to be quite interesting.

First, as noted earlier, findings from the CogUSA/CogEcon surveys have already been used to help select which new cognitive measures to add to the HRS for its 2010 wave. In particular, among all the components of intelligence measured in CogUSA the Number Series (NS) test, a measure of fluid intelligence, was found to be the most correlated with measures of financial knowledge, wealth, and portfolio choice. In CogUSA, this test contains 47 items of varying difficulty that call on individuals to recognize patterns in a series of numbers in order to fill in the missing number in a series such as 2, _, 6, 8.

Performance on this test is strongly related to wealth and portfolio choice, as illustrated in figure 13.2. The left-hand panel presents a lowess (locally weighted regression) plot of wealth versus the number series W-score for individuals who achieved between the 5th and 95th percentiles on this test. Over this range, household total wealth increases about threefold, from approximately $400,000 to $1,200,000. The panel plots the share of stocks in total wealth versus the NS score over the same range.[7] As can be seen, the share of stocks in wealth is also a sharply increasing function of NS. Clearly, one reason for the relationships depicted in figure 13.3 is that persons with high levels of *Gf* tend to obtain more schooling, enter higher-paying occupations, save more, and end up with greater wealth. Another reason is that people who anticipate accumulating more have a stronger incentive to acquire financial knowledge. Moreover, as discussed earlier, findings from CogEcon/CogUSA (Shapiro, 2010a) and from other bodies of data in other countries (Dohmen et al., 2010; Benjamin et al., 2006) suggest that risk tolerance and patience are both increasing functions of *Gf*.

Age profiles of the NS test from CogUSA and the financial knowledge test in CogEcon conform to the predictions of *Gf-Gc* theory shown in figure 13.1. The empirical profiles are presented in the left and right panels of figure 13.4 for persons between the ages of 50 and 85 with differing levels of educational attainment. The left panel shows large difference in the NS W-score by education group at any given age. However, there is a

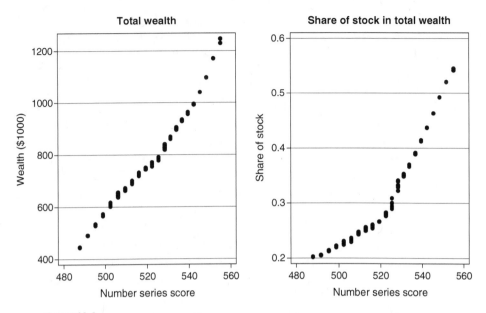

Figure 13.3
Total wealth and stock wealth by scores on the Number Series test. Source: *CogEcon 2008.*

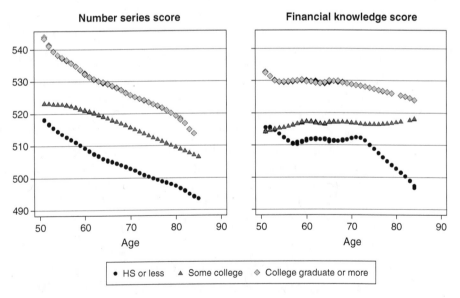

Figure 13.4
Age profiles of two cognitive measures. Sample: age 50–85, lowest plots. Source: *CogEcon* and *CogUSA* studies.

substantial decrease in scores by age within each education group that follows roughly linear and parallel trajectories. In contrast, the age profile of scores on the financial knowledge test, a component of crystallized intelligence, is much flatter at all educational levels, although the least-educated group shows a rather sharp decline beginning in the early 70s. If this decline is real (i.e., not an artifact of random noise), it would be consistent with findings in the medical literature suggesting that problems with money are an early signal of dementia, and that people with low education are most vulnerable to early loss of function (Marson, 2001; Pérès, Helmer, Amieva, Orgogozo et al., 2008; McArdle & Prescott, 2010).

13.10 Conclusion

This chapter has explored how two theories of cognitive development, one from psychology and one from economics, might be integrated into a unified theory that can help us understand how well or poorly individuals and households can deal with the complexities of preparing for retirement and managing their resources during retirement. The strength of neoclassical microeconomic theory is in describing the implications of optimal decision making subject to constraints. In recent years, behavioral economists have emphasized the importance of recognizing that cognitive capacity is a limited resource and that people may be unable to make optimal decisions. From its inception more than a half-century ago, human capital theory has emphasized that cognitive capacities and skills are scarce and, moreover, that they are distributed unequally in the society. Population heterogeneity in the supply of skills interacting with the influence of technological change in the demand for skills appear to be largely responsible for the dramatic growth in inequality of labor market earnings in the past quarter-century.

A key insight of human capital theory is that cognitive capacity is malleable. It is subject to purposive development through investment in human capital, subject to technological constraints embodied in a human capital production function. However, economics itself has little to say about these technological constraints, which involve the psychology and neurobiology of the brain as well as interactions of individuals with parents, family, schools, workplaces and the broader society and culture. The theoretical and empirical work of cognitive psychologists for more than a century has coalesced into a widely accepted theory of multiple aspects of intelligence known as the theory of fluid and crystallized intelligence (*Gf-Gc* theory).

This chapter shows how the developmental patterns implied by *Gf-Gc* theory parallel the processes emphasized in human capital theory. In particular, both disciplines stress the role of investment, a process by which a person applies his effort, ability to think and reason (*Gf*), and his accumulated knowledge (*Gc*) to learning new skills and knowledge that are durable and, therefore, accumulate. In addition, recent research on brain plasticity, which hypothesizes that the size and structure of the brain itself is influenced by these investment activities, is discussed briefly.

The final part of this chapter argues that the fruitfulness of an integrated psychological-economic theory of cognitive aging and human capital is greatly enhanced by the collection of new data that combines detailed measures of fluid and crystallized abilities together with economic measures of income, wealth, and portfolio composition. An ongoing project by the authors to collect such data is described. Evidence is presented suggesting that a particular measure of fluid intelligence known as the Number Series (NS) test—a test of the ability to recognize numerical patterns—is strongly related to economic outcomes such as total household wealth and the share of stocks in total wealth.

There are a number of potential pathways that might provide a link between measures of fluid intelligence and these economic outcomes. Higher values of NS are associated with higher educational attainment and being in occupations with higher lifetime income (Helppie, Kapinos, & Willis, 2010), which, in turn, would lead to higher rates of saving and higher wealth. NS is also positively related to components of crystallized intelligence that are important for financial decisions. For example, NS is strongly related to scores on the financial knowledge test, which measures the practical knowledge needed to manage financial matters (Delavande, Rohwedder, & Willis, 2008). In unpublished findings, NS is also related to a better knowledge of historical stock returns and to probabilistic beliefs that stock returns are higher than bond returns. NS is also positively related to risk tolerance (Shapiro, 2010b), which may help explain the positive relation between the NS and the share of stocks in total wealth. It remains for future research to attempt to extract causal pathways from these associations.

This evidence from the CogUSA/CogEcon surveys led to a decision to add a short adaptive version of NS to the Health and Retirement Study (HRS) beginning in 2010. The addition of improved cognitive measures to the extremely detailed data the HRS collects on economics, health, biomarkers and genetics, and social psychological measures is likely to stimulate further advances in the interdisciplinary approach to research on the economics and psychology of cognitive aging described in this chapter.

Acknowledgments

The authors gratefully acknowledge support by grants from the National Institute of Aging to the University of Southern California (AG07137) and to the University of Michigan (AG026571).

Notes

1. It should be noted that the effect on brain volume is quantitatively small. An anonymous referee suggests that there are "other dramatic effects of experience on the brain, particularly changes in synapse formation and connections, that may be much more important in explaining the learning and plasticity evidenced by changes in Gf-Gc."

2. An important assumption is that $\beta_1 + \beta_2 < 1$. This implies that there are diminishing returns to the amount of investment that can be produced by adding inputs in a given period, which, along with a constraint on the total available time, imply that it is not possible to achieve any desired stock of human capital within one period. Thus, the accumulation of human capital will be spread out over multiple periods of a person's life.

3. Becker (1964) shows that a worker will pay the cost of human capital that is acquired on the job if the training is "general," in the sense that the trained worker is at least as productive at other firms as she is at the firm from which she acquires the training. If the training is "firm-specific," the worker and firm will tend to share the cost of training.

4. See also Paglin and Rufolo (1990) and Murnane, Willett, and Levy (1995).

5. See Ofstedal, Fisher, and Herzog (2005) for descriptions of the cognitive measures that are available on the HRS.

6. See Fisher and Helppie (2010) and Helppie and Hsu (2010) for documentation reports on these two waves.

7. The share of total wealth is equal to total stock wealth divided by total wealth. Since the denominator in this ratio is measured with considerable error, plotting individual-level shares would produce badly biased estimates of the relationship between the stock share and NS. The plotted line in the left panel of figure 13.3 is the ratio of the lowess (locally weighted regression) estimate of stock wealth, conditional on NS, divided by the lowess estimate of total wealth, conditional on NS. Assuming that measurement error is mean zero for both totals, this procedure provides a consistent estimate of the stock share, conditional on NS.

References

Autor, D. H., Katz, L. F., & Kearney, M. S. (2008). Trends in U.S. wage inequality: Revising the revisionists. *Review of Economics and Statistics, 90*(2), 300–323.

Baltes, P. B., & Kliegl, R. (1992). Further testing of limits of cognitive plasticity: Negative age differences in a mnemonic skill are robust. *Developmental Psychology, 28*, 121–125.

Baltes, P. B., & Mayer, K. U. (Eds.). (1999). *The Berlin Aging Study: Aging from 70 to 100*. New York: Cambridge Univ. Press.

Banks, J., O'Dea, C., & Oldfield, Z. (2010). Cognitive function, numeracy and retirement savings trajectories. *Economic Journal, 120*(548), F381–F410.

Becker, G. S. (1962) Investment in human capital: A theoretical analysis. *Journal of Political Economy, 70*(5), Part 2: *Investment in Human Beings*, 9–49.

Becker, G. S. (1964). *Human capital*. New York: Columbia University Press for the National Bureau of Economic Research.

Becker, G. S. (1965). A theory of the allocation of time. *Economic Journal, LXXV*(299), 493–517.

Ben-Porath, Y. (1967). The production of human capital and the life cycle of earnings. *Journal of Political Economy, 75*(4), 352–365.

Benjamin, D.J., Brown, S.A, & Shapiro, J. M. (2006, May). Who is "Behavioral"? Cognitive ability and anomalous preferences. Harvard University mimeo, Cambridge, MA.

Binet, A., & Simon, T. (1905). Methodes novelles pour le diagnostic du niveau intellectuel des anormaux. *L'Année Psychologique, 11*, 191.

Borghans, L., Duckworth, A. L., Heckman, J. J., & ter Weel, B. (2008). The economics and psychology of personality traits. *Journal of Human Resources, 43*(4), 972–1059.

Bound, J., & Johnson, G. E. (1992). Changes in the structure of wages during the 1980's: An evaluation of alternative explanations. *American Economic Review, 82*, 371–392.

Camerer, C., Loewenstein, G., & Rabin, M. (Eds.). (2003). *Advances in behavioral economics*. Princeton, NJ: Princeton University Press.

Carneiro, P., Heckman, J. J., & Vytlacil, E. (forthcoming). Estimating marginal returns to education. *American Economic Review*.

Carroll, J. B. (1993). *Human cognitive abilities: A survey of factor-analytic studies*. Cambridge: Cambridge University Press.

Carroll, J. B. (1998). Human cognitive abilities: A critique. In J. J. McArdle & R. W. Woodcock (Eds.), *Human cognitive abilities in theory and practice* (pp. 5–24). Mahwah, NJ: Lawrence Erlbaum Associates, Publishers.

Cattell, R. B. (1941). Some theoretical issues in adult intelligence testing. [Abstract]. *Psychological Bulletin, 38,* 592.

Cattell, R. B. (1987). *Intelligence: Its structure, growth and action.* Amsterdam, The Netherlands: Elsevier.

Cattell, R. B. (1998). Where is intelligence? Some answers form the triadic theory. In J. J. McArdle & R. W. Woodcock (Eds.), *Human cognitive abilities in theory and practice* (pp. 29–38). Mahwah, NJ: Lawrence Erlbaum.

Cunha, F., & Heckman, J. J. (2008). Formulating, identifying and estimating the technology of cognitive and noncognitive skill formation. *Journal of Human Resources, 43*(4), 738–782.

Cunha, F., Heckman, J. J., & Schennach, S. M. (2010). Estimating the technology of cognitive and noncognitive skill. *Econometrica, 78*(3), 883–893.

Delavande, A., Rohwedder, S. & Willis, R. J. (2008). Preparation for retirement, financial literacy and cognitive resources (WP 2008–190). Michigan Retirement Research Center.

Dohmen, T., Falk, A., Huffman, D., & Sunde, U. (2010). Are risk aversion and impatience related to cognitive ability? *American Economic Review, 100*(3), 1238–1260.

Donaldson, G., & Horn, J. L. (1992). Age, cohort, and time developmental muddles: Easy in practice, hard in theory. *Experimental Aging Research, 18,* 213–222.

Ferrer, E., & McArdle, J. J. (2004). An experimental analysis of dynamic hypotheses about cognitive abilities and achievement from childhood to early adulthood. *Developmental Psychology, 40*(6), 935–952.

Fisher, G., & Helppie, B. (2010) Cognitive economics study, 2008 survey data: Background, data description and usage. Cognitive Economic Study Documentation Report, University of Michigan, September 30.

Fleiss, J. L., & Shrout, P. E. (1977). The effect of measurement error on some multivariate procedures. *American Journal of Public Health, 67,* 1184–1189.

Frederick, S. (2005). Cognitive reflection and decision making. *Journal of Economic Perspectives, 19*(4), 25–42.

Friedman, M. (1953). *Essays in positive economics.* Chicago: University of Chicago Press.

Grossman, M. (1972). On the concept of health capital and the demand for health. *Journal of Political Economy, 80*(2), 223–255.

Heckman, J. J. (1974). Shadow prices, market wages and labor supply. *Econometrica*, *42*(4), 679–694.

Heckman, J. J. (2007). The economics, technology and neuroscience of human capability formation. *Proceedings of the National Academy of Sciences of the United States of America*, *104*(33), 13250–13255.

Helppie, B., & Hsu, J. W. (2010) Cognitive economics study, 2009 post-crash survey data: background, data description and usage. Cognitive Economic Study Documentation Report, University of Michigan, September 30.

Helppie, B., Kapinos, K. & Willis, R. J. (2010) Occupational learning, financial knowledge, and the accumulation of retirement wealth. Michigan Retirement Research Center, WP 2010-237, October 2010.

Horn, J. L. (1988). Thinking about human abilities. In J. R. Nesselroade & R. B. Cattell (Eds.), *Handbook of multivariate psychology* (p. 5). New York: Academic Press.

Horn, J. L. (2003). Old and new theories of measurement. In A. Maydeu-Olivares & J. J. McArdle (Eds.), *Contemporary psychometrics* (pp. 201–224). Mahwah, NJ: Erlbaum.

Horn, J. L., & Cattell, R. B. (1967). Age differences in fluid and crystallized intelligence. *Acta Psychologica*, *26*, 107–129.

Horn, J. L., & McArdle, J. J. (2007). Understanding human intelligence since Spearman. In R. Cudeck & R. MacCallum (Eds.), *Factor analysis at 100 years* (pp. 205–247). Mahwah, NJ: Lawrence Erlbaum.

Horn, J. L., & Noll, J. (1997). Human cognitive capabilities: Gf-Gc theory. In D. P. Flanagan, J. L. Genshaft, & P. L. Harrison (Eds.), *Contemporary intellectual assessment: theories, tests, and issues* (pp. 53–91). New York: Guilford Press.

Hsu, J. W. (2011). Aging and strategic learning: The impact of spousal incentives on financial literacy (Unpublished dissertation chapter). Department of Economics, University of Michigan, Ann Arbor, MI.

Hsu, J. W., Fisher, G., & Willis, R. J. (2011) Internet access and cognitive ability: An analysis of the selectivity of internet interviews in the cognitive economics survey. In J. W. Hsu, *Essays on aging and human capital*, unpublished Ph.D. Dissertation (Economics), University of Michigan, chapter 4.

Hudomiet, P., Kézdi, G., & Willis, R. J. (2011). Stock market crash and expectations of American households. *Journal of Applied Econometrics*, *26*, 393–415.

Juhn, C., Murphy, K. M., & Pierce, B. (1993). Wage inequality and the rise in the returns to skill. *Journal of Political Economy*, *101*(3), 410–442.

Kaiser, H. F. (1960). The application of electronic computer to factor analysis. *Journ. Educ. Psych. Measurement*, *20*, 141–151.

Katz, L. F., & Murphy, K. M. (1992). Changes in relative wages, 1963–1987: Supply and demand factors. *Quarterly Journal of Economics, 107*(1), 35–78.

Kézdi, G. & Willis, R. J. (2008) Stock market expectations and portfolio choice of American households. (CogEcon WP-2008-5) University of Michigan, October 31.

Langa, K., Plassman, B., Wallace, R., Herzog, A. R., Heeringa, S., Ofstedal, M. B., et al. (2005). The Aging, Demographics and Memory study: Study design and methods. *Neuroepidemiology, 25*, 181–191.

Li, K. Z. H., & Lindenberger, U. (2002). Relations between aging sensory/sensorimotor and cognitive functions. *Neuroscience & Biobehavioral Reviews, 26*(7), 777–783.

Lillard, L. A., & Willis, R. J. (2001). Cognition and wealth: The importance of probabilistic thinking (WP 2001-007). Michigan Retirement Research Center.

Lindenberger, U., & Baltes, P. B. (1997). Intellectual functioning in the old and very old: Cross-sectional results from the Berlin Aging Study. *Psychology and Aging, 12*, 410–432.

Lövdén, M., Bäckman, L., Lindenberger, U., Schaefer, S. & Schmiedek, F. (2010). A theoretical framework for the study of adult cognitive plasticity. *Psychological Bulletin, 136*(4), 659–676.

Lustig, C., Shah, P., Seidler, R., & Reuter-Lorenz, P. A. (2009). Aging, training, and the brain: A review and future directions. *Neuropsychology Review, 19*, 504–522.

Maguire, E. A., Gadian, D. G., Johnsrude, I. S., Good, C. D., Ashburner, J., Frackowiak, R. S., et al. (2000). Navigation-related structural change in the hippocampi of taxi drivers. *Proceedings of the National Academy of Sciences of the United States of America, 97*(8), 4398–4403.

Marson, D. C. (2001). Loss of financial competency in dementia: Conceptual and empirical approaches. *Aging, Neuropsychology, and Cognition, 8*, 164.

McArdle, J. J. (2001). A latent difference score approach to longitudinal dynamic structural analysis. In R. Cudeck, S. du Toit, & D. Sorbom (Eds.). *Structural equation modeling: Present and future* (pp. 342–380). Lincolnwood, IL: Scientific Software International.

McArdle, J. J. (2010a). Dealing with attrition in the CogUSA experiment. Unpublished research study, Department of Psychology, University of Southern California, Los Angeles.

McArdle, J. J. (2010b). Some ethical issues in factor analysis. In A. Panter & S. Sterber (Eds.), *Quantitative methodology viewed through an ethical lens* (pp. 313–339). Washington, DC: American Psychological Association Press.

McArdle, J. J. (2010c). Contemporary challenges of longitudinal measurement using HRS data. In G. Walford, E. Tucker, & M. Viswanathan (Eds.), *The SAGE handbook of measurement* (pp. 509–536). London: SAGE Press.

McArdle, J. J. (2010d). What life-span data do we really need? In R. Lerner (Series Ed.), *The handbook of life-span development: Vol. 1. Biology, cognition and methods across the life-span*, W. F. Overton (Ed.) (pp. 36–55). Hoboken, NJ: Wiley.

McArdle, J. J. (2011). Longitudinal dynamic analysis of cognition in the health and retirement study. *AStA Advanced Statistical Analyses, 95*, 453–480.

McArdle, J. J., Ferrer-Caja, E., Hamagami, F., & Woodcock, R. W. (2002). Comparative longitudinal multilevel structural analyses of the growth and decline of multiple intellectual abilities over the life-span. *Developmental Psychology, 38*(1), 115–142.

McArdle, J. J., Fisher, G. G., & Kadlec, K. M. (2007). Latent variable analysis of age trends in tests of cognitive ability in the Health and Retirement Survey, 1992–2004. *Psychology and Aging, 22*(3), 525–545.

McArdle, J. J., Fisher, G., Kadlec, K., Prindle, J. P., & Woodcock, R. J. (2010). Unpublished cognitive tests based on WJ items. Unpublished manuscript, Department of Psychology, University of Southern California, Los Angeles, CA.

McArdle, J.J., Hamagami, F., Meredith, W., & Bradway, K.P. (2001). Modeling the dynamic hypotheses of Gf-Gc theory using longitudinal life-span data. *Learning and Individual Differences, 12*, 53–79.

McArdle, J. J., & Prescott, C. A. (1992). Age-based construct validation using structural equation modeling. *Experimental Aging Research, 18*(3), 87–115.

McArdle, J. J., & Prescott, C. A. (1996). Contemporary models for the biometric genetic analysis of intellectual abilities. In D. P. Flanagan, J. L. Genshaft, & P. L. Harrison (Eds.), *Contemporary intellectual assessment: Theories, tests and issues* (pp. 403–436). New York: Guilford Press.

McArdle, J. J., & Prescott, C. A. (2010). Contemporary modeling of gene-by-environment effects in randomized multivariate longitudinal studies. *Perspectives on Psychological Science, 5*, 606–621.

McArdle, J. J., Prescott, C. A., Hamagami, F., & Horn, J. L. (1998). A contemporary method for developmental-genetic analyses of age changes in intellectual abilities. *Developmental Neuropsychology, 14*(1), 69–114.

McArdle, J., Smith, J. P., & Willis, R. J. (2011). Cognition and economic outcomes. In D. Wise (Ed.), *Explorations in the economics of aging*. Chicago: University of Chicago Press.

McArdle, J. J., & Woodcock, J. R. (1997). Expanding test-rest designs to include developmental time-lag components. *Psychological Methods, 2*(4), 403–435.

McArdle, J. J., & Woodcock, J. R. (1998). *Human abilities in theory and practice.* Mahwah, NJ: Erlbaum.

McGrew, K., & Flanagan, D. (1998). *The intelligence test desk reference (ITDR): Gf-Gc cross-battery assessment.* New York: Allyn & Bacon.

Murnane, R. J., Willett, J. B., & Levy, F. (1995). The growing importance of cognitive skills in wage determination. *Review of Economics and Statistics, 77*(2), 251–266.

Murphy, K. M., & Welch, F. (1992). The structure of wages. *Quarterly Journal of Economics, 107*(1), 285–326.

Ofstedal, M. B., Fisher, G. G., & Herzog, A. R. (2005). Documentation of cognitive functioning measures in the health and retirement study (DR-006). HRS Documentation Report.

Paglin, M., & Rufolo, A. M. (1990). Heterogeneous human capital, occupational choice, and male-female earnings differences. *Journal of Labor Economics, 8,* 123–144.

Pérès, K., Helmer, C., Amieva, H., Orgogozo, J.-M., Rouch, I., Dartigues, J.-F., et al. (2008). Natural history of decline in instrumental activities of daily living performance over the 10 years preceding the clinical diagnosis of dementia: A prospective population-based study. *Journal of the American Geriatrics Society, 56,* 37–44.

Rae, J. 1905. *The sociological theory of capital.* New York: Macmillan. (Original work published 1834 as *The new principles of political economy*)

Rohwedder, S., & Willis, R. J. (2010). Mental retirement. *Journal of Economic Perspectives, 24*(1), 119–138.

Salthouse, T. A., Atkinson, T. M., & Berish, D. E. (2003). Executive functioning as a potential mediator of age-related ·cognitive decline in normal adults. *Journal of Experimental: General, 132,* 4.

Schaie, K. W. (1996). Intellectual development in adulthood. In J. E. Birren & K. W. Schaie (Eds.), *Handbook of the psychology of aging* (4th ed., pp. 266–286). San Diego: Academic Press.

Shapiro, M. (2010a). The effects of the financial crisis on the well-being of older Americans: Evidence from the Cognitive Economics Study (WP 2010-228). Michigan Retirement Research Center.

Shapiro, M. D. (2010b). *Risk tolerance, cognition, and the demand for risky assets.* In progress. University of Michigan.

Sjaastad, L.A. (1962) The costs and returns of human migration. *Journal of Political Economy,* 70 (5), Part 2: *Investment in Human Beings,* 80–93.

Smith, D. D. (1958). Abilities and interests: I. Factorial study. *Canadian Journal of Psychology/Revue canadienne de psychologie, 12*(3),191–201.

Smith, J. P., McArdle, J. J., & Willis, R. J. (2010). Financial decision making and cognition in a family context. *Economic Journal, 120,* F363–F380.

Spearman, C. E. (1904). "General intelligence" objectively determined and measured. *American Journal of Psychology, 15,* 201–293.

Stigler, G. J. (1962) Information in the labor market. *Journal of Political Economy, 70* (5), Part 2: *Investment in Human Beings,* 94–105.

Thurstone, L. L. (1935). *Vectors of mind: Multiple-factor analysis for the isolation of primary traits.* Chicago: University of Chicago Press.

Thurstone, L. L. (1938). *Primary mental abilities.* Chicago: University of Chicago Press.

Wechsler, D. (1997). *WAIS-III administration and scoring manual.* San Antonio, TX: The Psychological Corporation.

Willis, R. J. (1973). A new approach to the economic theory of fertility behavior. *Journal of Political Economy, 81,* S14–S64.

Willis, R. J. (1986). Wage determinants: A survey and reinterpretation of human capital earnings functions. In O. Ashenfelter & R. Layard (Eds.), *Handbook of labor economics* (pp. 525–602). Amsterdam: North-Holland.

Willis, R. J., & Rosen, S. (1979). Education and self-selection. *Journal of Political Economy, 87*(Suppl.), S7–S36.

Woodcock, R. W. (1990). Theoretical foundations of the WJ-R measures of cognitive ability. *Journal of Psychoeducational Assessment, 8,* 231–258.

Woodcock, R. W., McGrew, K. S., & Mather, N. (2001). *The Woodcock-Johnson-III tests of cognitive abilities.* Itasca, IL: Riverside Publishing Company.

Zimmerman, W. S. (1953). A revised orthogonal rotational solution for Thurstone's original primary mental abilities test battery. *Psychometrika, 18,* 77–93.

VI   UNIFYING PERSPECTIVES

14 Going Cognitive: Tools for Rebuilding the Social Sciences

Mathew D. McCubbins and Mark Turner

Most human beings have an absolute and infinite capacity for taking things for granted.

—Aldous Huxley

14.1   Introduction: The Cognitive Basis of the Social Sciences

Much of social science research is about how human beings interact. Much of cognitive science is about higher-order human cognition, such as communication, social cognition and interaction, and decision-making. Given how these subjects complement one another, it might seem that they would unify quickly.

But that has not happened. One cause may be that early cognitive science emphasized the thinking of the lone individual confronting a solitary task—in mathematics, logic, problem-solving, or syntax. Even in cases where early cognitive science did investigate the interaction of agents—as in games of chess, backgammon, or go—there was little consideration of agents or their thought, per se. For the most part, there was no consideration of joint attention, of how we understand other minds, of metacognition, or of social engagement.

But since the 1980s, attention to the social aspects of cognition has risen dramatically, and it is now a central feature of cognitive science. Social cognitive neuroscience is a vibrant field. Research into distributed cognition, joint attention, the social nature of learning, the cultural origins of conceptual systems, social interaction in child development, co-speech gesture, social cognitive aspects of computing, and other topics involving social interaction has brought social aspects of cognition firmly into the core of cognitive science. Cognitive scientists routinely consider the social aspects of cognition.

But the reverse is not yet true: social scientists do not routinely consider the cognitive aspects of how people interact. Graduate programs in the social sciences—economics, political science, law, management, anthropology, sociology—do not begin with foundation courses in the cognitive aspects of their subject matters. That is the reason for this chapter. In the space allowed, this chapter sketches some positive contributions that cognitive science might make to the social sciences.

The purpose of this chapter is not to replace something with nothing, to tear down the efforts of others, but to suggest how something that has some merits and some flaws can be replaced by something that has more merits and fewer flaws.

14.2   The End of Taking Things for Granted

as·sump·tion
–noun
1. something taken for granted; a supposition: a correct assumption.
2. the act of taking for granted or supposing.
—*Random House Dictionary* (2011)

The chief lesson of cognitive science over the last thirty years is that anything that seems to go without saying in human thought and behavior in fact never goes without saying. Ironically, what seems most obvious to us human beings is usually what most needs explanation, even though we are disposed not to see that. The concepts we take for granted about human thought, such as vision, object permanence, communication, and indeed the identity of ourselves and others, it turns out, while fine for conducting our lives, are not good foundations on which to build theories and empirical tests of human cognition. These are concepts that need to be explained, and no matter how easy or natural they feel to us, the scientist should be highly suspicious of them, especially if they are to form the foundations of our theories.

One concept that is almost always taken for granted is consciousness, by which we mean our experience of having things present to our minds. (The authors sidestep here the impressive debates over the nature and varieties of consciousness—though see van Gulick, 2011—and use only a simple distinction between what we take to be present in the mental "Cartesian theater" and the far greater range of cognition that is inaccessible there.) Consciousness has some specific powers, to be sure, especially for focusing our attention long enough to enable learning in backstage cognition, by which we mean thinking processes that take place behind consciousness (Fauconnier, 1999). But consciousness itself is extremely weak. What we see in the Cartesian theater of our consciousness—what we can see by introspection in our own thought—is not so much thought, but shadows on the cave wall. They are cartoons to keep us going. Advanced human cognition has the exceptional power of "taking things for granted," a power so comprehensive and strong that it looks more like an evolutionary design feature than a bug. Why would we take things for granted? Because attention is scarce. We need to behave effectively, efficiently, and in a fit manner in niches that call for decision and action. What on Earth would be the evolutionary advantage of having to calculate how to see during each instance of sight, or reaffirm object permanence each time we see an object? Taking most things about our own thought for granted is efficient, but it is absolutely wrong for building a science. We do not need to understand much about our thought if what we want to do is act. But it is exactly what we need to understand if what we want to do is explain why and how we act as we do.

Human performances that seem simplest to human beings (deciding between coffee or tea, choosing the currency with which to pay for the beverage, and engaging in chit-chat with the barista) are by now recognized in cognitive science as notoriously difficult to explain. It seems to us, in consciousness, that the things that must be difficult to explain are the things that feel difficult to us—like proving a theorem in geometry. But, instead, what we take for granted about human thought has proved in cognitive science to be unimaginably more complex than anyone had expected; to be profoundly misrepresented by our supposedly bedrock, commonsense, intuitive notions; and to be conducted almost entirely in the backstage of cognition, invisible to consciousness.

Consider vision. It seems to us that we know how we see: the world shoots right through our senses into our brain; we think we see the cup as one thing because it is one thing; we think we see color because color is directly in front of us, part of the object, which we recognize because a certain wavelength makes us see a corresponding color. Yet cognitive scientists know that this cartoon understanding of vision is altogether false. This blanket assertion of its falsity is not controversial among scientists of vision. Color, for example, is not in the world. It is something that the brain attributes to the world for various purposes. The same light coming from an object can be seen as two different colors, and different light coming from an object under different illuminations can be seen as the same color. It might be useful for us to think of the visual field as an *interface* for interacting with reality, rather than as a *representation* of reality. For a review, see standard works like Stephen Palmer's *Vision Science* (1999) or Semir Zeki's *A Vision of the Brain* (1993). It is important to emphasize that we cannot see how we see. All the work is done in backstage cognition. In consciousness, we are aware of only little products of vision, the cartoons that keep us going.

Next, consider language. How do we communicate? The answer might seem obvious: we were taught language and we know the rules we were taught and we use those rules as we speak, write, and read. But again, cognitive scientists know that almost everything we are doing in language happens backstage, in ways that we cannot see, and according to patterns that we cannot state, indeed patterns that we do not even know we are following. This is not a controversial statement among scientists, no matter which branch of linguistics one prefers, and it is true no matter who one is or how much education one has had.

This is the general story of cognitive science, and could be told for memory, for categorization, for mathematics, for thinking about one's self and others, for causal understanding, for understanding change, for hypothetical or counterfactual conception, for decision making, and for many other aspects of cognition. Indeed, at the core of most social sciences is an autistic theory of mind, self, and others. As with vision, we need to understand that our theory of mind is just an *interface* for interacting with reality, rather than a *representation* of reality itself. This is the basic disposition from cognitive science that might be integrated into the training of social scientists.

14.3   Obliviousness Is Bliss

From a scientific perspective, it is not a surprise that the human mind is so poor at looking into the human mind, so poor at seeing through the small cartoons of consciousness. The cartoons of consciousness are highly useful, and there is no evolutionary advantage in building consciousness so that it can see through them. It is better in this case, metaphorically, to drive without looking under the hood. What consciousness needs to do is to exploit these cartoons, not understand them.

Our inability to see into thought and our reliance on very superficial cartoons in consciousness are not surprising, considering the nature of the human brain. A human brain has perhaps 10 billion to 100 billion (10^{11})

neurons. The average number of synaptic connections per neuron is perhaps 10,000 (10^4). The total number of connections in the brain is therefore maybe 10 trillion to a quadrillion (10^{14}–10^{15}), or about 10,000 times as many stars as astronomers think might be in the entire Milky Way (10^{11}).

Ten thousand galaxies, all inside the human head. All those connections, in a system weighing about 1.4 kilograms, working. The timing and phases of firing in neuronal groups, the suites of neuronal development in the brain, the electrochemical effect of neurotransmitters on receptors, the scope and mechanisms of neurobiological plasticity—all going on in ways we cannot even begin to see directly. The only time we are likely even to sense that our mental system is so complicated is when something goes wrong—such as when someone's language degrades because of stroke, or our field of vision starts to swim because of food poisoning. We have also seen that decision making changes because of hunger, sleepiness, sexual arousal, or addiction.

Human beings are awesomely effective, but largely clueless about how they work. The human mind is not built to look into the human mind. In fact, it seems to be built to *prevent* looking into itself. In Ron Sun's terms, great ranges of human cognition are "implicit" and inaccessible rather than "explicit" and accessible (Sun, 2002, p. 3).

The problem we face is that social scientists, like all human beings, have a robust capacity for taking things for granted. Social scientists have naturally taken commonsense notions as a starting point, created deductive models from these notions, refined and corrected the models, and added epicycles, filters, and specificity so that we can keep taking for granted what we feel we should take for granted. But the first and most important lesson for the social sciences from cognitive science is that *what we take for granted about human beings is not a starting place for science*. It is rather part of what needs to be explained.

14.4   Some Rubrics for Cognitive Social Science Theory

14.4.1   Model a Person as Multiple Agents

Cognition depends upon conceptual compression: we are able to carry in our minds compressed, simplistic notions that we then expand as needed for use. One of our most natural compressions is that a human ontogenetic organism has a self. This assumption makes perfect sense to us. From the viewpoint of cognitive science, it is amazing that an organism that spans

scores of years, with the most dramatic changes, development, and transformations, from birth to death, should have a unitary sense of self. Evidently, this conscious cartoon of the self keeps us going. Cultures invest
in propping up this useful conception: for example, they give us persistent
proper names, providing culturally enforced linguistic continuity across
otherwise remarkable variation.

Individual human beings are, beneath the guise of professed belief,
smarter than that: whatever we say, the way in which we behave suggests
that we have keen knowledge that each individual is already a network of
not-entirely-coordinated agents. Although we may carry around a conception of a unitary self with a stable list of ranked preferences, this very useful
conception of a unitary self is only a compressed anchor in a network of
various possible selves, each depending on conditions.

Freed of disciplinary shibboleths, everyone knows this. In the present,
we are often concerned that a future self will forget some task, obligation,
promise, or principle. We fret that a future self will forget the correct
ranking of preference. We are not even sure that a future self will remember
all the preferences it ought to have. And we take decisive action to constrain and correct that future self, apprehensive that the future self will
look back with condescension on the error of the past self that tried to do
the constraining. The world's literatures, artistic representations, and
dramas revert constantly to the theme of the challenging variation in self,
moment to moment.

Quite a number of social scientists have made the point that a self needs
to be modeled as a network of agents, rather than a unit with consistent
preferences. Pronin, Olivola, and Kennedy (2008), for example, report
experiments showing how decisions people make for future selves and
other people are similar to each other and different from their decisions
for present selves. Other researchers report on the time-inconsistency of
self (Ainslie, 1992; Ainslie & Haslam, 1992; Elster, 1979; Schelling, 1984;
Thaler & Shefrin, 1981; Trope & Liberman, 2000, 2003; Monterosso &
Ainslie, 2007). These social scientists have not had nearly the comprehensive, foundational influence throughout social science that the authors,
from a cognitive perspective, think would be beneficial for the social
sciences.

One agenda for social science theory, then, is to recast models of human
decision making, judgment, choice, reaction, and planning so as to incorporate the variability of the self and of the choices and preferences of
our many selves. Just as economics has modeled different individuals as

different agents, so each individual human organism—a "person"—can be modeled, using the same methods, as a network of agents engaged in interdependent action.

Experiments can be designed to test for not only variation in self, but also the present self's awareness of this variability over agents in a network of time and space and conditions, where each of the nodes has an agent that, however different from the agents at other nodes, refers to the same human organism. We can test for regret, remorse, redemption, revenge, guilt, shame, and other functions of past selves. We can test for apprehension, hope, and strategizing with respect to future selves. We can test for ways in which the present self conceives of past selves as having tried to condition which present self will be active, and even ways in which past selves tried to arrange conditions so that future selves would moderate whatever activities might be engaged in by the present self.

Which self is operating at which moment depends upon what is active in the brain at that moment, and as cognitive science has shown repeatedly and in detail, activation in the brain is a highly variable phenomenon. The picture from cognitive science is the one that Sir Charles Sherrington famously expressed over a century ago. Sherrington referred to the brain and the central nervous system as an "enchanted loom" where "millions of flashing shuttles weave a *dissolving* pattern, always a meaningful pattern, though *never an abiding one*" (Sherrington, 1906; emphasis added).

14.4.2 Model Agency Accurately

Agency is a much more complicated theoretical notion in cognitive science than it is in economics. Cognitive scientists study the complex ways in which we project intentionality to create blended conceptions involving agency. We form concepts of quasi-agents, and these concepts can be highly useful, even indispensable, in our understanding. By no means are we obliged to believe that our conception of the quasi-agent refers to something exactly like itself in reality. Instead, our conception of the quasi-agent is an efficient compression, something congenial to the kinds of thinking human beings are evolved to do. Such mental compressions provide suitable cognitive platforms where we can, so to speak, get a foothold as we try to grasp, manipulate, and assess the mental network of information and relationships that this compressed quasi-agent grounds.

Quasi-agents include institutions, corporations, organizations, groups, and even statistical artifacts. A phrase like "the average Chicago trader" prompts us to construct a quasi-agent that can play a direct role in our choosing regardless of whether we think there exist any average traders. (Even if we do, it is not a specific one that we are thinking about.) We can say, "a Pennsylvania voter has a 42% likelihood of ..." without imagining that there is any specific Pennsylvania voter with that specific likelihood.

The construction, operation, and transformation of quasi-agents in thought are active topics of research in cognitive science—research that, at least in principle, could be imported to the social sciences to help improve its foundations. As far as we can see, almost every one of the central puzzles in social science, all of which currently impede progress, crucially involves the concept of agency. It is accordingly plausible, at least, that the joint reconsideration of the concept of agency might hold a key for moving ahead in the consideration of these puzzles.

14.4.3 Check for Variation in the Mechanisms of Reason Depending on Conceptual Domain

One of the most active areas of research in cognitive science over the last 25 years has been the exploration of the ways in which human reasoning might vary depending on domain. In cognitive science, "the century-old conviction that humans reason according to some content-independent logic was shattered by a number of factors ..." (Gigerenzer & Hug, 1992, p. 127). Although there is substantial disagreement in the field between strong evolutionary modularists and their critics on certain specific points (see, e.g., Elman, Bates, Johnson, Karmiloff-Smith et al., 1996), it is now always an open question, to be resolved empirically, whether human beings reason in a particular way in a particular domain. For any domain, we seek to know how reason operates *in that domain*. (For reviews, see Gigerenzer & Hug, 1992; Caramazza & Shelton, 1998; and Fiddick, Cosmides, & Tooby, 2000.)

The domain-generality of reason is part of the foundation of economics. Different games, for example, are *all* assumed to be manifestations of general principles of rationality, but those different games have content; they are easily understood by subjects as having different narrative structures, involving different domains. While game theorists would assume that decision processes are general, we often see that subjects' behavior in experiments is different for different narratives. To the extent that reason depends upon the particular narrative or the particular domain, we should not in principle, without empirical testing, expect uniform principles of reason across these domain-specific narratives. This is an area where cognitive science can offer a direct and quick improvement.

14.4.4 No Daydream Believing

One of the most vibrant areas of research in cognitive science in recent decades concerns the nature of hypothetical and counterfactual thinking. These kinds of thought are referred to as "decoupled," because they are not directly connected to the situation the thinker inhabits.

The reality is rather more complicated than this: even tightly "coupled" cognition, which directly addresses the situation one inhabits, routinely involves networks of mental spaces, some of which are hypothetical or counterfactual. Here is a simple example: suppose you look in the refrigerator and see what you see. Suppose someone points out to you that there is "no horseradish in the refrigerator." This expression prompts you to call up—in backstage cognition, of course, not focally in your visual imagination—another mental space in which there is horseradish in the refrigerator. These two mental spaces are counterfactual with respect to each other. Then you reconstrue your understanding of the refrigerator, so that now the counterfactual relationship is compressed into an absence: in the refrigerator, there is *absence of horseradish*, a new element that you previously may not have recognized, because it was not active.

The mental space with the absence of horseradish refers to the situation you actually inhabit; indeed, it can be causal for your real action: you may go to the store to purchase horseradish so as to change the situation, so as to get rid of the absence. Taking another step in this direction, let us now focus on those counterfactual and hypothetical mental spaces that we do not take as referring to the situation we inhabit. For example, "If the vote were held today, would you ... ?" or "If you won the super-lotto, what would you do?" These mental scenarios do not subtend actual present actions and actual present consequences. Hypothetical and counterfactual scenarios can be very helpful, useful, and influential for judgment, decision, and choice, but such decoupled scenarios, which do not refer to actual situations, *do not stand in for real scenarios*. These decoupled scenarios do not have to be plausible or even possible. They are never full, and they do not need to follow principles according to which we understand real scenarios. They do not need to apply to real situations in order to guide our thinking about real situations. We can use them to guide our thinking about the conceptual networks in which they are situated without their being anything more than mildly, sparsely, and tenuously connected to actual action. For a classic discussion of how vignettes do not stand in for actual situations, see LaPiere (1934).

One recommendation from cognitive scientific research would be that we never confuse decoupled cognition and coupled cognition. There is nothing wrong with investigating decoupled cognition and the amazing and powerful mechanisms of human imagination. But decoupled scenarios cannot be used in experiments as stand-ins for actual situations without demonstration of very high covariance. From the cognitive scientific perspective, it would be important to measure, test, and model hypothetical and counterfactual decoupled cognition, but also to make sure that it is not regarded as substituting for how cognition works in actual situations. Experiments must test causal relationships in actual situations rather than imagined situations.

14.4.5 Include Heuristics in Candidate Explanations

Since at least Simon (1957; see also Gigerenzer & Selten, 2002; Kahneman, 2003; and Rubinstein, 1998), economics has framed the heuristics that human beings evidently follow in judging, choosing, and deciding as makeshift cognitive shortcuts, expediencies for dealing with the fact that our cognition is limited and our rationality bounded. From the point of view of cognitive science, this is an unexamined assumption, open to empirical investigation. Perhaps it is wrong. Heuristics, which often run in backstage cognition in ways that make them unlikely to be noticed and articulated by people, may operate better and lead to more success than the explicit domain-general computational forms of reason proposed by economics (see, most notably, Becker, 1976, and Becker & Murphy, 2001). It may be that more information is not always better, that more computation is not always better, and that more time for deciding is not always better. These are empirical issues, open to empirical research. As a second consideration, the actual heuristics may vary depending on the conceptual domain of the action.

Cognitive science can help lead the way in introducing these kinds of empirical research to economics. As Gigerenzer and Brighton (2009) write:

Heuristics are efficient cognitive processes that ignore information. In contrast to the widely held view that less processing reduces accuracy, the study of heuristics shows that less information, computation, and time can in fact improve accuracy. We review the major progress made so far: (a) the discovery of less-is-more effects; (b) the study of the ecological rationality of heuristics, which examines in which environments a given strategy succeeds or fails, and why; (c) an advancement from vague labels to computational models of heuristics; (d) the development of a systematic theory of heuristics that identifies their building blocks and the evolved capacities they exploit, and views the cognitive system as relying on an "adaptive toolbox;" and (e) the development of an empirical methodology that accounts for individual differences, conducts competitive tests, and has provided evidence for people's adaptive use of heuristics. Homo heuristicus has a biased mind and ignores part of the available information, yet a biased mind can handle uncertainty more efficiently and robustly than an unbiased mind relying on more resource-intensive and general-purpose processing strategies.

The suggestion from cognitive science would then be that we do not assume that mechanisms of rational choice are best or even good; perhaps in a particular case they are bad; perhaps heuristics are in fact better. Social scientific explanations need always to include in the array of possible explanations for human behavior the relevant mechanisms of heuristics from the cognitive science literature, and to test which mechanisms work best in which situations.

14.5   Improving the Design of Experiments and Quasi-Experiments

14.5.1   Test for Compliance to Treatment

In most sciences, understanding is not an issue: we do not need to be concerned with the concepts used by the system we are studying. Electromagnetic waves, digestive systems, immune systems, and the weather are systems that do not think, consciously or unconsciously, about what they are doing. But in the study of human action, thinking is all-important for doing. Data from human action can tell us about human thinking only to the extent that we know what concepts underlie the behavior. Someone whose culture offers the stereotype that ghosts of dead relatives are sent by our enemies to defeat us are likely to react to a performance of *Hamlet* in ways that Shakespeare would not have predicted, and certainly in ways that we would misanalyze if we did not know that they were using this stereotype to make sense of the performance. Hamlet himself, on seeing the ghost of his father, if he held this stereotype, would have reactions we would misanalyze.

For empirical work in social science to carry weight, the research must be able to show that the researcher knows the concepts the actors are using. To assume that subjects are using this or that concept, such as *trade, offer, barter, collaboration, acceptance, rejection, play, decision, player, game, election, contract, bid, ask,* or *strike,* is unacceptable from the point of view of cognitive science. Do the subjects have the same understanding of the consequences of their actions as the researchers do? Do the subjects have the same understanding of the causes of those consequences as the researchers do? In short: is the subject understanding the situation in the way the researcher presupposes the subject is understanding the situation?

For example, it is typically assumed that subjects are motivated only by a payoff matrix. For starters, this is already an assumption that should not be made. Perhaps the subjects are not motivated by that presumed payoff matrix at all, having come to the experiment only to neutralize peer pressure or social obligation, and now, in the experiment room, having already accomplished that goal, are waiting only for time to pass to be released. We must establish first that they are motivated by a payoff matrix and second that they are motivated in the ways the researchers imagine. Articles reporting experiments with payoffs often present extremely complex and unintuitive payoff matrices. Absent evidence that the subject actually grasps the payoff matrix in the way the researcher imagines, the data cannot be interpreted as a record of behavior under these payoff conditions. Our experiments show that, in very sparse decision environments, subjects make decisions within one narrative—defined by whom they are playing against, the actions available to them, the information each player has, and the payoffs for each action combination—yet the decision rules implied by these decisions are routinely violated by the very same subjects within other narratives. Indeed, our experiments show that the decision rules implied by decisions a subject makes at one place in the game are routinely violated by the very same subject at a slightly different spot in the very same run of the very same game.

There are good, available methods for ascertaining compliance with the researcher's understanding of the experimental situation. Examples of experimental protocols that include thorough measures of compliance to protocol can be found in Boudreau (2009); Lupia and McCubbins (1998); Druckman (2001a, 2001b); McCubbins, Paturi, and Weller (2008); and Enemark, McCubbins, Paturi, and Weller (2011). We can test for compliance to protocol.

While commonplace in medical research, it is rare for economists or political scientists to test for "compliance to treatment." This is odd, since the changing frames that economists rely upon to create "treatment" and "control" conditions often merely change payoffs or change the "information or communication" frame, and are extremely small and subtle relative to, for example, a drug effect. Indeed, to combat the problems associated with the weak-effect size of their treatment and to save money, economists frequently train their subjects for 10 to 30 "training trials" before asking their subjects to comply to the treatment for but a few paid rounds (often as few as two or three actual paid trials).

Some scholars feel that testing for compliance is unnecessary, except in cases where experimental results do not support the theory under consideration. This would be a methodological mistake: data interpreted as supporting a position under the assumption of compliance may in fact undermine the position if subjects were known to be non-compliant; accordingly, in the absence of testing for compliance, the researcher will establish a position that should have been rejected. The view that we need not test for compliance in such cases is also at odds with any protocol that requires training rounds, for there can be no need to train subjects to a frame if we are allowed to assume compliance. When queried about the size and direction of their effects, some economists might want to respond, "How else could the demonstrated treatment effect come about when the only difference between the two groups studied was that one received a treatment and the other did not?" This very statement, of course, ignores the possibility that the result is the product of a differential compliance to treatment.

Let us consider a randomized control drug trial so that the compliance problem is easier to see. Suppose there are two groups, of thirty subjects each, randomly assigned to control and treatment conditions. Let us suppose that the drug is 100% effective and that the placebo is 86.67% effective. Suppose that all sixty subjects comply to their instructions. We will find a significant treatment effect with total cured of 30 versus 26 ($p<.05$). As the effectiveness of the control drops, the size and significance of the treatment effect increases. Now suppose 50% of subjects in the treatment group do not comply (do not take the drug), while only 20% of the subjects in the control group do not comply. The assumed rate of effectiveness for all noncompliers is zero. The comparison of treated to control is no longer significant with only 15 cured in the treated case versus 21 cured among the control cases. So, because of absence of compliance, we now reverse causation, and conclude that an effective treatment is actually harmful. Or, assume the reverse, that the subjects learn of the drug's effectiveness and 40% of the control group find and take the drug while all 100% of the subjects in the treated group comply to treatment, with comparable cure rates of 30 versus 28. In this case, we will also incorrectly reject the effectiveness of the treatment.

Now suppose that the treatment is 70% effective and the control is 30% effective. Perfect compliance will show the treatment to be effective. Suppose you have 50% noncompliance in the treated group and 100% in the control group. Now the control would appear to be significantly more effective (at the $p>.05$ level) than is the treatment. Of course, different examples can be designed that show similar outcomes where the true treatment and compliance effects vary.

14.5.2   Minimize Framing (Unless That's What You Mean to Test)

"Framing" refers to the use of relatively small conceptual organizing devices in thought and action. The frames for *mother, Sunday, player, win, competition, inquiry, restaurant, president,* or *traffic light* come with default conceptual structure, including relations, assessments, and patterns of standard action. A given conceptual frame may call upon many encyclopedic conceptual domains. Minimal and partial prompts of language, gesture, and situation lead us to activate various frames and to connect and blend them.

Different people can activate and deploy quite different frames under the identical prompts. Accordingly, one of the issues the social scientist always faces when inquiring into the behavior of subjects is what conceptual structure has been activated by the prompts encountered. This is a problem of framing.

The healthy but limited respect by economists for the importance of framing derives from the work of Kahneman and Tversky on "prospect theory," which sought to show that differences in the framing of a choice can bias the choice one way or another, despite the fact that the framing is immaterial to the consequences of the action with respect to the payoff matrix (Kahneman & Tversky, 1979, 2000). If we frame an action as a trade, then, since every trade is both a loss and a gain, it is possible for us to frame the action so as to emphasize loss or gain. Prospect theory proposes that there is a bias depending on this framing (Tversky & Kahneman, 1992; Tversky & Fox, 2000): it is assumed within expected utility theory that choosers are in general risk-averse, but, on the contrary, ceteris paribus, there is a four-fold pattern of risk attitudes: risk-seeking for gains of low probability, and for losses of high probability; risk-aversion for gains of high probability, and for losses of low probability (Tversky & Fox, 2000, p. 94). Accordingly, choosers will tend to make different choices depending on how the choice is framed—as loss or gain—despite the fact that the expected values of the alternative choices are identical. Kahneman and Tversky (1979) focus on framing effects in the decision-theoretic problem of choosing between alternative lotteries. Economists have since expanded this line of research into game-theoretic contexts, showing, for example, that framing affects the choice to contribute to a public good or impose externalities on others (see, e.g., Andreoni, 1995; Cookson, 2000). McDermott, Fowler, and Smirnov (2008) argue that "context-dependent" attitudes toward risk have a basis in evolutionary psychology. Post, Van den Assem, Baltussen, and Thaler (2008) show this same sensitivity to framing in the high-stakes choices of contestants on the game show "Deal or No Deal," a decision environment decidedly far-removed from the foraging of our evolutionary ancestors.[1]

However, all of this work tends to focus on how the description of payoffs influences subjects. This kind of framing is a mallet used to beat subjects into perceiving in some particular way the monetary benefits of their available alternatives. While many economists acknowledge the power of the mallet, the slightest feather of linguistic connotation can influence subjects. Even the suggestion that an experimental task is a game to play can affect behavior. Cognitive science has shown that framing, far from being a special epicycle for loss and gain, is endemic to human cognition and action. Sher and McKenzie (2008), after providing a review of framing effects, propose a basic correction to the understanding of framing in decision theory. In order for it to be rational for a subject to make the same decision under two framings, these framings would need to be not just logically equivalent but *informationally* equivalent. Given a fixed logical content, there are many different statements that can express it. But a speaker must choose one or another of these statements, and as Sher and McKenzie (2008) write,

a speaker who wishes to convey this logical content will not select a statement at random. Various factors will influence the speaker's selection, these factors varying in degree of intentionality and conscious accessibility. In general, the speaker's selection will vary as a function of the information that is available to the speaker, as well as the speaker's attitudes about the thing being described. But if the speaker's choice of frame varies as a function of the speaker's beliefs and attitudes, then it also potentially conveys information about those beliefs and attitudes. Surely rational actors would not be expected to artificially ignore such information, should it prove relevant to the choice at hand (p. 83.)

Consider, for example, perspective. Cognitive linguistics often studies the phenomenon of deixis, including ways in which the system of the language can prompt for taking one perspective or another. "I" can mean various things depending on the perspective. Deixis often goes unnoticed in consciousness: "go" and "come" can be used of the same event, but call for different perspectives. "Aquí" and "acá" both mean "here" in Mexican Spanish but call for quite different framings and perspectives. Linguistic framing can prompt for attitude, or stance. "If it rains tomorrow, we will eat inside" and "If it rained tomorrow, we would eat inside" refer to the same hypothetical situation, but suggest different stances by the speaker toward the likelihood of the situation (Fillmore, 1990; see also Dancygier & Sweetser, 2005.)

Cognitive science views framing as ubiquitous and powerful in human cognition, choice, and decision-making. Accordingly, in the social sciences, an experiment or inquiry that imposes one frame over another has not shown us anything general about human cognition but rather, at best, something about how human beings operate under the imposition of that particular frame. If the experimenter has used language having to do with frames for *play, game, competition, gain, loss, failure, success, winning, defeat, offer, acceptance, rejection, murder, cooperation, hypothetical candidate, hypothetical company, hypothetical benefit package,* or anything else, then the data cannot be interpreted as showing us how people behave in general, or that they, on their own, would use these frames. Showing as much would require experiments that demonstrate that even when the framing is neutral with respect to various possible framings, people nonetheless operate with this or that specific frame.

Certainly, different people do have different automatic framings, depending on their formation. A navigator, trained repeatedly to take a certain kind of situation as posing a certain kind of threat calling for a certain kind of action, is very likely to activate those frames immediately, without even imagining that alternatives could be chosen, just as the economist trained to frame certain interactions in certain ways, calling for certain arithmetical operations of assessment, will launch on that track straightaway, without reflection. But what this shows us is that people can be trained to use certain frames and mechanisms in certain situations, something cognitive science already knew.

From the perspective of cognitive science, it is important when working with subjects to root out specific coercive frames. This can be done by using the cognitive scientific toolkit for detecting and adjusting framing. For example, consider the following possible description in an experiment:

In this experiment, you are Player 1 in a game against Player 2. The experimenters have given each player $10. Here are the rules of the game: Player 1 must make a choice to send some, all, or none of their $10 to Player 2. Whatever Player 1 sends is subtracted from Player 1's account. The amount sent by Player 1 to Player 2 is tripled when it is moved to Player 2's account. Then Player 2 must make a choice: to send some, all, or none of the money in their account to Player 1. The money Player 2 chooses to send to Player 1 is subtracted from Player 2's account and is given to Player 1, but is not tripled. That is the end of the game.

Consider the framing. People interacting do not need to be framed as players in a game, and framing them as such can bias their actions, because the frame for playing a game can just lead subjects to think that the goal of having fun (playing games) is in some way a substitute for the incentives offered by the experimenters. Indeed, if this were not the case, Las Vegas casinos could not afford air conditioning. Nor do the people interacting need to be framed as playing *against* each other. Nor does the situation need to be framed as being driven by *rules*. The framing here includes forced choices, deductions, gifts, and experimenters in the role of players in a game—indeed, their role is limited in the game, but all-controlling: they launch it and they, arbitrarily, give cash gifts.

If we want to learn how subjects perform under forced frames of this sort, the data from this experiment could, in principle, be revealing. But the data could not—without further justification—be interpreted as showing anything about how subjects perform when they are freer to choose across alternative framings. The instructions might, for example, be written like this:

In this experiment, you are paired randomly with someone else participating in the experiment. The computer has established an account for you and an account for the other person, each containing $10. You may remove some, all, or none of the $10 in your account. The amount you remove will be multiplied by three, and that tripled amount will be added to the other person's account. Afterward, the other person may remove some, all, or none of the amount in his or her account. The computer will then credit that amount, untripled, to your account. The experiment ends at that point.

This rephrasing still comes with a host of framing issues: an *experiment*, with a *computer* that goes about keeping track of *money*, and *unknown* other people *randomly assigned*. But it does resist enforcing the frames of *play, game, competition, experimenter, deduction, sending*, and *gift*. We are often oblivious to forced framing, even when it has great consequences. Luckily, there is a direct remedy. Cognitive linguistics provides a toolkit for detecting and adjusting framing. Just as we might submit protocols for intended experiments to review for ethics, so we might submit them to a cognitive linguist for abatement of forced framing.

14.5.3 Remove the Curse of Knowledge

In cognitive science, "the curse of knowledge" refers to the way in which, once one has acquired knowledge and common cognitive practice, it is exceptionally difficult to imagine that others do not have this knowledge or follow these habits of framing and construal. These "others" can include one's former self. The term used in cognitive science was actually invented by three economists (Camerer, Loewenstein, & Weber, 1989).

The curse of knowledge is not rigid: when one uses what he or she recognizes as a specialized and sophisticated bit of knowledge, then the curse of knowledge naturally does not apply. We do not assume that the ten-year-old knows how to do integration by parts. But when we see an interaction and without conscious reflection frame it as a *trade*, it may be quite difficult for us to pay attention to the fact that the participants might be framing it in a quite different way. When we have designed an experiment and explained it to the subjects, we are unlikely to recognize that the subjects may operate under quite a different understanding of what is going on.

The curse of knowledge is not surprising, from an evolutionary point of view. Evolutionarily, it is for the most part more efficient and effective if our framings and understanding lie in the backstage of cognition, where the great powers of the human mind can work quickly and efficiently, without the limitations and frailty of human consciousness. Almost all cognition happens outside of consciousness. If we had to conduct in consciousness everything we are doing when we are reading or writing, we would never be able to read or write. What we have learned in our lives, we have learned mostly without being aware of the role of backstage cognition.

The curse of knowledge can be seen as a consequence of the powerful nature of our higher-order human cognition. But it poses a constant threat to the methodology of the social sciences, because it disposes experimenters and researchers to blindness: compliance and framing are fundamental methodological issues for the social sciences, but they are mostly invisible to the social scientist.

For any particular knowledge on which an empirical study depends, the social scientist should not assume that subjects in her experiment use that knowledge unless she can demonstrate by something much more reliable than intuition that everyone appropriately analogous to her subjects certainly possessed that knowledge several years before the age of the youngest subject in the study.

14.5.4 Don't Expect People to be Able to Predict Other People Very Well, or to Think They Can

Classical economics depends upon the assumption that a participant in interdependent decision-making is confident of his or her ability to predict what the other participants will do in a strip of activity, provided the participant has complete information and knows the payoff matrices of the other participants. It is this assumption that makes it possible to conclude that "players" will "eliminate" "dominated strategies," for example.

At first blush, it might seem as if cognitive science encourages classical economics to assume confidence on the part of the participant to predict the thoughts of others, since work on the curse of knowledge shows that calculations, assessments, judgments, and interpretations someone has achieved seep into her conceptions of what other people are thinking, so we judge that other people share our knowledge and perspective more than they do. We are clearly overconfident regarding our ability to understand other minds.

But is there in fact evidence that economics should assume people are confident to the degree that economics assumes? Since economics is dealing with the realities of human cognition rather than axioms of a mathematical system, evidence must come from demonstration that human beings in fact are possessed of such confidence. When a mathematician is inventing a mathematical system, the mathematician is permitted to choose any axioms he or she prefers, because the test of these preferences is (1) the imperviousness of the system to attempts to demonstrate that the axioms are inconsistent, and (2) evidence of the power of the mathematical system to contribute to mathematics we already care about, or to provide applications to real situations we are motivated to explain. The warrant for classical microeconomic theory is not its power to explain real systems—it is uncontroversial that it fails systematically on this point. Neither is its warrant that it contributes to mathematical knowledge we care about—we know of no mathematics used in economics that goes beyond mid-college level. The warrant for the axioms in economics must accordingly be that they have been established by research on human cognition, which is to say, the warrant for economics is its standing as a branch of cognitive science.

If participants in interdependent decision-making lack confidence in their ability to predict the behavior of other participants, then it can be rational for these participants to make choices that would be judged irrational by the principles of classical economics.

Should participants be confident? No.

Consider recognizing a Nash equilibrium. As Daskalakis, Goldberg, and Papadimitriou (2009) review, von Neumann (1928) proved that any two-person zero-sum game has an equilibrium. It was later shown that such problems are computationally tractable. In 1951, Nash showed that if mixed strategies are allowed, then every game with a finite number of players in which each player can choose from finitely many strategies has at least one equilibrium. But are these equilibria computationally tractable?

In computational complexity theory, a problem X is called "hard" for a complexity class C if any problem in C can be reduced to X, which implies that no problem in C is harder than X, since a solution to X provides a solution to any problem in C. If X is both hard for C and in C, then X is called "C-complete." A problem that is C-complete is the hardest problem, computationally, in C, or rather, there is no harder problem in C. In a breakthrough discovery, Daskalakis, Goldberg, and Papadimitriou (2005) proved that four-player Nash is PPAD-complete, and then that three-player Nash is PPAD-complete. Then, a few weeks later (Chen & Deng, 2006) proved that even *two-player* Nash is PPAD-complete.

What does this mean? The complexity class PPAD is considered computationally infeasible. As the MIT news summary puts it (Hardesty, 2009), a solution to such a problem is in general "so hard to calculate that all the computers in the world couldn't find it in the lifetime of the universe. And in those cases, Daskalakis believes, human beings playing the game probably haven't found it either." The summary continues: "By showing that some common game-theoretical problems are so hard that they'd take the lifetime of the universe to solve, Daskalakis is suggesting that they can't accurately represent what happens in the real world." Of course, in reality, people might use heuristics, or find approximation strategies, or work by analogy with previous similar games where they felt they had found a Nash equilibrium, or try to find some subset of games where the Nash equilibrium happens not to be so hard to calculate and operate adaptively to move their lives toward those games. In any event, participants' lack of general confidence in their ability to calculate Nash equilibria would be justified, and accordingly, alternative kinds of play would be rational.

Indeed, equilibrium strategies are quite commonly irrational, given a perfectly reasonable set of beliefs about the other player. This is because one player's equilibrium action is only a best response to the other player's equilibrium action. If the first player deviates from equilibrium (or fails to find it), the second player can likely improve his own payoff by abandoning the action prescribed by equilibrium analysis. In zero-sum games, deviations from equilibrium can give clever players a strategic advantage (see, for example, Rubinstein & Tversky's, 1993, hide-and-seek games). In coordination games, everyone can benefit by deviating from the mixed-strategy equilibria in a coordinated manner, as subjects do in Schelling's

(1960) famous focal-point games by responding intelligently to supposedly irrelevant framing.

Camerer (2003), in a review of the experimental economics literature, shows that across a wide range of settings, subjects do not follow equilibrium strategies when those strategies require a certain amount of iterative or forward thinking to uncover. Chou, McConnell, Nagel, and Plott (2009) show that subjects are unable to find the equilibrium to games that involve even mildly complex reasoning. In a recent experiment, we found that even when the game is simple and transparent, with a dominant strategy that is easy to find, subjects choose repeatedly not to play Nash equilibrium strategies, even to the extent that they lose a substantial fraction of their potential earnings. (We know that our protocols are simple to understand because we measure compliance by quizzing subjects on the instructions.)

In our experiment, we have each subject play both roles in the Trust Game (the Truster and the Trustee), and then, after a large battery of intervening tasks, we have each subject play the role of Dictator in a Dictator Game, in which the relative endowments are exactly the same as when the subject played the second, reciprocative, Trustee role in the Trust Game.[2] Thus, each subject faces exactly the same strategic decision in the Dictator Game and as player 2 in the Trust Game, with the only difference being that in the Trust Game, any disparity in endowment is due to a "gift" given by player 1, which may come with an implicit expectation of reciprocity.

In both games, the equilibrium strategy is to give nothing. In our experiment, all 26 subjects had $14 in the role of Reciprocator in the Trust game, while their partners had $2, and, of course, all 26 subjects had $14 in the role of Dictator in the Dictator game, while their partners had $2. Eight of those 26 subjects (31%) chose to give nothing when they were in the role of Reciprocator in the Trust Game, but 13 of those 26 subjects (50%) chose to give nothing when they were in the role of Dictator in the Dictator Game, despite the fact that the payoffs and action possibilities were identical in the two situations. The average transfer was $2.73 by Reciprocators in the Trust Game, which is 26% more than the $2.15 by Dictators in the Dictator Game. This shows that the same subjects can make significantly different choices even in situations that have identical structure and payoffs going forward. Perhaps the framing of reciprocal trustworthiness induces at least some subjects to deviate further from equilibrium. In any case, we observe high rates of deviation from equilibrium in both the Trustee and Dictator conditions.

Now let us consider whether it is rational to be confident in the choice of one's strategies in even a basic 2-person zero-sum game. Let us consider one such game often used in economics: Centipede, with 100 rounds (two turns per round), beginning with $1 on the table, and under the rule that the pot increases by a factor of 10 at the end of every round. The player who picks up the pot keeps 95% and the other player receives 5%. The algebra works like this: the player who picks up the pot in round n receives $.95(10^{n-1})$ and the other player receives $.05(10^{n-1})$. Player 2 loses money if he or she picks up the pot but has underguessed by even one round the round in which Player 1 would have picked up the pot. Player 1 loses money if he or she picks up the pot but has underguessed by even two rounds the round in which Player 2 would have picked up the pot.[3]

Accordingly, unless economists can prove that participants have strong confidence in their guesses of the other participant's actions over 100 rounds, then the players should not behave as the classical economists predict even in a game where every round is just the same except for the amount of money in the pot; it would be irrational for the players to do what the economists think they should, given the participants' doubt that they are perfect mind-readers. And of course, these economics games are the merest cartoons compared to the complexity of interdependent decision-making in the real world. We should expect that actors are much less confident in ecologically valid situations than they are in playing these cartoon games about their ability to predict the choices of others.

To the extent that participants lack confidence, it is rational for them to take a number of actions that would otherwise appear to be irrational. Among the possible strategies would be to

1. exit the strip of interdependent decision-making with whatever guaranteed positive consequence is available, or at least the smallest cost.
2. prune the tree of possible paths of interdependent decision-making so as to move from a part of the tree where the player has less confidence to an area of the tree where the player has greater confidence.
3. relinquish choosing and let things take their course, especially in cases where the current situation seems analogous to remembered scenarios in which the actor made choices that, it turned out, were less than optimal, leaving the actor with the conclusion at that time that he or she had, by acting precipitously and overconfidently, cheated himself or herself of a better reward that life would have brought except for the meddling of the actor.
4. obtain more information of the sort that might raise one's degree of confidence, including information from cheap talk.

Trying to imagine what other people are thinking and what they will do, and making this attempt to imagine repeatedly across agents, is such a standard activity of the human mind that it must be taken into account when trying to understand strategies of choosing and deciding. Where confidence in predicting the actions of others is not perfect, such considerations are part of the analysis of rationality, not extraneous to it.

We can test for degree of confidence whenever we place participants into a situation of choice. We can offer them a choice of exiting with a low payoff and see whether they take it. We can offer them a choice of pruning the tree. We can ask them to rate their ability to predict against the abilities of others in the experiment, on the understanding that they will be significantly rewarded if they rate their comparative ability accurately. These and related measures can measure confidence experimentally rather than assuming it unscientifically. The agenda here is to replace an unexamined assumption with a scientific investigation. Confidence in the ability to predict the choices and actions of others is something we need to discover, and then build our models of human choosing according to what we discover.

The suggestion from cognitive science is that social science researchers should never assume that subjects are confident of their ability to recognize or predict best response. Instead, always test for confidence, not just by asking subjects if they are confident (that would be mere self-report), but rather by putting them in situations where they must bet, or consider taking insurance, or in some other way make consequential choices on the basis of their confidence.

14.6 Conclusion

Human beings make choices, decisions, judgments, and assessments by using their available cognitive processes in situations. Accordingly, models, theories, and experiments in the social sciences are cognitive at their core, at least to the extent that they make claims about the behavior of actual individuals. And yet, the cognitive foundations of human behavior are often ignored or skipped over in the social sciences. Explaining behavior by conducting an equilibrium analysis for all strategic situations sweeps the cognitive bases of that behavior under the rug, as if cognitive operations provide something fixed across all behavior that can logically be factored out when we are looking for causal connections, because, if they are fixed across different behavior, they cannot contribute to causing the difference. But equilibrium analysis factors out differences that are any-thing but fixed, differences that are all-important for causal explanation.

There are social science models that in principle are justified in abstract-ing away from the thought of individuals or even groups. For example, a precise predictive model of Gross National Product following the discovery of new oil reserves may be quite useful in macroeconomics and political science, even though it makes no claims about thought. But for the micro-level in the social sciences, social science modulo cognitive science seems to be self-contradictory and unlikely to succeed. Equilibrium analysis in particular needs to be reexamined for the ways in which choice depends upon actual cognition in actual ecologically valid situations.

This chapter is an attempt to provide a few examples of the ways in which routine methods of cognitive science could be applied to improve social scientific investigations concerned with judgment, decision, reason, and choice. Since judgment, decision, reason, and choice are cognitive and conceptual operations, these foundations should draw as much as possible on the best research from cognitive science.

Notes

1. Unlike experimental studies of framing, Post et al. (2008) rely on observational data, in which the frame (previous earnings) is not controlled by an experimenter but generated endogenously by the subject.

2. In our Trust Game, players 1 and 2 are endowed with $5, and player 1 can choose to give any integer amount of this $5 to his anonymous partner. The experimenter triples the amount that player 1 gave and adds this to player 2's initial endowment. Player 2 can choose to give any integer amount from his updated total back to player 1, although this second gift is not tripled. Player 2 has no incentive to give player 1 any money in the second round, so player 1 has no incentive to give player 2 any money in the first round. The equilibrium, therefore, is for neither player to give any money to his partner. Our Dictator Game is identical in framing to the Trust Game, except that it does not include the Trust Game's first round.

3. Here are the details, taken from Turner (2001), pages 114–115: Consider P1, who guesses that the opponent will pick up the money in round n, so picks it up first in round n. If P1 underguesses by as little as two turns, meaning that P2 would actually have picked up the money in round $n+2$, then P1 would have been considerably better off to let P2 do it. Specifically, by not picking up the money, P1 would win $\$.05(10^{n+1})$, which is much more than the $\$.95(10^{n-1})$ P1 wins by picking it up in round n. For example, if P1 guesses that P2 will pick up the pot in round four, and so P1 picks it up in round four, it is true that P1 wins $950. Yet if P1 underguessed by just two turns, and P2 actually would have picked it up in round six, P1 would

have won \$5,000 by letting P2 pick up the pot. So P1 loses \$4,050 by taking the money off the table. (In fact, if P1 underguessed by only one turn, P1 would win \$500 by keeping hands off the pot, which, true, isn't \$950, but it's still more than half of the amount P1 wins by picking up the pot in round four.) Unless P1 thinks he can guess almost perfectly over 100 rounds when P2 will pick up the pot, it is better for P1 to pass. Now consider P2, who guesses that P1 will pick up the money in round n, where n is greater than 1, and P2 therefore picks up the pot right before P1, in round n-1. If P2 underguessed by even only one turn, so P1 would actually have picked it up in round n+1, then P2 would have been considerably better off to let P1 do it. Specifically, with P1 picking it up in round n+1, P2 would have won \$.05(10^n) by keeping hands off the pot, but P2 wins only \$.95(10^{n-2}) by picking it up in round n-1. For example, if P2 guesses that P1 will pick up the pot in round four, and this leads P2 to pick it up right before P1—that is, on the second turn of round three—it is true that P2 wins \$95. Yet if P2 underguessed by just one turn, so P1 actually would have picked up the pot in round five, P2 would have won \$500—that is \$.05(10^{5-1})—by letting P1 pick up the pot. So unless P2 thinks he will not underguess at all, P2 is better off passing every time.

References

Ainslie, G. (1992). *Picoeconomics: The interaction of successive motivational states within the person*. New York: Cambridge University Press.

Ainslie, G., & Haslam, N. (1992). Hyperbolic discounting. In G. F. Loewenstein & J. Elster (Eds.), *Choice over time* (pp. 57–92). New York: Russell Sage.

Andreoni, J. (1995). Warm-glow versus cold-prickle: The effects of positive and negative framing on cooperation in experiments. *Quarterly Journal of Economics, 110*(1), 1–21.

Becker, G. S. (1976). *The economic approach to human behavior*. Chicago, IL: University of Chicago Press.

Becker, G. S., & Murphy, K. M. (2001). *Social economics: Market behavior in a social environment*. Cambridge, MA: Harvard University Press.

Boudreau, C. (2009). Closing the gap: When do cues eliminate differences between sophisticated and unsophisticated citizens? *Journal of Politics, 71*(3), 1–13.

Camerer, C. F. (2003). *Behavioral game theory: Experiments on strategic interaction*. Princeton, NJ: Princeton University Press.

Camerer, C. F., Loewenstein, G. F., & Weber, M. (1989). The curse of knowledge in economic settings: An experimental analysis. *Journal of Political Economy, 97*(5), 1232–1254.

Caramazza, A., & Shelton, J. R. (1998). Domain-specific knowledge systems in the brain: The animate-inanimate distinction. *Journal of Cognitive Neuroscience, 10,* 1–34.

Chen, X., & Deng, X. (2006). Settling the complexity of 2-player Nash-equilibrium. In *Proceedings of the 47th Annual IEEE symposium on foundations of computer science* (FOCS'06; pp. 261–272).

Chou, E., McConnell, M., Nagel, R., & Plott, C. (2009). The control of game form recognition in experiments: Understanding dominant strategy failures in a simple two person 'guessing' game. *Experimental Economics, 12*(2), 159–179.

Cookson, R. (2000). Framing effects in public goods experiments. *Experimental Economics, 3,* 55–79.

Dancygier, B., & Sweetser, E. (2005). *Mental spaces in grammar: Conditional constructions* (Cambridge Studies in Linguistics, Vol. 108.) Cambridge, MA: Cambridge University Press.

Daskalakis, C., Goldberg, P. W., & Papadimitriou, C. H. (2005). The complexity of computing a Nash equilibrium (TR05–115). Electronic Colloquium on Computational Complexity. Retrieved from http://eccc.hpi-web.de/eccc-reports/2005/TR05-115

Daskalakis, C., Goldberg, P. W., & Papadimitriou, C. H. (2009). The complexity of computing a Nash equilibrium. *Communications of the ACM, 52*(2), 89–97.

Druckman, J. (2001a). The implications of framing effects for citizen competence. *Political Behavior, 23,* 225–256.

Druckman, J. (2001b). Evaluating framing effects. *Journal of Economic Psychology, 22,* 91–101.

Elman, J. L., Bates, E. A., Johnson, M. H., Karmiloff-Smith, A., Parisi, D., & Plunkett, K. (1996). *Rethinking innateness: A connectionist perspective on development.* Cambridge, MA: MIT Press.

Elster, J. (1979). *Ulysses and the sirens.* New York: Cambridge University Press.

Enemark, D., McCubbins, M., Paturi, R., & Weller, N. (2011). Does more connectivity help groups to solve social problems? In *Proceedings of the 12th ACM conference on electronic commerce* (pp. 21–26). New York: Association for Computing Machinery.

Fauconnier, G. (1999). Methods and generalizations. In T. Janssen & G. Redeker (Eds.), *Scope and foundations of cognitive linguistics* (pp. 95–128). The Hague, The Netherlands: Mouton De Gruyter.

Fiddick, L., Cosmides, L., & Tooby, J. (2000). No interpretation without representation: The role of domain-specific representations and inferences in the Wason selection task. *Cognition, 77,* 1–79.

Fillmore, C. (1990). Epistemic stance and grammatical form in English conditional sentences. In *Papers from the 26th Regional Meeting of the Chicago Linguistic Society. Vol. 1: Main Session* (pp. 137–162).

Gigerenzer, G., & Brighton, H. (2009). Homo heuristicus: Why biased minds make better inferences. *Topics in Cognitive Science, 1*(1), 107–143.

Gigerenzer, G., & Hug, K. (1992). Domain-specific reasoning: Social contracts, cheating, and perspective change. *Cognition, 43*, 127–171.

Gigerenzer, G., & Selten, R. (2002). *Bounded rationality: The adaptive toolbox*. Cambridge, MA: MIT Press.

Hardesty, L. (2009, November 9). What computer science can teach economics. *MIT News*. Retrieved from http://web.mit.edu/newsoffice/2009/game-theory.html

Kahneman, D. (2003). Maps of bounded rationality: Psychology for behavioral economics. *American Economic Review, 93*(5), 1449–1475.

Kahneman, D., & Tversky, A. (1979). Prospect theory: An analysis of decision under risk. *Econometrica, 47*, 263–291.

Kahneman, D., & Tversky, A. (Eds.). (2000). *Choices, values, and frames*. Cambridge: Cambridge University Press.

LaPiere, R. T. (1934). Attitudes versus actions. *Social Forces, 3*(2), 230–237.

Lupia, A., & McCubbins, M. (1998). The democratic dilemma: Can citizens learn what they need to know? Cambridge: Cambridge University Press.

McCubbins, M., Paturi, R., & Weller, N. (2008). Connected coordination: Network structure and group coordination. *American Politics Research, 37*, 899–920.

McDermott, R., Fowler, J. H., & Smirnov, O. (2008). On the evolutionary origin of prospect theory preferences. *Journal of Politics, 70*(2), 335–350.

Monterosso, J., & Ainslie, G. (2007). The behavioral economics of will in recovery from addiction. *Drug and Alcohol Dependence, 90*, S100–S111.

Palmer, S. (1999). *Vision science: Photons to phenomenology*. Cambridge, MA: The MIT Press.

Post, T., Van den Assem, M. J., Baltussen, G., & Thaler, R. H. (2008). Deal or no deal? Decision making under risk in a large-payoff game show. *American Economic Review, 98*(1), 38–71.

Pronin, E., Olivola, C., & Kennedy, K. (2008). Doing unto future selves as you would do unto others: Psychological distance and decision making. *Personality and Social Psychology Bulletin, 34*(2), 224–236.

Rubinstein, A. (1998). *Modeling bounded rationality*. Cambridge, MA: MIT Press.

Rubinstein, A., & Tversky, A. (1993). Naïve strategies in zero-sum games (working paper 17–93). Sackler Institute of Economic Studies, Tel Aviv University.

Schelling, T. (1960). *The strategy of conflict*. Cambridge, MA: Harvard University Press.

Schelling, T. (1984). Self-command in practice, in policy, and in a theory of rational choice. *American Economic Review, 74*, 1–11.

Sher, S., & McKenzie, C. (2008). Framing effects and rationality. In N. Chater & M. Oaksford (Eds.), *The probabilistic mind: Prospects for Bayesian cognitive science* (pp. 79–96). Oxford: Oxford University Press.

Sherrington, C. S. (1906). *The integrative action of the nervous system*. New York: Charles Scribner's Sons.

Simon, H. (1957). A behavioral model of rational choice. In *Models of man, social and rational: Mathematical essays on rational human behavior in a social setting* (pp. 241–260). New York: Wiley.

Sun, R. (2002). *Duality of the mind: A bottom-up approach toward cognition*. Mahwah, NJ: Lawrence Erlbaum.

Thaler, R. H., & Shefrin, H. M. (1981). An economic theory of self-control. *Journal of Political Economy, 89*, 392–406.

Trope, Y., & Liberman, N. (2000). Time-dependent changes in preferences. *Journal of Personality and Social Psychology, 79*, 876–889.

Trope, Y., & Liberman, N. (2003). Temporal construal. *Psychological Review, 110*, 403–421.

Turner, M. (2001). *Cognitive dimensions of social science: The way we think about politics, economics, law, and society*. New York: Oxford University Press.

Tversky, A., & Fox, C. R. (2000). Weighing risk and uncertainty. In Kahneman, D., & Tversky, A. (Eds.), *Choices, values, and frames* (pp. 93–117). Cambridge: Cambridge University Press.

Tversky, A., & Kahneman, D. (1992). Advances in prospect theory: Cumulative representation of uncertainty. *Journal of Risk and Uncertainty, 5*, 297–323.

van Gulick, R. (2011). Consciousness. In E. N. Zalta (Ed.). *The Stanford encyclopedia of philosophy* (Summer 2011 ed.). Retrieved from http://plato.stanford.edu/archives/sum2011/entries/consciousness

von Neumann, J. (1928). Zur theorie der gesellshaftsspiele. *Mathematische Annalen, 100*, 295–320.

Zeki, S. (1993). *A vision of the brain*. Oxford: Blackwell.

15 Role of Cognitive Processes in Unifying the Behavioral Sciences

Herbert Gintis

15.1 Introduction

The behavioral sciences include economics, anthropology, sociology, psychology, and political science, as well as biology insofar as it deals with animal and human behavior. These disciplines have distinct research foci and include three conflicting models of decision-making, as determined by what is taught in the graduate curriculum and what is accepted in journal articles without reviewer objection. The three models are economic, sociological, and biological. A more sustained appreciation of the contributions of cognitive psychology to our understanding of decision-making will help adjudicate among these models.

These three models are not only different, which is to be expected given their distinct explanatory goals, but *incompatible*. This means, of course, that at least two of the three are certainly incorrect. However, this chapter will argue that in fact all three are flawed, but can be modified to produce a unified framework for modeling choice and strategic interaction for all of the behavioral sciences through a considered appreciation of the evolved nature of human psychology. Such a framework can then be enriched in different ways to meet the particular needs of each discipline.

In a summary to be amplified below, we may characterize the *economic model* as rational choice theory, which takes the individual as maximizing a self-regarding preference subject to an unanalyzed and pre-given set of beliefs, called subjective priors. The *sociological model* is that of the pliant individual who internalizes the norms and values of society and behaves according to the dictates of the social roles he occupies. The *biological model* is that of the fitness maximizer, who is the product of a long process of Darwinian evolution. Each of these models is deeply insightful but equally deeply one-sided and flawed. The analysis presented here, which is a development of the extended argument in Gintis (2009a), shows a way forward in conserving the major insights of these three models while rejecting their weakness—weaknesses that account for their mutual incompatibility.

This framework for unification includes five conceptual units: (a) gene-culture coevolution; (b) the socio-psychological theory of norms; (c) game theory; (d) the rational actor model; and (e) complexity theory. Gene-culture coevolution comes from the biological theory of social organization (sociobiology), and is foundational because *Homo sapiens* is an evolved, biological, highly social species. The socio-psychological theory of norms includes fundamental insights from sociology that apply to all forms of human social organization, from hunter-gatherer to advanced technological societies.

The role of these five principles (and others that might emerge alongside or replace them as the process of unification progresses) is to render models of human behavior from different disciplines *compatible where they overlap*. Of course, unification will not and should not collapse all disciplines into one, or cause some disciplines to become subdisciplines of others. The behavioral disciplines will and should retain their constitutive explanatory goals. Rather, the above five principles, supplemented by additional perspectives, will ensure that where the objects of inquiry in distinct disciplines coincide, the explanations from the various disciplines are concordant (Levine & Palfrey, 2007; Goeree et al., 2007).

Complexity theory is relevant because human society has emergent properties that cannot be derived analytically from lower-level constructs. This is why agent-based modeling (Sun, 2006, 2008) as well as historical and ethnographic studies of human social dynamics (Geertz, 1963) are needed to supplement analytical models (see chapter 1 in this volume by Ron Sun).

For the sake of clarity, we note that the emergent properties of a complex system are not *caused* by its higher-level structure. Rather, the lower-level causes of emergent properties are too complex to be analytically modeled. For instance, water is "caused" by its constituents (hydrogen, oxygen, and their pair-bonds), but the causation is so complex that we cannot predict some of the most important properties of water by simply solving the Schrödinger equation with the appropriate potential function. Similarly, a digital computer has emergent properties with respect to its solid-state and electronic components, in the sense that we must develop wholly novel ideas, including the concepts of hardware, software, memory, and algorithms, to understand the computer, rather than explaining its operation in terms of the laws of quantum mechanics.

Human cognition is an emergent property of human evolution because the emergence of language entailed a transformation of individual cognition into *social cognition*, in which cognitive processes are distributed across and between brains via material cognitive tools (Dunbar, Gamble, & Gowlett, 2010). For this reason, human cognition cannot be completely understood by studying individual brains in social isolation. Similarly, human morality and ethics, including the role of social norms in regulating human behavior, cannot be understood by investigating the content and operation of individual brains: morality is a distributed social cognition.

Game theory includes three branches related to cognitive psychology: behavioral, epistemic, and evolutionary. Behavioral game theory, which uses classical game-theoretic methodology to reveal and quantify the social psychology of human strategic interaction, reveals that rational behavior involves deep structural psychological principles, including other-regarding preferences and the human propensity to value virtuous behavior for its own sake. Chapter 11 in this volume by Don Ross explains the cognitive basis of behavioral game theory and its relationship to neuroeconomics. Epistemic game theory is the application of the modal logic of knowledge and belief to strategic interaction, and fosters the integration of the rational actor model with the socio-psychological theory of norms (Gintis, 2009a). Epistemic game theory reveals dimensions of cognitive processes that are key to understanding human cooperation, but are obscured in classical game theory, which equates rationality with "maximization." The cognitive foundations of epistemic game theory, which deals with the epistemological structure of knowledge and belief, are increasingly revealed in neuroscientific studies, which reveal how the brain represents other agents during game-theoretic interactions (Hampton, Bossaerts, & O'Doherty, 2006, 2008). Finally, evolutionary game theory is a macro-level analytical apparatus allowing the insights of biological and cultural evolution to be analytically modeled.

The rational actor model is the single most important analytical construct in the behavioral sciences operating at the level of the individual. While gene-culture coevolutionary theory is a form of "ultimate" explanation that does not predict, the rational actor model provides a "proximate" description of behavior that can be tested in the laboratory and real-world situations, and is the basis of the explanatory success of economic theory. Game theory makes no sense without the rational actor model, and behavioral disciplines, like sociology and psychology, that have abandoned this model have fallen into theoretical disarray. Cognitive psychology without the rational actor model is a seriously crippled enterprise. This conclusion holds for social psychology as well. For examples of socio-psychological analyses based on behavioral economics principles, see Keiser, Lindenburg, and Steg (2008), as well as Stapel and Lindenberg (2011) and references therein.

However, the rational actor model has an obvious shortcoming that must be dealt with before the model can fit harmoniously with a sophisticated cognitive psychology of human decision making. The rational actor model describes how individuals make decisions *in social isolation*, using an agent's "subjective prior," or informed opinion prior to the collection of data, to represent his beliefs as to how actions link to real-world outcomes and thence to personal payoffs. Given the social nature of the human brain, however, it is inaccurate to equate subjective priors, which are constituted in social networks, with "beliefs," as the latter become constituted and evolve in social life. Beliefs are the product of rational social interaction, a fact that undermines the rather naive methodological individualism choice theorists. Thus, rational decision making inexorably involves *imitation*, sometimes exhibits *conformist bias*, and generally entails the constitution of networks of mutual beliefs characteristic of a distributed mind. Chapter 10 in this book by Harvey Whitehouse on the instrumental opacity of ritualistic activity illustrates rather well the connection between rationality and tradition found in virtually every society that has been studied, as does chapter 9 by Ilkka Pyysiäinen, which argues that analytical treatments of religion are ineluctably reductivist, purportedly "explaining" religion fully in terms of lower-level cognitive and social variables. In fact, religion is an emergent property of human societies, giving meaning to human life that cannot be deduced from more mundane aspects of human social existence and cognitive organization.

Complexity theory is needed because human society is a complex adaptive system with emergent properties that cannot now, and perhaps never will be, explained starting with more basic units of analysis. The hypothetico-deductive methods of game theory and the rational actor model, and even gene-culture coevolutionary theory, must therefore be complemented by the work of behavioral scientists who deal with society in more interpretive terms and develop holistic schemas that shed light where analytical models cannot penetrate. Anthropological and historical studies fall into this category, as well as macroeconomic policy and comparative economic systems. Agent-based modeling of complex adaptive dynamical systems is also useful in dealing with emergent properties of complex systems (see Ron Son's contribution to this volume, as well as Sun, 2006, 2008).

15.2 Gene-Culture Coevolution

Because culture is key to the evolutionary success of *Homo sapiens*, individual fitness in humans depends on the structure of social life. It follows that human cognitive, affective, and moral capacities are the product of an evolutionary dynamic involving the interaction of genes and culture. This dynamic is known as *gene-culture coevolution* (Cavalli-Sforza & Feldman, 1982; Boyd & Richerson, 1985; Dunbar, 1993; Richerson & Boyd, 2004). This coevolutionary process has endowed us with preferences that go beyond the egoistic motives emphasized in traditional economic and biological theory, and embrace a social epistemology facilitating the sharing of intentionality across minds, as well as such non-self-regarding values as a taste for cooperation, fairness, and retribution, the capacity to empathize, and the ability to value honesty, hard work, piety, toleration of diversity, and loyalty to one's reference group.

The genome encodes information that is used both to construct a new organism and to endow it with instructions for transforming sensory inputs into decision outputs. Because learning is costly and error-prone, efficient information transmission occurs when the genome encodes all aspects of the organism's environment that are constant, or that change only slowly through time and space. By contrast, environmental conditions that vary rapidly, or require a complex series of coordinated actions, can be dealt with by providing the organism with the capacity to *learn*.

There is an intermediate case, however, that is efficiently handled neither by genetic encoding nor learning. When environmental conditions are positively correlated across generations, each generation acquires valuable information through learning that it cannot transmit genetically to the succeeding generation, because such information is not encoded in the germ line. Indeed, there is no plausible mechanism whereby complex learned behaviors can be incorporated in the genome. Hence, there is a fitness benefit to the transmission of *epigenetic* information concerning the current state of the environment. Such epigenetic information is quite common in the natural world (Jablonka & Lamb, 1995), but achieves its highest and most flexible form in *cultural transmission* in humans and to a considerably lesser extent in other primates (Bonner, 1984; Richerson & Boyd, 1998).

Cultural transmission occurs via various modes: vertical (parents to children), horizontal (peer to peer), and oblique (elder to younger), as in Cavalli-Sforza and Feldman (1981); prestige (higher influencing lower status), as in Henrich and Gil-White (2001); popularity-related, as in Newman, Barabasi, and Watts (2006); and even random population-dynamic transmission, as in Shennan (1997) and Skibo and Bentley (2003).

The parallel between cultural and biological evolution goes back to Huxley (1955), Popper (1979), and James (1880) (see Mesoudi, Whiten, & Laland, 2006, for details). The idea of treating culture as a form of epigenetic transmission was pioneered by Richard Dawkins, who coined the term "meme" in *The Selfish Gene* (1976) to represent an integral unit of information that could be transmitted phenotypically. There quickly followed several major contributions to a biological approach to culture, all based on the notion that culture, like genes, could evolve through replication (intergenerational transmission), mutation, and selection.

Cultural elements reproduce themselves from brain to brain and across time, mutate, and are subject to selection according to their effects on the fitness of their carriers (Parsons, 1964; Cavalli-Sforza & Feldman, 1982). Moreover, there are strong interactions between genetic and epigenetic elements in human evolution, ranging from basic physiology (e.g., the transformation of the organs of speech with the evolution of language) to sophisticated social emotions, including empathy, shame, guilt, and revenge-seeking (Zajonc, 1980, 1984; Krajbich, Adolphs, Tranel, Denburg et al., 2009).

Because of their common informational and evolutionary character, there are strong parallels between genetic and cultural modeling (Mesoudi et al., 2006). Like biological transmission, culture is transmitted from parents to offspring, and like cultural transmission, which is transmitted horizontally to unrelated individuals, so in microbes and many plant species, genes are regularly transferred across lineage boundaries (Jablonka & Lamb, 1995; Rivera & Lake, 2004; Abbott, James, Milne, & Gillies, 2003). Moreover, anthropologists reconstruct the history of social groups by analyzing homologous and analogous cultural traits, much as biologists reconstruct the evolution of species by the analysis of shared characters and homologous DNA (Mace & Pagel, 1994). Indeed, the same computer programs developed by biological systematists are used by cultural anthropologists (Holden, 2002; Holden & Mace, 2003). In addition, archeologists who study cultural evolution have a similar *modus operandi* as paleobiologists who study genetic evolution (Mesoudi et al., 2006). Both attempt to reconstruct lineages of artifacts and their carriers. Like paleobiology, archeology assumes that when analogy can be ruled out, similarity implies causal connection by inheritance (O'Brien & Lyman, 2000). Like biogeography's study of the spatial distribution of organisms (Brown & Lomolino, 1998), behavioral ecology studies the interaction of ecological, historical, and geographical factors that determine distribution of cultural forms across space and time (Smith & Winterhalder, 1992).

Dawkins added a fundamental mechanism of epigenetic information transmission in *The Extended Phenotype* (1982), noting that organisms can directly transmit environmental artifacts to the next generation in the form of such constructs as beaver dams, bee hives, and even social structures (e.g., mating and hunting practices). The phenomenon of a species creating an important aspect of its environment and stably transmitting this environment across generations, known as *niche construction*, is a widespread form of epigenetic transmission (Odling-Smee, Laland, & Feldman, 2003). Moreover, niche construction gives rise to what might be called a *gene-environment coevolutionary process*, since a genetically induced environmental regularity becomes the basis for genetic selection, and genetic mutations that give rise to mutant niches will survive if they are fitness-enhancing for their constructors.

An excellent example of gene-environment coevolution is the honey bee: its eusociality probably arose from the high degree of relatedness fostered by haplodiploidy, but persists in modern species despite the fact that relatedness in the hive is generally quite low as a result of multiple queen matings, multiple queens, queen deaths, and the like (Gadagkar, 1991; Seeley, 1997; Wilson & Holldobler, 2005). The social structure of the hive is transmitted epigenetically across generations, and the honey bee genome is an adaptation to the social structure laid down in the distant past.

Gene-culture coevolution in humans is a special case of gene-environment coevolution, in which the environment is culturally constituted and transmitted (Feldman & Zhivotovsky, 1992). The key to the success of our species in the framework of the hunter-gatherer social structure in which we evolved is the capacity of unrelated, or only loosely related, individuals to cooperate in relatively large egalitarian groups in hunting and territorial acquisition and defense (Boehm, 2000; Richerson & Boyd, 2004). While contemporary biological and economic theory have attempted to show that such cooperation can be effected by self-regarding rational agents (Trivers, 1971; Alexander, 1987; Fudenberg, Levine, & Maskin, 1994), the conditions under which this is the case are highly implausible even for small groups (Boyd & Richerson, 1988; Gintis, 2005;

Bowles & Gintis, 2011). Rather, the social environment of early humans was conducive to the development of prosocial traits, such as empathy, shame, pride, embarrassment, and reciprocity, without which social cooperation would be impossible.

Neuroscientific studies exhibit clearly the genetically based neural structures that are specialized to support moral cognition and action. Brain regions involved in moral judgments and behavior include the prefrontal cortex, especially the orbitalfrontal cortex, and the superior temporal sulcus (Moll, Zahn, di Oliveira-Souza, Krueger et al., 2005). These brain structures are most highly developed in humans and are doubtless evolutionary adaptations (Schulkin, 2000). The evolution of the human prefrontal cortex is closely tied to the emergence of human morality (Allman, Hakeem, & Watson, 2002). Patients with focal damage to one or more of these areas exhibit sociopathic behavior (Miller, Darby, Benson, Cummings et al., 1997) and a variety of antisocial behaviors, including the absence of embarrassment, pride, and regret (Beer, Heerey, Keltner, Skabini et al., 2003; Camille, 2004). There is a likely genetic predisposition underlying sociopathy, and sociopaths comprise 3%–4% of the male population, but they account for between 33% and 80% of the population of chronic criminal offenders in the United States (Mednick, Kirkegaard-Sorenson, Hutchins, Knop et al., 1977).

The author argues on the basis of this body of empirical information that culture is directly encoded into the human brain and achieves its force by virtue of the intimate interactions of socially distributed cognition, which of course is the central claim of gene-culture coevolutionary theory.

The evolution of the physiology of speech and facial communication is a dramatic example of gene-culture coevolution. The increased social importance of communication in human society rewarded genetic changes that facilitate speech. Regions in the motor cortex expanded in early humans to facilitate speech production. Concurrently, nerves and muscles to the mouth, larynx, and tongue became more numerous to handle the complexities of speech (Jurmain, Nelson, Kilgore, & Travathan, 1997). Parts of the cerebral cortex—the Broca's and Wernicke's areas, which do not exist or are relatively small in other primates—are large in humans and permit grammatical speech and comprehension (Binder, Frost, Hammeke, Cox et al., 1997; Belin, Zatorre, Lafaille, Ahad et al., 2000).

Adult modern humans have a larynx low in the throat, a position that allows the throat to serve as a resonating chamber capable of a great number of sounds (Relethford, 2007). The first hominids that have skeletal structures supporting this laryngeal placement are the *Homo heidelbergensis*, who lived from 800,000 to 100,000 years ago. In addition, the production of consonants requires a short oral cavity, whereas our nearest primate relatives have much too long an oral cavity for this purpose. The position of the hyoid bone, which is a point of attachment for a tongue muscle, developed in *Homo sapiens* in a manner permitting highly precise and flexible tongue movements.

Another indication that the tongue has evolved in hominids to facilitate speech is the size of the hypoglossal canal, an aperture that permits the hypoglossal nerve to reach the tongue muscles. This aperture is much larger in Neanderthals and humans than in early hominids and nonhuman primates (Dunbar, 1996). Human facial nerves and musculature have also evolved to facilitate communication. This musculature is present in all vertebrates, but for all except mammals, it serves feeding and respiratory functions alone (Burrows, 2008). In mammals, this mimetic musculature attaches to the skin of the face, thus permitting the facial communication of such emotions as fear, surprise, disgust, and anger. In most mammals, however, a few wide sheet-like muscles are involved, rendering fine information differentiation impossible, whereas in primates, this musculature divides into many independent muscles with distinct points of attachment to the epidermis, thus permitting higher bandwidth facial communication. Humans have the most highly developed facial musculature by far of any primate species, with a degree of involvement of lips and eyes that is not present in any other species. This example is quite a dramatic and concrete illustration of the intimate interaction of genes and culture in the evolution of our species.

15.3   Rational Decision Theory

General evolutionary principles suggest that individual decision making for members of a species can be modeled as optimizing a preference function. Natural selection leads the content of preferences to reflect biological fitness. The principle of expected utility extends this optimization to stochastic outcomes. The resulting model is called the *rational actor model* or *rational decision theory* in economics.

For every constellation of sensory inputs, each decision taken by an organism generates a probability distribution over outcomes, the expected value of which is the *fitness* associated with that decision. Since fitness is a scalar variable, for each constellation of sensory inputs, each possible action the organism might take has a specific fitness value, and organisms whose decision mechanisms are optimized for this environment will choose the available action that maximizes this value. This argument was presented by Darwin (1872) and is implicit in the standard notion of "survival of the fittest," but formal proof is recent (Grafen, 1999, 2000, 2002). The case with frequency-dependent (non-additive genetic) fitness has yet to be formally demonstrated, but the informal arguments are compelling.

Given the state of its sensory inputs, if an organism with an optimized brain chooses action A over action B when both are available, and chooses action B over action C when both are available, then it will also choose action A over action C when both are available. Thus choice consistency follows from basic evolutionary dynamics.

The so-called *rational actor model* was developed in the twentieth century by John von Neumann, Leonard Savage and many others. The model is often presented as though it applies only when actors possess extremely strong information-processing capacities. In fact, the model depends only on *choice consistency* (Gintis, 2009a). When preferences are consistent, they can be represented by a numerical function, often called a *utility function*, which the individual maximizes subject to his subjective beliefs. When consistency extends over lotteries (actions with probabilistic outcomes), then agents act as though they were maximizing expected utility subject to their subjective priors (beliefs concerning the effect of actions on the probability of diverse outcomes).

Four caveats are in order. First, individuals do not *consciously* maximize something called "utility," or anything else. Rather, consistent preferences imply that there is an objective function such that the individual's choices can be predicted by maximizing the expected value of the objective function (much as a physical system can be understood by solving a set of Hamiltonian equations, with no implication that physical systems carry out such a set of mathematical operations). Second, individual choices, even if they are self-regarding (e.g., personal consumption) are not necessarily welfare-enhancing. For instance, a rational decision-maker may smoke cigarettes knowing full well their harmful effects, and even while wishing that he did not smoke cigarettes. It is fully possible for a rational decision-maker to wish that he had preferences other than those he actually has, and even to transform his preferences accordingly. Third, preferences must have some stability across time to be theoretically useful, but preferences are ineluctably a function of an individual's *current state*, which includes both his physiological state and his current beliefs. Because beliefs can change dramatically and rapidly in response to sensory experience, preferences may be subject to discontinuous change. Finally, beliefs need not be correct, nor need they be updated correctly in the face of new evidence, although Bayesian assumptions concerning updating can be made part of consistency in elegant and compelling ways (Jaynes, 2003).

The rational actor model is the cornerstone of contemporary economic theory, and in the past few decades has also become the heart of the biological modeling of animal behavior (Real, 1991; Alcock, 1993; Real & Caraco, 1986). Economic and biological theory thus have a natural affinity: the choice consistency on which the rational actor model of economic theory depends is rendered plausible by evolutionary theory, and the optimization techniques pioneered in economics are routinely applied and extended by biologists in modeling the behavior of nonhuman organisms.

In a stochastic environment, natural selection will ensure that the brain makes choices that, at least roughly, maximize expected fitness, and hence satisfy the expected utility principle (Cooper, 1987). To see this, suppose an organism must choose from action set X, where each $x \in X$ determines a lottery that pays i offspring with probability $p_i(x)$, for $i=0,1, \ldots ,n$. Then the expected number of offspring from this lottery is $\psi(x) = \sum_{j=1}^{n} j P_j(x)$. Let L be a lottery on X that delivers $x_i \in X$ with probability q_i for $i=1, \ldots ,k$. The probability of j offspring given L is then $\sum_{i=1}^{k} q_i p_j(x_i)$, so the expected number of offspring given L is

$$\sum_{j=1}^{n} j \sum_{i=1}^{k} q_i p_j(x_i) = \sum_{i=1}^{k} q_i \sum_{i=1}^{k} j p_j(x_i) = \sum_{i=1}^{k} q_i \psi(x_i) , \qquad (15.1)$$

which is the expected value theorem with utility function $\psi(\cdot)$.

Evidence from contemporary neuroscience suggests that expected utility maximization is not simply an "as if" story. In fact, the brain's neural circuitry actually makes choices by internally representing the payoffs of various alternatives as neural firing rates, and choosing a maximal such rate (Glimcher, 2003; Dorris & Glimcher, 2003; Glimcher, Dorris, & Bayer, 2005; Kable & Glimcher, 2009). Neuroscientists increasingly find that an aggregate decision-making process in the brain synthesizes all available information into a single, unitary value (Parker & Newsome, 1998; Schall & Thompson, 1999; Glimcher, 2003). Indeed, when animals are tested in a repeated trial setting with variable reward, dopamine neurons appear to encode the difference between the reward that an animal expected to receive and the reward that an animal actually received on a particular trial (Schultz, Dayan, & Montague, 1997; Sutton & Barto, 2000), an evaluation mechanism that enhances the environmental sensitivity of the animal's decision-making system. This error-prediction mechanism has the drawback of only seeking local optima (Sugrue, Corrado, & Newsome, 2005). Montague and Berns (2002) and Niv and Montague (2009) have shown, however, that often error-prediction algorithms locate global optima. Montague and Berns (2002), for instance, show that the orbito-frontal cortex and striatum contain mechanisms for global predictions that include making assessment of risk and the discounting of future rewards. Their data suggest a decision model analogous to the famous Black-Scholes options-pricing equation (Black & Scholes, 1973).

Perhaps the most pervasive critique of the rational decision model is that put forward by Herbert Simon (1982), holding that because informa-tion processing is costly and humans have finite information-processing capacity, individuals *satisfice* rather than *maximize*, and hence are only *boundedly rational*. There is much substance to Simon's premises, especially that of including information-processing costs and limited information in modeling choice behavior, and that of recognizing that the decision on how much information to collect depends on unanalyzed subjective priors at some level (Winter, 1971; Heiner, 1983). Indeed, from basic information theory and quantum mechanics, it follows that *all rationality is bounded*.

However, the popular message taken from Simon's work is that we should reject the rational actor model. For instance, the mathematical psychologist D. H. Krantz (1991) asserts, "The normative assumption that individuals *should* maximize *some* quantity may be wrong. People do and should act as *problem solvers*, not *maximizers*." This is incorrect. In fact, as long as individuals are involved in routine choice and hence have consis-tent preferences, they can be modeled as maximizing an objective function subject to constraints. This point is lost on even such capable researchers as Gigerenzer and Selten (2001), who reject the "optimization subject to constraints" method on the grounds that individuals do not in fact solve optimization problems. However, just as the billiards players do not solve differential equations in choosing their shots, so decision makers do not solve Lagrangian equations, even though in both cases we may use such optimization models to describe their behavior.

15.4 From Rational Action to Bayesian Cognition

Bayesian models of cognitive inference that generalize the rational actor model are increasingly prominent is several areas of cognitive psychology, including animal and human learning (Courville, Daw, & Toureztsky, 2006; Tenenbaum, Griffiths, & Kemp, 2006; Steyvers, Tenenbaum, Wagenmakers, & Blum, 2003; Griffiths & Tenenbaum, 2008); visual perception and motor control (Yuille & Kersten, 2006; Kording & Wolpert, 2006); semantic memory and language processing (Steyvers, Griffiths, & Dennis, 2006; Chater & Manning, 2006; Xu & Tenenbaum, 2007); and social cognition (Baker, Tenenbaum, & Saxe, 2007). For a recent overview of Bayesian models of cognition, see Griffiths, Kemp, and Tenenbaum (2008). These models are especially satisfying because they bridge the gap between traditional cognitive models that stress symbolic representations and their equally traditional adversaries that stress statistical testing. Bayesian models are symbolic in that they are predicated upon a repertoire of pre-existing models that can be tested, as well as upon statistical techniques that carry out the testing and provide the feedback through which the underlying models can be chosen and modified.

Bayesian information-processing models may solve the problem of how humans acquire complex understandings of the world given severely underdetermining data. For instance, the spectrum of light waves received in the eye depends both on the color spectrum of the object being observed and the way the object is illuminated. Therefore, inferring the object's color is severely underdetermined, yet we manage to consider most objects to have a constant color even as background illumination changes. Brainard and Freeman (1997) show that a Bayesian model solves this problem fairly well, given reasonable subjective priors as to the object's color and the effects of the illuminating spectra on the object's surface.

Several students of developmental learning have stressed that children's learning is similar to scientific hypothesis testing (Carey, 1985; Gopnik & Meltzoff, 1997), but without offering specific suggestions regarding the calculation mechanisms involved. Recent studies suggest that these mechanisms include causal Bayesian networks and related algorithmic processes (Glymour, 2001; Gopnik & Schultz, 2007; Gopnik & Tenenbaum, 2007; Sun, 2008). One schema, known as constraint-based learning, uses observed patterns of independence and dependence among a set of observational variables experienced under different conditions to work backward in determining the set of causal structures compatible with the set of observations (Pearl, 2000; Spirtes, Glymour, & Scheines, 2001). Eight-month-old babies can calculate elementary conditional independence relations well enough to make accurate predictions (Sobel & Kirkham, 2007). Two-year-olds can combine conditional independence and hands-on information to isolate causes of an effect, and four-year-olds can design purposive interventions to gain relevant information (Gopnik, Sobel, Schulz, & Glymour, 2001; Schultz & Gopnik, 2004). "By age four," observe Gopnik and Tenenbaum (2007), "children appear able to combine prior knowledge about hypotheses and new evidence in a Bayesian fashion." (p. 284). Moreover, neuroscientists have begun studying how Bayesian updating is implemented in neural circuitry (Knill & Pouget, 2004). For instance, suppose an individual wishes to evaluate an hypothesis h about the natural world given observed data x and under the constraints of a background repertoire T. The value of h may be measured by the Bayesian formula

$$P(h \mid x, T) = \frac{P(x \mid h, T)P(h \mid T)}{\sum_{h' \in T} P(x \mid h', T)P(h' \mid T)} \tag{15.2}$$

Here, $P(x|h,T)$ is the likelihood of the observed data x, given h and the background theory T, and $P(h|T)$ gives the likelihood of h in the agent's repertoire T. The constitution of T is an area of active research. In language acquisition, it will include predispositions to recognize certain forms as grammatical and not others. In other cases, T might include different models of folk-physics, folk-biology, or natural theology. Recent theoretical work in cognitive neuroscience has shown how neural networks could implement Bayesian inference (Ma, Beck, Latham, & Pouget, 2006).

15.5 Evolutionary Game Theory

The analysis of living systems includes one concept that is not analytically represented in the natural sciences: that of a *strategic interaction*, in which the behavior of agents is derived by assuming that each is choosing a *best response* to the actions of other agents. The study of systems in which agents choose best responses, and in which such responses evolve dynamically, is called *evolutionary game theory*.

A *replicator* is a physical system capable of drawing energy and chemical building blocks from its environment to make copies of itself. Chemical crystals, such as salt, have this property, but biological replicators have the additional ability to assume a myriad of physical forms based on the highly variable sequencing of chemical building blocks. Biology studies the dynamics of such complex replicators using the evolutionary concepts of replication, variation, mutation, and selection (Lewontin, 1974).

Biology plays a role in the behavioral sciences much like that of physics in the natural sciences. Just as physics studies the elementary processes that underlie all natural systems, so biology studies the general characteristics of survivors of the process of natural selection. In particular, genetic replicators, the epigenetic environments to which they give rise, and the effect of these environments on gene frequencies account for the characteristics of species, including the development of individual traits and the nature of intraspecific interaction. This does not mean, of course, that behavioral science in any sense *reduces* to biological laws. Just as one cannot deduce the character of natural systems (e.g., the principles of inorganic and organic chemistry, the structure and history of the universe, robotics, plate tectonics) from the basic laws of physics, similarly one cannot deduce the structure and dynamics of complex life forms from basic biological principles. But, just as physical principles inform model creation in the natural sciences, so must biological principles inform all the behavioral sciences.

Within population biology, evolutionary game theory has become a fundamental tool. Indeed, evolutionary game theory is basically population biology with frequency-dependent fitnesses. Throughout much of the twentieth century, classical population biology did not employ a game-theoretic framework (Fisher, 1930; Haldane, 1932; Wright, 1931). However, Moran (1964) showed that Fisher's fundamental theorem, which states that as long as there is positive genetic variance in a population, fitness increases over time, is false when more than one genetic locus is involved. Eshel and Feldman (1984) identified the problem with the population genetic model in its abstraction from mutation. But how do we attach a fitness value to a mutant? Eshel and Feldman (1984) suggested that payoffs be modeled game-theoretically on the phenotypic level, and a mutant gene be associated with a strategy in the resulting game. With this assumption, they showed that under some restrictive conditions, Fisher's fundamental theorem could be restored. Their results were generalized by Liberman (1988), Hammerstein and Selten (1994), Hammerstein (1996), Eshel, Feldman, and Bergman (1998), and others.

The most natural setting for genetic and cultural dynamics is game theoretic. Replicators (genetic, cultural, or both) endow copies of themselves with a repertoire of strategic responses to environmental conditions, including information concerning the conditions under which each is to be deployed in response to the character and density of competing replicators. Genetic replicators have been well understood since the rediscovery of Mendel's laws in the early twentieth century. Cultural transmission also apparently occurs at the neuronal level in the brain, in part through the action of *mirror neurons* (Williams, Whiten, Suddendorf, & Perrett, 2001; Rizzolatti, Fadiga, Fogassi, & Gallese, 2002; Meltzhoff & Decety, 2003). Mutations include replacement of strategies by modified strategies, and the "survival of the fittest" dynamic (formally called a *replicator dynamic*) ensures that replicators with more successful strategies replace those with less successful strategies (Taylor & Jonker, 1978).

Cultural dynamics, however, do not reduce to replicator dynamics. For one thing, the process of switching from lower- to higher-payoff cultural norms is subject to error, and with some positive frequency, lower-payoff forms can displace higher-payoff forms (Edgerton, 1992). Moreover, cultural evolution can involve conformist bias (Henrich & Boyd, 1998, 2001; Guzman, Rodriguez-Sickert, & Rowthorn, 2007), as well as oblique and horizontal transmission (Lumsden & Wilson, 1981; Cavalli-Sforza & Feldman, 1981; Gintis, 2003).

In rational choice theory, choices give rise to probability distributions over outcomes, the expected values of which are the payoffs to the choice from which they arose. Game theory extends this analysis to cases where there are multiple decision makers. In the language of game theory, *players* (or *agents*) are endowed with a set of available *strategies* and have certain *information* concerning the rules of the game, the nature of the other players and their available strategies, as well as the structure of payoffs. Finally, for each combination of strategy choices by the players, the game specifies a distribution of *payoffs* to the players. Game theory predicts the behavior of the players by assuming each maximizes its preference function subject to its information, beliefs, and constraints (Kreps, 1990).

Game theory is a logical extension of evolutionary theory. To see this, suppose there is only one replicator, deriving its nutrients and energy from non-living sources. The replicator population will then grow at a geometric rate, until it presses upon its environmental inputs. At that point, mutants that exploit the environment more efficiently will out-compete their less efficient conspecifics, and with input scarcity, mutants will emerge that "steal" from conspecifics who have amassed valuable resources. With the rapid growth of such predators, mutant prey will devise means of avoiding predation, and predators will counter with their own novel predatory capacities. In this manner, strategic interaction is born from elemental evolutionary forces. It is only a conceptual short step from this point to cooperation and competition among cells in a multicellular body, among conspecifics who cooperate in social production, between males and females in a sexual species, between parents and offspring, and among groups competing for territorial control.

Historically, game theory did not emerge from biological considerations, but rather from strategic concerns in World War II (von Neumann & Morgenstern, 1944; Poundstone, 1992). This led to the widespread caricature of game theory as applicable only to static confrontations of rational self-regarding agents possessed of formidable reasoning and information-processing capacity. Developments within game theory in recent years, however, render this caricature inaccurate.

Game theory has become the basic framework for modeling animal behavior (Smith, 1982; Alcock, 1993; Krebs & Davies, 1997), and thus has shed its static and hyperrationalistic character in the form of evolutionary game theory (Gintis, 2009b). Evolutionary and behavioral game theory do not require the formidable information-processing capacities of classical game theory, so disciplines that recognize cognition is scarce and costly can make use of game-theoretic models (Young, 1998; Gintis, 2009b; Gigerenzer & Selten, 2001). Thus, agents may consider only a restricted subset of strategies (Winter, 1971; Simon, 1972), and they may use rule-of-thumb heuristics rather than maximization techniques (Gigerenzer & Selten, 2001). Game theory is thus a generalized schema that permits the precise framing of meaningful empirical assertions, but imposes no particular structure on the predicted behavior.

15.6 Socio-psychological Theory of Norms

Complex social systems generally have a division of labor, with distinct social positions occupied by individuals specially prepared for their roles. For instance, a bee hive has workers, drones, and queens, and workers can be nurses, foragers, or scouts. Preparation for roles is by gender and larval nutrition. Modern human society has a division of labor characterized by dozens of specialized *roles*, appropriate behavior within which is given by *social norms*, and individuals are *actors* who are motivated to fulfill these roles through a combination of *material incentives* and *normative commitments*.

The centrality of culture in the social division of labor was clearly expressed by Emile Durkheim (1902/1933), who stressed that the great multiplicity of roles (which he called *organic solidarity*) required a commonality of beliefs (which he called *collective consciousness*) that would permit the smooth coordination of actions by distinct individuals. This theme was developed by Talcott Parsons (1937), who used his knowledge of economics to articulate a sophisticated model of the interaction between the situation (role) and its inhabitant (actor). The actor/role approach to social norms was filled out by Erving Goffman (1959), among others.

The social role has both normative and positive aspects. On the positive side, the payoffs—rewards and penalties—associated with a social role must provide the appropriate incentives for actors to carry out the duties associated with the role. This requirement is most easily satisfied when these payoffs are independent from the behavior of agents occupying other roles. However, this is rarely the case. In general, social roles are deeply interdependent, and can be modeled as the strategy sets of players in an epistemic game, the payoffs to which are precisely these rewards and penalties, the choices of actors then forming a *correlated equilibrium*, for which the required commonality of beliefs is provided by a society's common culture (Gintis, 2009a). This argument provides an analytical link uniting the actor/role framework in sociological theory with game theoretic models of cooperation in economic theory.

Appropriate behavior in a social role is given by a *social norm* that specifies the duties, privileges, and normal behavior associated with the role. In the first instance, the complex of social norms has an instrumental character devoid of normative content, serving merely as an informational device that coordinates the behavior of rational agents (Lewis, 1969; Gauthier, 1986; Binmore, 2005; Bicchieri, 2006). However, in most cases, high-level performance in a social role requires that the actor have a *personal commitment* to role performance that cannot be captured by the self-regarding "public" payoffs associated with the role (Conte & Castelfranchi, 1999; Gintis, 2009a). This is because (a) actors may have private payoffs that conflict with the public payoffs, inducing them to behave counter to proper role-performance (e.g., corruption, favoritism, aversion to specific tasks); (b) the signal used to determine the public payoffs may be inaccurate and unreliable (e.g., the performance of a teacher or physician); and (c) the public payoffs required to gain compliance by self-regarding actors may be higher than those required when there is at least partial reliance upon the personal commitment of role incumbents (e.g., it may be less costly to use personally committed rather than purely materially motivated physicians and teachers). In such cases, self-regarding actors who treat social norms purely instrumentally will behave in a socially inefficient manner. This point is stressed in chapter 6 in this volume by Peter Bull and Ofer Feldman in their discussion of the social skills model of political discourse.

The normative aspect of social roles flows from these considerations. First, to the extent that social roles are considered legitimate by incumbents, they will place an intrinsic positive ethical value on role-performance. We may call this the *normative bias* associated with role-occupancy (Gintis, 2009a). This phenomenon has strong support in the cognitive neuroscience literature. For instance, social norms appear to affect the same neural regions as those involved in reward learning and valuation

(Klucharev, Hytonen, Rijpkema, Smidts et al., 2009), showing that morality is not a constraint on behavior, but rather a positive motivator of behavior. Second, human ethical predispositions include *character virtues*, such as honesty, trustworthiness, promise-keeping, and obedience, that may increase the value of conforming to the duties associated with role-incumbency (Gneezy, 2005). Third, humans are also predisposed to care about the esteem of others even when there can be no future reputational repercussions (Masclet, Noussair, Tucker, & Villeval, 2003), and take pleasure in punishing others who have violated social norms (Fehr & Fischbacher, 2004). These normative traits by no means contradict rationality (section 3), because individuals trade off these values against material reward, and against each other, just as described in the economic theory of the rational actor (Andreoni & Miller, 2002; Gneezy & Rustichini, 2000).

The socio-psychological model of norms can thus resolve the contradictions between the sociological and economic approaches to social cooperation, retaining the analytical clarity of game theory and the rational actor model, while incorporating the normative and cultural considerations stressed in psycho-social models of norm compliance.

15.7 Conclusion

This chapter has presented five analytical tools that together help adjudicate the conflicting disciplinary models of human action in the behavioral disciplines, with an emphasis on the role of cognitive processes in weaving social principles into a coherent conceptual web. While there are doubtless formidable scientific issues involved in providing the precise articulations of these tools and the major conceptual tools of the various disciplines, these scientific issues are likely to be dwarfed by the sociological issues surrounding the semifeudal nature of the modern behavioral disciplines.

Note

This chapter is based on ideas more fully elaborated in Gintis (2009a).

References

Abbott, R. J., James, J. K., Milne, R. I., & Gillies, A. C. M. (2003). Plant introductions, hybridization and gene flow. *Philosophical Transactions of the Royal Society B: Biological Sciences, 358*, 1123–1132.

Alcock, J. (1993). *Animal behavior: An evolutionary approach*. Sunderland, MA: Sinauer.

Alexander, R. D. (1987). *The biology of moral systems.* New York: Aldine.

Allman, J., Hakeem, A., & Watson, K. (2002). Two phylogenetic specializations in the human brain. *Neuroscientist, 8,* 335–346.

Andreoni, J., & Miller, J. H. (2002). Giving according to GARP: An experimental test of the consistency of preferences for altruism. *Econometrica, 70*(2), 737–753.

Baker, C. L., Tenenbaum, J. B., & Saxe, R. R. (2007). Goal inference as inverse planning. In *Proceedings of the 29th annual conference of the Cognitive Science Society* (pp. 779–784).

Beer, J. S., Heerey, E. A., Keltner, D., Skabini, D., & Knight, R. T. (2003). The regulatory function of self-conscious emotion: Insights from patients with orbitofrontal damage. *Journal of Personality and Social Psychology, 65,* 594–604.

Belin, P., Zatorre, R. J., Lafaille, P., Ahad, P., & Pike, B. (2000). Voice-selective areas in human auditory cortex. *Nature, 403,* 309–312.

Bicchieri, C. (2006). *The grammar of society: The nature and dynamics of social norms.* Cambridge: Cambridge University Press.

Binder, J. R., Frost, J. A., Hammeke, T. A., Cox, R. W., Rao, S. M., & Prieto, T. (1997). Human brain language areas identified by functional magnetic resonance imaging. *Journal of Neuroscience, 17,* 353–362.

Binmore, K. G. (2005). *Natural justice.* Oxford: Oxford University Press.

Black, F., & Scholes, M. (1973). The pricing of options and corporate liabilities. *Journal of Political Economy, 81,* 637–654.

Boehm, C. (2000). *Hierarchy in the forest: The evolution of egalitarian behavior.* Cambridge, MA: Harvard University Press.

Bonner, J. T. (1984). *The evolution of culture in animals.* Princeton, NJ: Princeton University Press.

Bowles, S., & Gintis, H. (2011). *A cooperative species: Human reciprocity and its evolution.* Princeton, NJ: Princeton University Press.

Boyd, R., & Richerson, P. J. (1985). *Culture and the evolutionary process.* Chicago: University of Chicago Press.

Boyd, R., & Richerson, P. J. (1988). The evolution of reciprocity in sizable groups. *Journal of Theoretical Biology, 132,* 337–356.

Brainard, D. H., & Freeman, W. T. (1997). Bayesian color constancy. *Journal of the Optical Society of America A: Optics, Image Science, and Vision, 14,* 1393–1411.

Brown, J. H., & Lomolino, M. V. (1998). *Biogeography.* Sunderland, MA: Sinauer.

Burrows, A. M. (2008). The facial expression musculature in primates and its evolutionary significance. *BioEssays, 30*(3), 212–225.

Camille, N. (2004). The involvement of the orbitofrontal cortex in the experience of regret. *Science, 304,* 1167–1170.

Carey, S. (1985). *Conceptual change in childhood.* Cambridge, MA: MIT Press.

Cavalli-Sforza, L. L., & Feldman, M. W. (1981). *Cultural transmission and evolution: A quantitative approach.* Princeton, NJ: Princeton University Press.

Cavalli-Sforza, L. L., & Feldman, M. W. (1982). Theory and observation in cultural transmission. *Science, 218,* 19–27.

Chater, N., & Manning, C. (2006). Probabilistic models of language processing and acquisition. *Trends in Cognitive Sciences, 10,* 335–344.

Conte, R., & Castelfranchi, C. (1999). From conventions to prescriptions: Towards an integrated view of norms. *Artificial Intelligence and Law, 7,* 323–340.

Cooper, W. S. (1987). Decision theory as a branch of evolutionary theory. *Psychological Review, 4,* 395–411.

Courville, A. C., Daw, N. D., & Touretzky, D. S. (2006). Bayesian theories of conditioning in a changing world. *Trends in Cognitive Sciences, 10,* 294–300.

Darwin, C. (1872). *The origin of species by means of natural selection* (6th ed.). London: John Murray.

Dawkins, R. (1976). *The selfish gene.* Oxford: Oxford University Press.

Dawkins, R. (1982). *The extended phenotype: The gene as the unit of selection.* Oxford: Freeman.

Dorris, M. C. and Glimcher, P. W. (2003). Monkeys as an animal model of human decision making during strategic interactions. Unpublished manuscript.

Dunbar, R. M. (1993). Coevolution of neocortical size, group size and language in humans. *Behavioral and Brain Sciences, 16*(4), 681–735.

Dunbar, R. M. (1996). *Grooming, gossip, and the evolution of language.* Cambridge, MA: Harvard University Press.

Dunbar, R. M. Gamble, C., & Gowlett, J. (2010). *Social brain, distributed mind.* Proceedings of the British Academy (158). Oxford University Press.

Durkheim, E. (1933). *The division of labor in society.* New York: The Free Press. (Original work published 1902)

Edgerton, R. B. (1992). *Sick societies: Challenging the myth of primitive harmony.* New York: Free Press.

Eshel, I., & Feldman, M. W. (1984). Initial increase of new mutants and some continuity properties of ESS in two locus systems. *American Naturalist, 124,* 631–640.

Eshel, I., Feldman, M. W., & Bergman, A. (1998). Long-term evolution, short-term evolution, and population genetic theory. *Journal of Theoretical Biology, 191,* 391–396.

Fehr, E., & Fischbacher, U. (2004). Third-party punishment and social norms. *Evolution and Human Behavior, 25,* 63–87.

Feldman, M. W., & Zhivotovsky, L. A. (1992). Gene-culture coevolution: Toward a general theory of vertical transmission. *Proceedings of the National Academy of Sciences of the United States of America, 89,* 11935–11938.

Fisher, R. A. (1930). *The genetical theory of natural selection.* Oxford: Clarendon Press. (Original work published 1930)

Fudenberg, D., Levine, D. K., & Maskin, E. (1994). The folk theorem with imperfect public information. *Econometrica, 62,* 997–1039.

Gadagkar, R. (1991). On testing the role of genetic asymmetries created by haplodiploidy in the evolution of eusociality in the hymenoptera. *Journal of Genetics, 70*(1), 1–31.

Gauthier, D. (1986). *Morals by agreement.* Oxford: Clarendon Press.

Geertz, C. (1963). *Peddlers and princes: Social change and economic modernization in two Indonesian towns.* Chicago: University of Chicago Press.

Gigerenzer, G., & Selten, R. (2001). *Bounded rationality.* Cambridge, MA: MIT Press.

Gintis, H. (2003). Solving the puzzle of human prosociality. *Rationality and Society, 15*(2), 155–187.

Gintis, H. (2005). Behavioral game theory and contemporary economic theory. *Analyze Kritik 27*(1), 48–72.

Gintis, H. (2009a). *The bounds of reason: Game theory and the unification of the behavioral sciences.* Princeton, NJ: Princeton University Press.

Gintis, H. (2009b). *Game theory evolving* (2nd ed.). Princeton, NJ: Princeton University Press.

Glimcher, P. W. (2003). *Decisions, uncertainty, and the brain: The science of neuroeconomics.* Cambridge, MA: MIT Press.

Glimcher, P. W., Dorris, M. C., & Bayer, H. M. (2005). Physiological utility theory and the neuroeconomics of choice. *Games and Economic Behavior* (52), 213–256.

Glymour, C. (2001). *The mind's arrows: Bayes nets and graphical causal models in psychology.* Cambridge, MA: MIT Press.

Gneezy, U. (2005). Deception: The role of consequences. *American Economic Review*, *95*(1), 384–394.

Gneezy, U., & Rustichini, A. (2000). A fine is a price. *Journal of Legal Studies, 29*, 1–17.

Goeree, J. K., Palfrey, T., Rogers, B., & McKelvey, R. (2007). Self-correcting information cascades. *Review of Economic Studies, 74*(3), 733–762.

Goffman, E. (1959). *The presentation of self in everyday life.* New York, NY: Anchor.

Gopnik, A., & Meltzoff, A. (1997). *Words, thoughts, and theories.* Cambridge, MA: MIT Press.

Gopnik, A., & Schultz, L. (2007). *Causal learning, psychology, philosophy, and computation.* Oxford: Oxford University Press.

Gopnik, A., Sobel, D. M., Schulz, L. E., Glymour, C. (2001). Causal learning mechanisms in very young children: Two-, three-, and four-year-olds infer causal relations from patterns of variation and covariation. *Developmental Psychology, 37*(50), 620–629.

Gopnik, A., & Tenenbaum, J. B. (2007). Bayesian networks, Bayesian learning and cognitive development. *Developmental Studies, 10*(3), 281–287.

Grafen, A. (1999). Formal Darwinism, the individual-as-maximizing-agent: analogy, and bet-hedging. *Proceedings of the Royal Society B: Biological Sciences, 266*, 799–803.

Grafen, A. (2000). Developments of Price's equation and natural selection under uncertainty. *Proceedings of the Royal Society B: Biological Sciences, 267*, 1223–1227.

Grafen, A. (2002). A first formal link between the Price equation and an optimization program. *Journal of Theoretical Biology, 217*, 75–91.

Griffiths, T. L., Kemp, C., & Tenenbaum, J. B. (2008). *Bayesian models of cognition.* Berkeley, CA: University of California.

Griffiths, T. L., & Tenenbaum, J. B. (2008). From mere coincidences to meaningful discoveries. *Cognition, 103*, 180–226.

Guzman, R. A., Rodriguez-Sickert, C., & Rowthorn, R. (2007). When in Rome, do as the Romans do: The coevolution of altruistic punishment, conformist learning, and cooperation. *Evolution and Human Behavior, 28*, 112–117.

Haldane, J. B. S. (1932). *The causes of evolution.* London: Longmans, Green.

Hammerstein, P. (1996). Darwinian adaptation, population genetics and the streetcar theory of evolution. *Journal of Mathematical Biology, 34*, 511–532.

Hammerstein, P., & Selten, R. (1994). Game theory and evolutionary biology. In R. J. Aumann & S. Hart (Eds.), *Handbook of game theory with economic applications* (pp. 929–993). Amsterdam, The Netherlands: Elsevier.

Hampton, A. N., Bossaerts, P., & O'Doherty, J. P. (2006). The role of the ventromedial prefrontal cortex in abstract state-based inference during decision making in humans. *Journal of Neuroscience, 26*(32), 8360–8367.

Hampton, A. N., Bossaerts, P., & O'Doherty, J. P. (2008). Neural correlates of mentalizing-related computations during strategic interactions in humans. *Proceedings of the National Academy of Sciences of the United States of America, 105*(18), 6741–6746.

Heiner, R. A. (1983). The origin of predictable behavior. *American Economic Review, 73*(4), 560–595.

Henrich, J., & Boyd, R. (1998). The evolution of conformist transmission and the emergence of between-group differences. *Evolution and Human Behavior, 19,* 215–242.

Henrich, J., & Boyd, R. (2001). Why people punish defectors: Weak conformist transmission can stabilize costly enforcement of norms in cooperative dilemmas. *Journal of Theoretical Biology, 208,* 79–89.

Henrich, J., & Gil-White, F. (2001). The evolution of prestige: Freely conferred status as a mechanism for enhancing the benefits of cultural transmission. *Evolution and Human Behavior, 22,* 165–196.

Holden, C. J. (2002). Bantu language trees reflect the spread of farming across sub-Saharan Africa: A maximum-parsimony analysis. *Proceedings of the Royal Society B: Biological Sciences, 269,* 793–799.

Holden, C. J., & Mace, R. (2003). Spread of cattle led to the loss of matrilineal descent in Africa: A coevolutionary analysis. *Proceedings of the Royal Society B: Biological Sciences, 270,* 2425–2433.

Huxley, J. S. (1955). *Evolution, cultural and biological* (pp. 2–25). Yearbook of Anthropology.

Jablonka, E., & Lamb, M. J. (1995). *Epigenetic inheritance and evolution: The Lamarckian case.* Oxford: Oxford University Press.

James, W. (1880). Great men, great thoughts, and the environment. *Atlantic Monthly, 46,* 441–459.

Jaynes, E. T. (2003). *Probability theory: The logic of science.* Cambridge, MA: Cambridge University Press.

Jurmain, R., Nelson, H., Kilgore, L., & Travathan, W. (1997). *Introduction to physical anthropology.* Cincinnati, OH: Wadsworth Publishing.

Kable, J. W., & Glimcher, P. W. (2009). The neurobiology of decision: Consensus and controversy. *Neuron, 63*(6), 733–745.

Keiser, K., Lindenberg, S., & Steg, L. (2008). The spreading of disorder. *Science, 322*, 1681–1685.

Klucharev, V., Hytonen, K., Rijpkema, M., Smidts, A., & Fernandez, G. (2009). Reinforcement learning signal predicts social conformity. *Neuron, 61*(1), 140–151.

Knill, D., & Pouget, A. (2004). The Bayesian brain: The role of uncertainty in neural coding and computation. *Trends in Cognitive Psychology, 27*(12), 712–719.

Kording, K. P., & Wolpert, D. M. (2006). Bayesian decision theory in sensorimotor control. *Trends in Cognitive Sciences, 10*, 319–326.

Krajbich, I., Adolphs, A., Tranel, D., Denburg, N. L., & Camerer, C. (2009). Economic games quantify diminished sense of guilt in patients with damage to the prefrontal cortex. *Journal of Neuroscience, 29*(7), 2188–2192.

Krantz, D. H. (1991). From indices to mappings: The representational approach to measurement. In D. Brown & J. Smith (Eds.), *Frontiers of mathematical psychology* (pp. 1–52). Cambridge: Cambridge University Press.

Krebs, J. R., & Davies, N. B. (1997). *Behavioral ecology: An evolutionary approach* (4th ed.). Oxford: Blackwell Science.

Kreps, D. M. (1990). *A course in microeconomic theory.* Princeton, NJ: Princeton University Press.

Levine, D. K., & Palfrey, T. R. (2007). The paradox of voter participation: A laboratory study. *American Political Science Review, 101*, 143–158.

Lewis, D. (1969). *Conventions: A philosophical study.* Cambridge, MA: Harvard University Press.

Lewontin, R. C. (1974). *The genetic basis of evolutionary change.* New York: Columbia University Press.

Liberman, U. (1988). External stability and ESS criteria for initial increase of a new mutant allele. *Journal of Mathematical Biology, 26*, 477–485.

Lumsden, C. J., & Wilson, E. O. (1981). *Genes, mind, and culture: The coevolutionary process.* Cambridge, MA: Harvard University Press.

Ma, W. J., Beck, J. M., Latham, P. E., & Pouget, A. (2006). Bayesian inference with probabilistic population codes. *Nature Neuroscience, 9*(11), 1432–1438.

Mace, R., & Pagel, M. (1994). The comparative method in anthropology. *Current Anthropology, 35*, 549–564.

Masclet, D., Noussair, C., Tucker, S., & Villeval, M.-C. (2003). Monetary and non-monetary punishment in the voluntary contributions mechanism. *American Economic Review, 93*(1), 366–380.

Mednick, S. A., Kirkegaard-Sorenson, L., Hutchings, B., Knop, J., Rosenberg, R., & Schulsinger, F. (1977). An example of bio-social interaction research: The interplay of socio-environmental and individual factors in the etiology of criminal behavior. In S. A. Mednick & K. O. Christiansen (Eds.), *Biosocial bases of criminal behavior* (pp. 9–24). New York: Gardner Press.

Meltzhoff, A. N., & Decety, J. (2003). What imitation tells us about social cognition: A rapprochement between developmental psychology and cognitive neuroscience. *Philosophical Transactions of the Royal Society B: Biological Sciences, 358*, 491–500.

Mesoudi, A., Whiten, A., & Laland, K. N. (2006). Towards a unified science of cultural evolution. *Behavioral and Brain Sciences, 29*, 329–383.

Miller, B. L., Darby, A., Benson, D. F., Cummings, J. L., & Miller, M. H. (1997). Aggressive, socially disruptive and antisocial behaviour associated with fronto-temporal dementia. *British Journal of Psychiatry, 170*, 150–154.

Moll, J., Zahn, R., di Oliveira-Souza, R., Krueger, F., & Grafman, J. (2005). The neural basis of human moral cognition. *Nature Neuroscience, 6*, 799–809.

Montague, P. R., & Berns, G. S. (2002). Neural economics and the biological substrates of valuation. *Neuron, 36*, 265–284.

Moran, P. A. P. (1964). On the nonexistence of adaptive topographies. *Annals of Human Genetics, 27*, 338–343.

Newman, M., Barabasi, A.-L., & Watts, D. J. (2006). *The structure and dynamics of networks*. Princeton, NJ: Princeton University Press.

Niv, Y., & Montague, P. R. (2009). Theoretical and empirical studies of learning. In P. W. Glimcher, C. Camerer, E. Fehr, & R. A. Poldrack (Eds.), *Neuroeconomics: Decision making and the brain* (pp. 329–349). New York: Academic Press.

O'Brien, M. J., & Lyman, R. L. (2000). *Applying evolutionary archaeology*. New York: Kluwer Academic.

Odling-Smee, F. Laland, K. N., & Feldman, M. W. (2003) *Niche construction: The neglected process in evolution*. Princeton, NJ: Princeton University Press.

Parker, A. J., & Newsome, W. T. (1998). Sense and the single neuron: Probing the physiology of perception. *Annual Review of Neuroscience, 21*, 227–277.

Parsons, T. (1937). *The structure of social action*. New York: McGraw-Hill.

Parsons, T. (1964). Evolutionary universals in society. *American Sociological Review, 29*(3), 339–357.

Pearl, J. (2000). *Causality*. New York: Oxford University Press.

Popper, K. (1979). *Objective knowledge: An evolutionary approach*. Oxford: Clarendon Press.

Poundstone, W. (1992). *Prisoner's dilemma*. New York, NY: Doubleday.

Real, L. A. (1991). Animal choice behavior and the evolution of cognitive architecture. *Science, 253*, 980–986.

Real, L. A., & Caraco, T. (1986). Risk and foraging in stochastic environments. *Annual Review of Ecology and Systematics, 17*, 371–390.

Relethford, J. H. (2007). *The human species: An introduction to biological anthropology*. New York: McGraw-Hill.

Richerson, P. J., & Boyd, R. (1998). The evolution of ultrasociality. In I. Eibl-Eibesfeldt & F. K. Salter (Eds.), *Indoctrinability, ideology and warfare* (pp. 71–96). New York: Berghahn Books.

Richerson, P. J., & Boyd, R. (2004). *Not by genes alone*. Chicago: University of Chicago Press.

Rivera, M. C., & Lake, J. A. (2004). The ring of life provides evidence for a genome fusion origin of eukaryotes. *Nature, 431*, 152–155.

Rizzolatti, G., Fadiga, L., Fogassi, L., & Gallese, V. (2002). From mirror neurons to imitation: Facts and speculations. In A. N. Meltzhoff & W. Prinz (Eds.), *The imitative mind: Development, evolution and brain bases* (pp. 247–266). Cambridge: Cambridge University Press.

Schall, J. D., & Thompson, K. G. (1999). Neural selection and control of visually guided eye movements. *Annual Review of Neuroscience, 22*, 241–259.

Schulkin, J. (2000). *Roots of social sensitivity and neural function*. Cambridge, MA: MIT Press.

Schultz, W., Dayan, P., & Montague, P. R. (1997). A neural substrate of prediction and reward. *Science, 275*, 1593–1599.

Schultz, L., & Gopnik, A. (2004). Causal learning across domains. *Developmental Psychology, 40*, 162–176.

Seeley, T. D. (1997). Honey bee colonies are group-level adaptive units. *American Naturalist, 150*, S22–S41.

Shennan, S. (1997). *Quantifying archaeology*. Edinburgh, Scotland: Edinburgh University Press.

Simon, H. (1972). Theories of Bounded Rationality. In C. B. McGuire & R. Radner (Eds.), *Decision and organization* (pp. 161–176). New York, NY: Elsevier.

Simon, H. (1982). *Models of bounded rationality.* Cambridge, MA: MIT Press.

Skibo, J. M., & Bentley, R. A. (2003). *Complex systems and archaeology.* Salt Lake City: University of Utah Press.

Smith, J. M. (1982). *Evolution and the theory of games.* Cambridge, England: Cambridge University Press.

Smith, E. A., & Winterhalder, B. (1992). *Evolutionary ecology and human behavior.* New York, NY: Aldine de Gruyter.

Sobel, D. M., & Kirkham, N. Z. (2007). Bayes nets and babies: Infants' developing statistical reasoning abilities and their representations of causal knowledge. *Developmental Science, 10*(3), 298–306.

Spirtes, P., Glymour, C., & Scheines, R. (2001). *Causation, prediction, and search.* Cambridge, MA: MIT Press.

Stapel, D. A., and Lindenberg, S. (2011). Coping with chaos: How disordered contexts promote stereotyping and discrimination. *Science 332*(8),: 251–253.

Steyvers, M., Griffiths, T. L., & Dennis, S. (2006). Probabilistic inference in human semantic memory. *Trends in Cognitive Sciences, 10*, 327–334.

Steyvers, M., & Tenenbaum, J. B., Wagenmakers, E. J., & Blum, B. (2003). Inferring causal networks from observations and interventions. *Cognitive Science, 27*, 453–489.

Sugrue, L. P., Corrado, G. S., & Newsome, W. T. (2005). Choosing the greater of two goods: Neural currencies for valuation and decision making. *Nature Reviews. Neuroscience, 6*, 363–375.

Sun, R. (2006). *Cognition and multi-agent interaction: From cognitive modeling to social simulation.* Cambridge: Cambridge University Press.

Sun, R. (2008). *The Cambridge handbook of computational psychology.* Cambridge: Cambridge University Press.

Sutton, R., & Barto, A. G. (2000). *Reinforcement learning.* Cambridge, MA: MIT Press.

Taylor, P., & Jonker, L. (1978). Evolutionarily stable strategies and game dynamics. *Mathematical Biosciences, 40*, 145–156.

Tenenbaum, J. B., Griffiths, T. L., & Kemp, C. (2006). Bayesian models of inductive learning and reasoning. *Trends in Cognitive Sciences, 10*, 309–318.

Trivers, R. L. (1971). The evolution of reciprocal altruism. *Quarterly Review of Biology, 46*, 35–57.

Von Neumann, J., & Morgenstern, O. (1944). *Theory of games and economic behavior.* Princeton, NJ: Princeton University Press.

Williams, J. H. G., & Whiten, A., Suddendorf, T., & Perrett, D. I. (2001). Imitation, mirror neurons and autism. *Neuroscience and Biobehavioral Reviews, 25,* 287–295.

Wilson, E. O., & Holldobler, B. (2005). Eusociality: Origin and consequences. *Proceedings of the National Academy of Sciences of the United States of America, 102*(38), 13367–13371.

Winter, S. G. (1971). Satisficing, selection and the innovating remnant. *Quarterly Journal of Economics, 85,* 237–261.

Wright, S. (1931). Evolution in Mendelian populations. *Genetics, 6,* 111–178.

Xu, F., & Tenenbaum, J. B. (2007). Word learning as Bayesian inference. *Psychological Review 114,* 245–272.

Young, H. P. (1998). *Individual strategy and social structure: An evolutionary theory of institutions.* Princeton, NJ: Princeton University Press.

Yuille, A. & Kersten, D. (2006). Vision as Bayesian inference: Analysis by synthesis? *Trends in Cognitive Sciences, 10*(7), 301–308.

Zajonc, R. B. (1980). Feeling and thinking: Preferences need no inferences. *American Psychologist, 35*(2), 151–175.

Zajonc, R. B. (1984). On the primacy of affect. *American Psychologist, 39,* 117–123.

Contributors

Scott Atran Institute for Social Research, University of Michigan, Ann Arbor, MI

Peter Bull Department of Psychology, University of York, Heslington, York, United Kingdom

Erin Cassese Department of Political Science, West Virginia University, Morgantown, WV

Ofer Feldman Faculty of Policy Studies, Doshisha University, Kyoto, Japan

Stanley Feldman Department of Political Science, Stony Brook University, Stony Brook, NY

Leonie Huddy Department of Political Science, Stony Brook University, Stony Brook, NY

Herbert Gintis Santa Fe Institute and Central European University, Northampton, MA

Joseph W. Kable Department of Psychology, University of Pennsylvania, Philadelphia, PA

John J. McArdle Department of Psychology, University of Southern California, Los Angeles, CA

Mathew D. McCubbins School of Business, School of Law, and Department of Political Science, University of Southern California, Los Angeles, CA

Kristen Renwick Monroe Interdisciplinary Center for the Scientific Study of Ethics and Morality, University of California, Irvine, Irvine, CA

Ilkka Pyysiäinen Faculty of Theology, Helsinki University, Finland

Don Ross School of Economics, University of Cape Town, Cape Town, South Africa and Center for Economic Analysis of Risk, Georgia State University, Atlanta, GA

Norbert Ross Department of Anthropology, Vanderbilt University, Nashville, TN

Bradd Shore Department of Anthropology, Emory University, Atlanta, GA

Ron Sun Department of Cognitive Sciences, Rensselaer Polytechnic Institute, Troy, NY

Paul Thagard Department of Philosophy, University of Waterloo, Waterloo, Canada

Mark Turner Department of Cognitive Science, Case Western Reserve University, Cleveland, OH

Harvey Whitehouse School of Anthropology, University of Oxford, Oxford, United Kingdom

Robert J. Willis Department of Economics and Institute for Social Research, University of Michigan, Ann Arbor, MI

Index